Mössbauer Effect
Methodology

Volume 10

MÖSSBAUER EFFECT METHODOLOGY

Proceedings of Symposia Sponsored by
the New England Nuclear Corporation, Boston

Edited by Irwin J. Gruverman

A Continuation Order Plan is available for this series. A continuation order will bring delivery of each new volume immediately upon publication. Volumes are billed only upon actual shipment. For further information please contact the publisher.

A Publication of the New England Nuclear Corporation

Mössbauer Effect Methodology

Volume 10

Edited by

Irwin J. Gruverman
Nuclear Medicine and Technology Center
New England Nuclear Corporation
Billerica, Massachusetts

and

Carl W. Seidel
Nuclides and Sources Division
New England Nuclear Corporation
Billerica, Massachusetts

SPRINGER SCIENCE+BUSINESS MEDIA, LLC

The Library of Congress cataloged the first volume of this title as follows:

Symposium on Mössbauer Effect Methodology.
Mössbauer effect methodology; proceedings. 1st–
; 1965–
₍Boston?₎ New England Nuclear Corporation; distributed
by Plenum Press, New York.

v. illus. 24 cm.

Symposia for 1965–   sponsored by the New England
Nuclear Corporation and the Technical Measurement Corporation.

1. Mössbauer effect—Addresses, essays, lectures.   I. New England Nuclear Corporation, Boston.   II. Technical Measurement Corporation. III. Title.

QC490.S94   65—21188

Proceedings of the Tenth Symposium on Mossbauer
Effect Methodology held in New York City,
February 1, 1976

Library of Congress Catalog Card Number 65-21188
ISBN 978-1-4684-8075-7   ISBN 978-1-4684-8073-3 (eBook)
DOI 10.1007/978-1-4684-8073-3

© 1976 Springer Science+Business Media New York
Originally published by New England Nuclear Corporation in 1976
Softcover reprint of the hardcover 1st edition 1976

Preface

This is the tenth volume of a continuing series intended to provide a forum for publication of developments in Mössbauer Effect Methodology and in Spectroscopy and its applications.

Mössbauer Effect Methodology, Volume 10, records the proceedings of the Tenth Symposium on Mössbauer Effect Methodology. The Symposium was sponsored by the New England Nuclear Corporation, with special emphasis on applications in catalysis and in biology. The Symposium was held in the Mercury Ballroom of the New York Hilton on February 1, 1976. Dr. M. Good presided over the meeting.

More than one hundred participants were involved in the technical sessions and the exhibit of Mössbauer effect instruments, equipment and materials by Elscint, Inc., Ranger Engineering and New England Nuclear. Continued evolution and improvement was the keynote of the exhibit.

As has been our experience in recent Symposia, many more papers were submitted than could be accommodated. The Selection Committee was hard-pressed to limit the number of papers, and the sessions were lengthy, despite their efforts.

The revelation of the breadth of catalytic applications work was impressive to most participants, and may represent the first such broad application to a commercially-viable area. Spectroscopy in biological systems was the subject of a session. Methodology papers included discussions of electron-re-emission spectroscopy and parallel-plate avalanche detection. Studies of ^{125}Te, ^{151}Eu, ^{181}Ta and ^{83}Kr were presented. Applications in studies of treated alloys and surface properties of metals were also presented.

As always, the editors are indebted to their colleagues for the organization, solicitation, selection and performance of the Symposium functions. Dr. Good continued the tradition of gentle-firm conduct of the meeting.

It is anticipated that the series will continue, and we look forward to perpetuating this forum and to its next edition, probably in 1978.

Billerica, Massachusetts   Irwin Gruverman

May, 1976   Carl W. Seidel

Contents

STUDY OF CATALYSTS

BIOLOGICAL APPLICATIONS

CONTENTS

MÖSSBAUER SPECTROSCOPY IN HETEROGENEOUS CATALYSIS

W. N. Delgass

School of Chemical Engineering
Purdue University
West Lafayette, Indiana, 47907

ABSTRACT

A typical solid catalyst consists of 1-10 nm particles
of the active components dispersed on a high-surface-area
($\sim$300 m^2/g), relatively inert support. When one of the
elements in the active component phase has a Mössbauer effect,
a variety of catalytically important, and otherwise inacces-
sible, chemical properties of the catalyst can be measured.
The detailed chemical information contained in the recoil
free fraction, isomer shift, quadrupole splitting and magnetic
dipole splitting derived from the Mössbauer spectrum can
elucidate structure, bonding, composition and particle size
of the active component phase. When particle size is small,
a high fraction of the Mössbauer atoms are surface atoms, and
their interactions with adsorbed gases make a strong contribution to the spectrum. The technique is illustrated by
discussion of studies of supported iron, iron-exchanged
zeolites, the ammonia synthesis catalyst and oxidation/
reduction catalysts. The emphasis is on the nature of catalytic problems susceptible to investigation by Mössbauer
spectroscopy.

INTRODUCTION

Since the first application of the Mössbauer effect to the study of surface chemistry by Flinn, Ruby, and Kehl in 1964 (1), the number of papers dealing with Mössbauer spectroscopic studies related to heterogeneous catalysis has grown past one hundred. The progress of the field from feasibility studies to detailed probing of important catalytic phenomena has been monitored in numerous reviews, including two very recent and complete ones by Gager and Hobson (2) and by Dumesic and Topsøe (3). This paper assumes a knowledge of the Mössbauer effect on the part of the reader and focuses attention on the nature of catalysis and heterogeneous catalysts and the classes of catalytic problems to which Mössbauer spectroscopy is likely to contribute solutions. An exhaustive review of the literature has not been attempted, but a variety of studies is presented to illustrate successful applications of the technique.

THE NATURE OF CATALYSIS

A catalyst may be defined as a substance which alters the rate of a chemical reaction but does not appear in the reaction products. In heterogeneous catalysis the catalyst is usually a solid, the reactants and products are often gases, and the function of the catalyst is to selectively enhance the rate of reaction to the desired product. Since the rate of a reaction is controlled by the free energy of activation, the role of a catalyst may be viewed as providing a new reaction path which has lower free energy barriers and therefore a higher rate for the selected reaction. The seat of catalytic action is clearly the solid surface since only gas/solid contact can alter the reaction path. In this view, important questions that arise concerning a catalytic reaction are: What is the reaction path? How and why does that particular catalyst surface allow that reaction path? Determination of the elementary kinetic steps along the reaction path is the realm of chemical kinetics and.is often made without reference to specific gas/surface interactions. Questions of how and why certain kinetic steps occur under the influence of the catalyst surface cannot be answered, however, without detailed knowledge of the chemistry and physics of the surface. It is for this knowledge that we turn to spectroscopic techniques such as the Mössbauer effect. As we focus attention on the surface, it is important to remember that the complete picture of a catalytic reaction will usually require information from several sources and must also include chemical kinetics.

THE NATURE OF CATALYSTS

In order to be more specific about the kinds of questions that are important to ask about catalysts, we must first consider their physical and chemical nature. Because the gas/solid interface is the source of catalytic activity, efficient utilization of both catalytic material and reactor volume requires catalysts to have high specific surface area, which in turn implies small particle size and/or high porosity. To prevent agglomeration, or sintering, the small particles are usually distributed on a high specific surface area support. Typical supports, Al_2O_3, SiO_2, SiO_2/Al_2O_3,

MgO, TiO$_2$, or carbon, have surface areas ranging from
100 m^2/g to 1000 m^2/g and must maintain their high surface
areas in the reaction medium (i.e., carbon is not usually
used in oxidation reactions). Supported metal catalysts
may contain 0.1 - 50% metal by weight but typical loadings
of noble metals are in the 1 wt % region. Metal particle
sizes range from 1 to 20 nanometers (nm) average diameter
and are often discussed in terms of dispersion, the ratio of
surface metal atoms to total metal atoms in the catalyst.
The relation between particle size and dispersion is demonstrated by Poltorok and Boronin who have estimated that a
regularly faceted Pt crystallite 0.9 nm on an edge will be
comprised of 19 atoms and have a fraction of atoms on the
surface equal to 0.95, while a crystallite 5 nm on an edge
will contain nearly 4000 atoms and have only 30% of those
atoms on the surface (4). Thus, the active component of a
catalyst is often found in particles less than 5 nm in size,
distributed on a support with surface density less than 1
particle per 100 square nanometers. Characterization of
such a catalyst should include study of the size distribution of the metal particles and their location within the
pore structure of the support. Such detail is seldom available, but specific metal surface areas are essential for
quantitative interpretation of catalyst behavior and are
readily measured for many systems by selective chemisorption
(5). When more than one active component is present on a
supported catalyst, however, questions of particle composition and component distribution can be very difficult. The
Mössbauer effect can be used to unique advantage in this case,
as has been shown by Bartholomew and Boudart (6), Garten and
Ollis (7), Garten and Vannice (8), and Garten (9).

Several very useful catalysts have also been developed without the need for a support. Many such catalysts, of which the ammonia synthesis catalyst is a notable example, have a number of added components, called promoters, which improve catalytic performance. The presence of additional components again raises the need for composition and component distribution information. Study of the bulk properties of unsupported catalysts can elucidate the chemical and physical properties of the foundation on which the catalytic surface rests. Direct observation of the surface is generally more difficult, except for the crystalline solids known as zeolites or molecular sieves, which have a very high internal surface area accessible to small molecules.

Perhaps the most important information needed to answer questions about the origins and control of catalytic behavior is that concerning surface composition. Clearly, chemical and physical theory cannot be built in the absence of information concerning surface stoichiometry; and yet the complexity of catalytic materials, particularly multicomponent and supported catalysts, makes even this fundamental information elusive. The problem is complicated further by the fact, anticipated by Gibbs (10), that the composition of the surface of a multicomponent solid depends on its environment. Sachtler and coworkers report, for example, that a Pd/Ag alloy is enriched in Pd at the surface when CO is adsorbed (11). Williams and Boudart found Ni/Au alloys to be Ni-rich in an oxidizing atmosphere and Au rich in vacuo (12). These results suggest that a catalyst surface must be considered as dynamic and that the true working catalyst can be created only in the reaction medium.

Assuming that the composition and structure of the surface are known, the next problem to consider is the nature
of the gas/surface interactions which provide the new, catalytic reaction path. Investigation of this area is often
divided into geometric and electronic considerations. Such
a division is misleading because the two factors are clearly
related, but the notion that geometry can contribute to re-
activity can be a useful one, particularly for comparison of
the catalytic activity of different crystal planes (13) and
analysis of catalyst selectivity and poisoning (14). The
electronic factors in gas/surface interactions have been
described by both localized chemical bonding and band
structure pictures (15,16). Each approach has lead to useful
concepts, but considerably more theoretical work is needed.

Particles in the size range 1-5 nm have surfaces with
special geometric and electronic properties. Such small
crystallites cannot be terminated entirely by the most stable
low Miller index faces and thus present unusual geometric
arrangements of surface atoms (17). Furthermore, the number
of corner and edge atoms increases as particle size decreases.
These atoms are coordinatively unsaturated because they do
not have the full complement of ligands and might be expected
to interact more readily with adsorbing atoms. Catalytic
reactions sensitive to crystallite size are termed structure
sensitive (18), but somewhat surprisingly not all reactions
exhibit this phenomenon. An interesting electronic question
associated with small particle size is that of how big a
metal atom cluster must be for atomic properties to give way
to metallic properties. Calculations suggest that clusters
of up to 50 atoms do not yet have the exact electronic

structure of the bulk (19,20). In this size region, perturbation of the electronic structure of the metal particle by
the bonding between the particle and the support can also be
significant, signalling the need to consider the influence
of support interactions on the behavior of supported metals.
Thus, though the support is frequently considered as an inert
carrier, it can have intrinsic activity, often due to surface
acidity, or a secondary effect on activity through its interaction with small particles.

To this point, the discussion has centered on characterization of the actual catalyst or a close percursor
thereof. There is, in addition, a rich chemistry which
governs the preparation of a catalyst with the properties
described above. Factors including pH, counter-ion, order
of addition of chemicals, and exact dehydration procedure,
to mention a few, play a decisive role in determining surface
area and component distribution in a finished catalyst. Control of these factors is largely the art of catalysis, but
tools such as the Mössbauer effect that can follow chemical
and physical changes at each stage in the genesis of a catalysts can bring improved understanding and precision to the
production of these complex solids.

The complexity of catalytic materials provides both
fascination and frustration as we strive for more detailed
knowledge. Why not dissect the problem and study simpler
systems that can be controlled and understood in greater
depth? Indeed, vigorous research in surface chemistry and
physics is providing important fundamental knowledge on the
interactions of a variety of gas molecules with well defined
surfaces. It should be remembered, however, that the

chemistry of a flat single crystal metal surface cannot be
expected a priori to be the same as that of a 1 nm metal
crystallite on an alumina support. Somorjai (21) and co-
workers and others have made remarkable progress in elucidat-
ing well characterized surface features with catalytic impli-
cations, but there remains a pressing need to unravel the
complexities of catalysis in situ. The unique opportunities
offered by the Mössbauer effect for filling this need are
taken up in the next section.

CATALYTIC INFORMATION FROM THE MÖSSBAUER EFFECT

Experimental Considerations

Some of the advantages and limitations of the Mössbauer
effect in catalysis are prescribed by the physical requirements of the Mössbauer experiment. The Mössbauer effect is
primarily a solid state, bulk phenomenon. The relatively
high penetrtaing power of the γ-radiation allows the spectroscopy to "see" into catalyst pores, where most of the
useful surface of a catalyst support is. On the other hand,
the Mössbauer effect in standard transmission geometry records surface chemistry, the prime interest in catalysis,
only when a high fraction of the Mössbauer atoms are surface
atoms. Fortunately, this is exactly the condition sought
for the active component of a good supported catalyst. Surface chemistry can also be followed when Mössbauer atoms are
adsorbed on a surface from gas phase (22,23) or from solution
by ion exchange (24).

As in any transmission experiment there is an optimal
sample thickness for good signal without saturation, and

thus a restriction on sample size or concentration. Mass attenuation of γ rays due to the support or matrix can be an experimental limitation when the Mössbauer atom is present in low concentration. For ^{57}Fe experiments, however, use of enriched iron easily extends useful range from 1% to below 0.1 wt % Fe on a support such as silica or alumina. Studies of flat or single crystal surfaces cannot be done by absorber experiments in transmission geometry. Source experiments, however, have sufficient sensitivity for such studies (25). A surface layer 10 to 300 nm thick can be probed by absorber experiments using conversion electron detection in backscattering geometry (26).

Perhaps the most important experimental requirement for successful Mössbauer investigation of catalysts is careful control of the environment of the absorber. Catalyst surfaces are usually highly reactive and adjust quickly to changes in the surrounding gases. Trace impurities of O_2 or H_2O, for example, can drastically change the chemical state of a catalyst surface. An absorber cell must be capable of good, hydrocarbon-free vacuum ($<10^{-6}$ Torr) and allow chemical reaction or pretreatment of the catalyst in a variety of gases at 1 atm pressure and temperatures of at least 673 K. To take full advantage of the ability of the Mössbauer effect to probe catalyst chemistry of isotopes such as ^{57}Fe, ^{119}Sn, and ^{151}Eu _in situ_, it should be possible to record spectra _during_ treatment. A variety of glass and metal cells have been designed for this purpose. The example shown in Figure 1 is a stainless steel cell with a Cu gasket seal, Be windows, and capability for absorber wafer temperatures of 800 K in 1 atm of gas (27). The entering gas is preheated in the

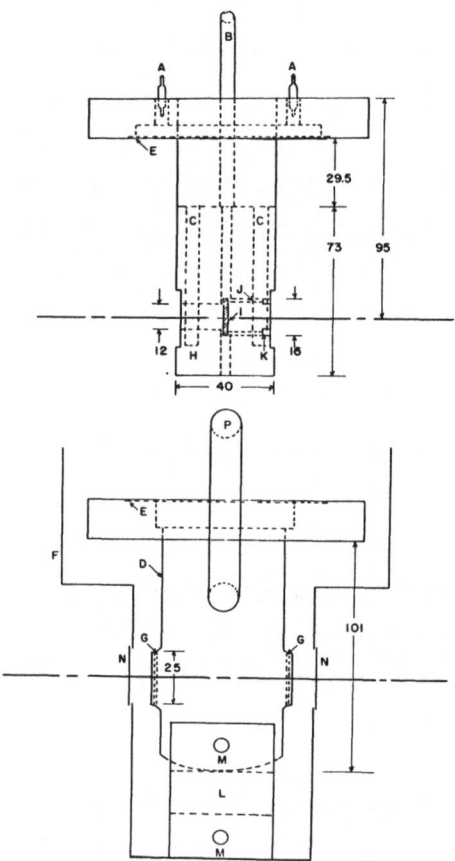

Figure 1. Mössbauer absorber cell with the sample holder
assembly raised and thermocouples and flange bolts and holes
omitted. Dimensions in mm. A - Thermocouple feed throughs,
B - Gas inlet tubing, C - Heater wells, D - Outer shell, E -
Gasket for Conflat flange, F - Copper heat shield, G - Be-
window, H - Sample holder block, I - Sample, J - Retaining
ring, K - Hollow boron nitride screw, L - Cell mount, M -
Screws to secure cell and mount, N - Be-window, P - Gas
outlet. From (27).

heater block and flows directly over the catalyst wafer to
insure good gas/solid contact during chemical reaction. An
external heat shield protects the radiation source and detector and an 80 mm minimum path length permits good count
rates for weak sources. Isotopes such as ^{99}Ru which require
liquid helium temperatures for source and absorber while
Mössbauer data is being collected pose special problems. In
situ pretreatment of the sample at 773 K is not easily recon-
ciled with the cryogenic requirements but can be accomplished
(28).

Spectral Parameters

The characteristic parameters of the Mössbauer spectrum
of a catalyst with a controlled chemical history provide
several kinds of information and will be discussed here in
turn. The isomer shift (IS) results from differences in
electron density at the nuclei of the Mössbauer atoms in the
absorber as compared to those in the source. Through this
parameter the Mössbauer effect probes changes in oxidation
state, covalency, coordination number, and kinds of ligands
for Mössbauer atoms in catalysts. The isomer shift can
indicate, for example, that impregnation of Al_2O_3 with an
$Eu(NO_3)_3$ solution and subsequent drying at 413 K produces a
surface complex which no longer resembles $Eu(NO_3)_3$, but is
probably a hydroxy-oxide (29). Such a result is not surprising in view of the cation-surface interactions anticipated during the genesis ·of the catalyst, but cannot be expected a priori. Impregnation of SiO_2 with a ruthenium
chloride solution, followed by drying at 383 K, still leaves
a Ru surface complex which closely resembles $RuCl_3 \cdot XH_2O$

(28). Other studies presented in the applications section will document the use of the isomer shift to follow oxidation state changes in catalytic materials. Its use in composition analysis is probably best exemplified in the identification of the Fe-Pt (6,8,9) and Fe-Pd (7) supported clusters already mentioned. While complete quantitative interpretation of isomer shifts due to surface atoms is not usually attempted, it should be remembered, as was pointed out ten years ago (25), that since the mean square velocity of a surface atom will usually be lower than that of an atom in the bulk, interpretation should include consideration of the second order Doppler shift.

The next higher order nuclear hyperfine interaction is the quadrupole splitting (QS). The dependence of this parameter on the electric field gradient tensor at the nucleus makes it an effective monitor of the symmetry of the environment of a Mössbauer atom in a catalyst. Since a surface represents a discontinuity with respect to the bulk, marked changes in symmetry and strong electric field gradients can be expected there. Quadrupole splittings in excess of 2 mm/sec for high spin Fe^{3+}, a $^6S_{5/2}$ state ion, dispersed on silica (30) confirm this expectation. Use of this parameter in catalysis has ranged from correlation of the quadrupole splitting of Fe in mixed oxides with their catalytic activity for ammoxidation of propylene (31) to qualitative indications of gas-surface interaction by changes in IS and QS after adsorption (32). The asymmetry of bonding at the surface which leads to large electric field gradients can also produce a directional dependence of the mean square displacement and, therefore, a significant Gol'danskii-Karayagin effect

(33). Quantitative interpretation of Gol'danskii-Karayagin effects have been reported for supported Fe (1), Sn (34) and Co (25).

The highest order nuclear hyperfine interaction is the magnetic dipole splitting due to the presence of a magnetic field at the nucleus of a Mössbauer atom. This interaction is most important for ^{57}Fe, which exhibits ferro-, antiferro- and ferrimagnetic ordering. Since the size of the magnetic field at the nucleus of a Mössbauer atom depends on the structure and composition of the solid in which it resides, magnetic dipole splittings can often provide phase identi- fication and composition information. Topsøe and Boudart (35) for example, have carefully analyzed the effect of adding 5 atomic percent Pb to a Cr promoted Fe_3O_4 water-gas shift catalyst to show that lead enters the tetrahedral sites as Pb^{4+} causing a change in structure from inverse to a partly normal spinel. Measurement of the Neel or Curie temperature at which magnetic ordering occurs can also aid in phase identification and determination of particle size. The latter determination is possible because of the dependence of the Neel or Curie temperature on particle volume. The phenomenon of superparamagnetism will be discussed further in the next section. Another kind of magnetic effect occurs when electron spins are not ordered but remained fixed in space for a time on the order of the Larmor precession period of the nucleus. The slow electron relaxation times required for this magnetic broadening are characteristic of S state ions in low concen- tration. Analysis of relaxation broadening in Eu^{2+} exchanged zeolites confirmed true exchange of the zeolites and reflected differences in the ionicity of the Eu bonding (36).

The final parameter to consider is the percent effect
or spectral area. Quantitative analysis of spectral areas
requires detailed knowledge of the lattice dynamics of the
solid in question and is generally not possible for the
complex, heterogeneous solids typical of catalysts. Changes
in spectral area as a function of catalyst treatment, however,
or the temperature dependence of spectral area for one catalyst compared to another can yield a qualitative measure of
the strength of bonding of the Mössbauer atom to the solid
matrix. Comparisons of peak areas showed stronger support
interaction between Eu^{3+} (37) or Ru (28) and Al_2O_3 as compared to SiO_2. The dependence of spectral area on lattice
dynamics can complicate composition analysis, particularly
in view of that fact that the Debye temperature of surface
atoms can be as low as 1/2 that of bulk atoms (38), but
measurement of the dependence of spectral area on temperature
can provide relative, effective Debye temperatures for the
different species identified in the spectrum. Broadening of
lines due to diffusion (39) has also been observed for highly
dispersed Mössbauer atoms but is restricted to cases where
the diffusion coefficient is of order 10^{-8} cm^2/sec (40).

APPLICATIONS TO CATALYSIS

In the previous section the overlap between Mössbauer
spectroscopy and heterogeneous catalysis was discussed from
the spectroscopic point of view. This section illustrates
four areas in which Mössbauer spectroscopy has contributed
to the understanding of catalysts. It stresses, from the
catalytic point of view, a few of many accomplishments which
demonstrate specific use and development of the method.

Supported Iron

Since this catalytic material was the first and is probably the most studied by the Mössbauer effect, it provides a logical starting point. Two important questions about supported iron concern the particle sizes of the iron-containing phases and the reducibility of the iron to the metal. In the limit of high surface concentration of iron, corresponding to 5 - 15 wt % Fe on most silicas and aluminas, iron tends to form particles of αFe_2O_3 in an oxidizing environment. This assignment is made on the basis of a magnetic field at the nucleus of the order 510 kOe, as measured by the splitting of the outer lines of the six line Mössbauer spectrum. In the limit of low surface concentration, < 1% Fe by weight carefully distributed on the support, the room temperature Mössbauer spectrum of the oxidized sample is typical of paramagnetic, high spin Fe^{3+}. At room temperature the critical particle diameter for antiferromagnetic ordering in αFe_2O_3 is 13.5 nm (41,42). Thus high loading of iron on a support leads to large particles of αFe_2O_3 on oxidation. The exact state of the iron at low loading is less clear. For particles of αFe_2O_3 of the order of 5.0 nm - 7.5 nm in diameter, the electron spins flip faster than every 10^{-8} seconds and the iron appears to be paramagnetic, or superparamagnetic, in the Mössbauer spectrum. Thus a quadrupole doublet can be indicative of small αFe_2O_3 particles or of Fe^{3+} monatomically dispersed on the support surface. Kundig et al. (41) have observed that the quadrupole splitting increases for small particles and that this parameter can help to identify the state of highly dispersed iron. Flinn, Ruby and Kehl, in their pioneering work on the Fe/Al_2O_3 system (1),

showed that the asymmetry of the Fe^{3+} quadrupole doublet
could be attributed to a Gol'danskii-Karayagin effect, demonstrating the strong influence of surface Fe^{3+} ions on the
Mössbauer spectrum. Spectral changes due to chemisorption
of gases confirm that much of the iron is on the surface (32).
Thus, while Mössbauer spectroscopy leaves some uncertainty in
the chemical state of Fe^{3+} highly dispersed on a support, the
technique readily identifies poor dispersion (particle radius
> 13 nm). Study of the temperature dependence of the anti-
ferromagnetic _versus_ the superparamagnetic fraction of Fe^{3+}
at decreasing temperatures can extend particle size information to the 5 nm region (41). Use of the Mössbauer effect
to monitor particle size has suggested that high dispersion
is more easily accomplished on Al_2O_3 compared to SiO_2 (43)
and with iron nitrate as compared to iron chloride as an
impregnation salt (44).

For many catalytic applications it is iron metal, not
the oxide, which is important. In spite of the fact that
αFe_2O_3 readily reduces to Fe metal in H_2 at 673 K (45), the
degree of reduction of supported iron at similar conditions
depends on the particle size of the oxidized iron phase and
on the support. Gager and Hobson (2) suggest that the antiferromagnetic αFe_2O_3 phase indicated in the room temperature
Mössbauer spectrum of an oxidized catalyst reduces to iron
metal while the paramagnetic or superparamagnetic phase does
not. This rule of thumb is not without exception (46) but
is generally consistent with published results. A picture
that emerges from the data taken so far is that at low surface concentration Fe^{3+} occupies sites in the surface that
provide intimate bonding of the iron to the support to form

a complex akin to a surface silicate or aluminate. When
these sites are saturated, Fe^{3+} is less strongly bonded to
the support and αFe_2O_3 particles grow as the impregnation
complex is decomposed. Thus, at low loading, support interactions produce a stable, dispersed iron oxide which can be
reduced to Fe^{2+} in H_2 at 673 - 773 K but not to iron metal.
Most of the iron in the larger Fe_2O_3 particles formed at
higher surface concentration is not susceptible to the influence of the support and reduces readily. These conclusions suggest that supported catalysts having small particles
of Fe must be prepared by alternative means. Two such methods
developed to date employ "alloying" the iron with noble metals
to improve reducibility (6,9) and carefully controlling
deposition of iron on MgO (47). The unique ability of
Mössbauer spectroscopy to indicate Fe-noble metal clustering
and details of the surface chemistry of small iron crystallites
is discussed by Garten and Dumesic in following papers in this
symposium.

Fe - Y Zeolite

Although the Mössbauer effect is quite sensitive to
subtle changes in chemical bonding, it is often difficult
to extract an accurate model of the environment of a Mössbauer
atom from Mössbauer spectroscopic data alone. The study of
the reversible oxidation of $Fe^{2+}Y$ zeolite (48) is a good
example of a case where a combination of structure information,
chemisorption measurements, and Mössbauer effect have led to
a detailed analysis of a specific gas/surface interaction.
Y-zeolite is a crystalline alumino-silicate. The structure
is built from SiO_4 and AlO_4 tetrahedra which join by sharing

corner oxygens to produce truncated octrahedra called sodalite
cages. The sodalite cages join through oxygen bridges, termed
hexagonal prisms, to form a tetrahedrally bonded array with
the diamond structure. As shown in Figure 2, the structure
encloses voids called supercages. These large cavities are
interconnected through windows with ∿0.9 nm free diameter,
thus offering a high internal surface area to small gas molecules. Molecules such as water can also enter the sodalite
cages through the 0.22 nm free-diameter, hexagonal windows.
By virtue of the charge deficiency of Al^{3+} compared to Si^{4+},
charge compensating cations must be present in the lattice.
A variety of cations can be ion-exchanged into the Y-zeolite
lattice and their most common crystallographic positions are
designated SI, SI', SII, and SII' in Figure 2. A sequence
of room temperature Mössbauer spectra of $Fe^{2+}Y$ dehydrated at
673 K, exposed to O_2 at 673 K, reduced in H_2 at 673 K and
hydrated at room temperature comprise Figure 3. The isomer
shift of the broadened quadrupole doublet in spectrum 3b is
0.57 mm/sec with respect to the ^{57}Co-Cr source and the quadrupole splitting was only slightly decreased in a spectrum
recorded at 673 K. Thus, not surprisingly, exposure of $Fe^{2+}Y$
to O_2 at 673 K oxidized the iron to a high spin Fe^{3+} state.
Spectrum 3c shows that the oxidation could be completely re-
versed by H_2 exposure at 673 K. Analysis of the parameters
of the $Fe^{2+}Y$ spectra, 3a and 3c, indicated the presence of
two ferrous species, one with IS = 1.40 mm/sec with respect
to ^{57}Co-Cr and QS = 2.36 mm/sec and the other with IS = 1.04
mm/sec and QS = 0.62 mm/sec. Chemisorption of gases that
could enter the supercage but not the sodalite cage strongly
perturbed the species with IS = 1.04 mm/sec but not the other
one. On the basis of this behavior and a correlation of IS

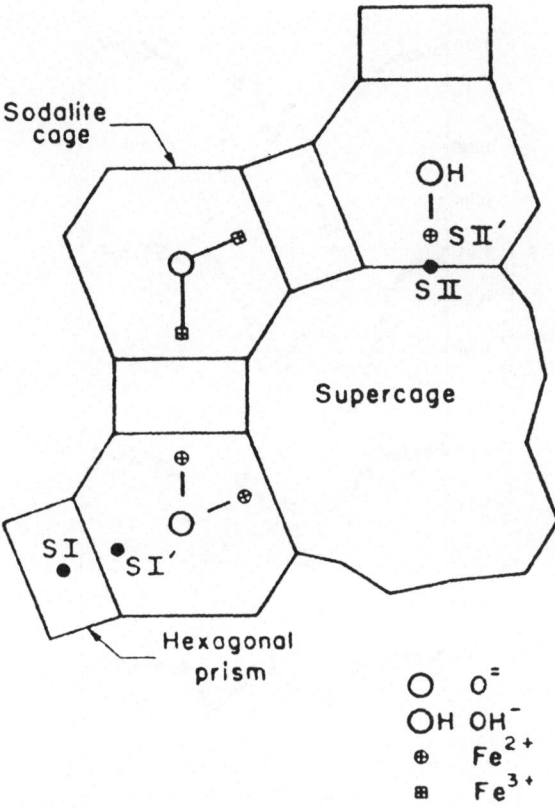

Figure 2. Schematic diagram of the structure and cation sites in Y zeolite. From (49).

with coordination number in ferrous oxides, the species with IS = 1.40 mm/sec was assigned to site SI while that with IS = 1.04 mm/sec was assigned to site SII'. The choice of SII' was dictated by the need for four-fold coordination rather than the six-fold coordination present at SII, the center of the puckered hexagonal ring, and by analysis of the heat of adsorption of Ar which indicated that gases in the

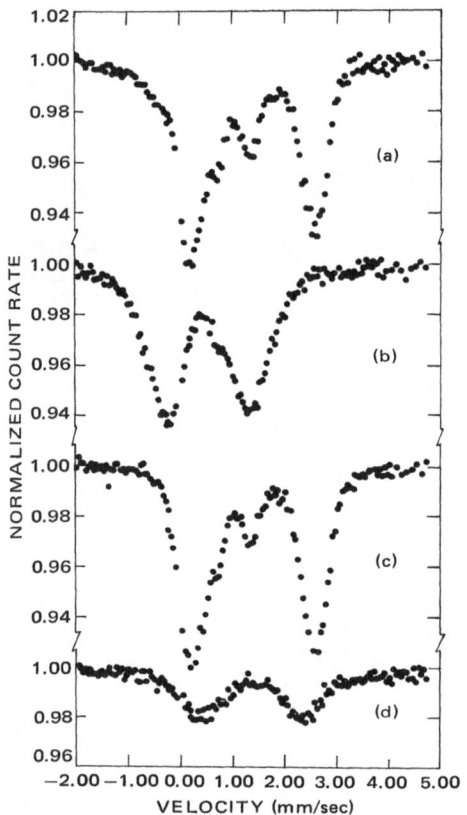

Figure 3. Oxidation/reduction of iron Y zeolite: (a) Fe^{2+}-Y dehydrated at 673 K; (b) oxidized at 673 K; (c) reduced in H_2 at 673 K; (d) exposed to 15 Torr H_2O. All spectra at room temperature and on the same sample. From (48).

supercage were not directly exposed to Fe^{2+} cations (50). The structure corresponding to the ferric species in spectrum 3b was assigned to an Fe^{3+}-O-Fe^{3+} bridge when volumetric chemisorption data showed the stoichiometry of adsorption to be one oxygen atom for each pair of Fe ions in the zeolite.

A similar bridge structure, slightly more accessible to small gas molecules, has been found for iron-exchanged mordenite (FeM), another crystalline alumino-silicate (51). FeM has significant catalytic activity for the reverse water-gas shift and ammoxidation of propylene reactions (51), but the possible catalytic role of the Fe-O-Fe bridge structure has not yet been examined in detail.

Additional information on the nature of ferrous ions in zeolites comes from spectrum 3d. Hydration markedly reduced the spectral area. This is due to solvation of Fe^{2+} in the intercrystalline water and shows how spectral areas reflect relative Debye temperatures. In the case of Fe^{2+}-Y the effect is reversible and subsequent dehydration regenerates spectrum 3a. In Eu^{3+} exchanged Y-zeolite a large increase in spectral area after the first dehydration of the freshly exchanged zeolite is irreversible and has been interpreted as being indicative of irreversible movement of Eu^{3+} from the super-cage to the sodalite cage or hexagonal prism (52).

Finally, we observe that ion exchanged zeolites provide further evidence for the influence of the cation environment on its oxidation/reduction chemistry. Iron in Fe-Y or M could not be reduced to Fe metal in H_2 at 773 K, in keeping with the effect of high dispersion discussed for supported iron. It is interesting to note, however, that dispersion enhances the reducibility of Eu^{3+}. The bulk oxide Eu_2O_3 cannot be reduced to EuO in H_2 at temperatures below 1473 K, but Eu^{3+} is readily reduced to Eu^{2+} in H_2 at 773 K when it is supported on Al_2O_3 or SiO_2 (29) or ion-exchanged in a zeolite (52).

The Ammonia Synthesis Catalyst

Detailed study of the workings of a catalyst often focuses on the surface, but careful examination of the bulk properties of catalytic solids can also be enlightening. This point is well demonstrated by the work of Topsøe, Dumesic and Boudart on the singly-promoted iron ammonia synthesis catalyst (53). Based on the classic work of Emmett and Brunauer (54), alumina, added at the ~1% level has long been considered to cover roughly half of the surface of αFe in the reduced catalyst and in so doing to stabilize the catalyst surface area. Although several subsequent studies confirmed these conclusions, the question of the position and role of Al_2O_3 was reopened when careful x-ray diffraction analysis of a singly-promoted catalyst (3 atomic % Al_2O_3) suggested that most of the alumina was in the bulk of the αFe in the form of very small $FeAl_2O_4$ clusters (55,56,57). Mössbauer spectra presented in reference 53 for the reduced catalyst showed the bulk to be pure αFe. Since $FeAl_2O_4$, finely dispersed in αFe, can be expected to lower the hyperfine field, the measured value of 330.6 kOe for the catalyst ruled out the presence of occluded clusters of 10 molecules of $FeAl_2O_4$ suggested by the x-ray analysis and showed that $FeAl_2O_4$ particles, if present, must be at least 3 nm in diameter. Careful analysis of spectral areas after background correction showed that, at the reduction conditions used in the x-ray diffraction experiments, the alumina could be present as >3 nm particles of $FeAl_2O_4$, but no Fe^{2+} was observed after more severe reduction. Thus these findings ruled out a random distribution of $FeAl_2O_4$ molecules in αFe, and the authors suggest that the strain observed in the x-ray diffraction patterns could be produced by large occluded clusters which are probably pure Al_2O_3 after severe reduction.

Oxidation/Reduction Catalysts

Since the Mössbauer effect can readily follow changes in oxidation state of Fe, Sn, and Eu, studies of electron exchange with these elements in catalysts for oxidation/ reduction reactions is a natural application for Mössbauer spectroscopy. Firsova et al. (58), for example, have used the Mössbauer effect in ^{119}Sn to study the behavior of supported tin molybdate (Sn/Mo = 1) as a catalyst for partial oxidation of propylene. The room temperature spectrum, Figure 4a, of the catalyst aged in vacuum at 723 K was similar to that of SnO_2, indicating that the tin was present as Sn^{4+}. Adsorption of propylene at 473K did not alter the Mössbauer spectrum of the catalyst, e.g. did not reduce the Sn^{4+}. If the catalyst was first exposed to oxygen at 473 K and then to propylene at the same temperature, however, the spectrum shown in Figure 4b resulted. The doublet with IS = 3.33 mm/sec with respect to SnO_2 and QS = 2.45 mm/sec shows clearly that in this case some of the tin was reduced to Sn^{2+}. Comparison with model compounds suggested that Sn^{2+}-O-C bonds were formed. In contrast to the behavior of the catalyst after 473 K adsorption, exposure of the catalyst to propylene at 673 K, without preadsorbed oxygen, gave a spectrum similar to Figure 4b. These results are particularly interesting in view of the accompanying finding that supported tin, in the absence of Mo, was not reduced by similar treatment. The direct observation of reduction of Sn^{4+} to Sn^{2+} in this work supports the proposal reviewed by Margolis (59) that the

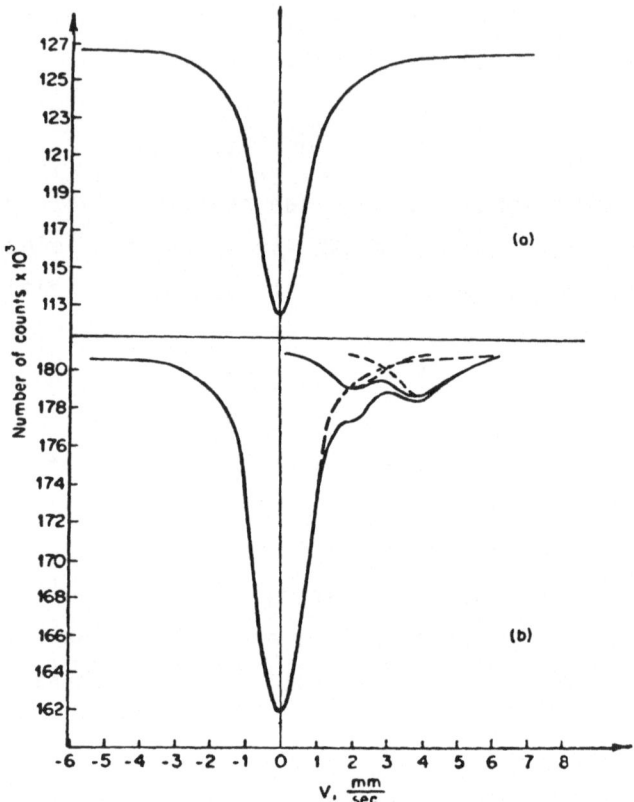

Figure 4.   Room temperature ^{119}Sn Mössbauer spectra of a
supported tin molybdate catalyst: (a) initial sample; (b)
after adsorption of oxygen, then propylene at 473 K.
From (59).

second cation in molybdate partial oxidation catalysts under-
goes synergistic oxidation/reduction with Mo cations during
catalysis. For Sn-Sb oxide on the other hand, Mössbauer
spectra of both ^{119}Sn and ^{121}Sb show that the presence of
Sn^{4+} in the oxide matrix enhances Sb^{5+}/Sb^{3+} ratio and that

adsorption of propylene at 473 K, with or without oxygen, reduces the Sb^{5+}/Sb^{3+} ratio but does not affect the oxidation state of the tin (60).

Experiments such as these illuminate important facets of catalyst chemistry not easily discovered by other techniques. One can move still closer to catalytically important chemistry by collecting Mössbauer data during chemical reaction, as shown recently for iron-doped cobalt molybdate (61). When oxidation/reduction reactions proceed by a regenerative sequence of elementary kinetic steps involving sequential oxidation/reduction of cations on the surface (62), the composition of the reaction mixture controls the degree of oxidation of the surface at steady state. Comparison of direct Mössbauer measurements of the oxidation state of the surface as a function of gas composition to values predicted from quantitative analysis of the kinetics then provides a consistency check for a reaction model. Such agreement has not been found between measured and predicted values for Eu^{2+}/Eu^{3+} ratios during the $CO_2 + H_2 \rightarrow CO + H_2O$ reaction over Eu/Al_2O_3 and Eu-M, indicating that if this reaction occurs by a regenerative sequence over these catalysts the number of active sites is a small fraction of the total number of Eu atoms in the catalyst (63,64). Ultimately, proof that spectroscopic data has true catalytic significance requires direct coupling of the spectroscopic result to the kinetics of the reaction, a job often done best at unsteady state (65,66).

CONCLUSIONS

A combination of techniques is usually required to un-

ravel the complexities of a catalytic problem. Through its
chemical parameters, the Mössbauer effect can contribute to
the solution of selected problems by revealing catalyst
chemistry and structure. The unique advantages of this tool
lie in its ability to monitor one element in a complex mixture
and to provide its wealth of information in situ, at or near
reaction conditions. The limited number of Mössbauer isotopes
restricts application, but studies of ^{57}Fe, ^{119}Sn, ^{121}Sb,
^{151}Eu, and more recently ^{99}Ru (28) catalysts have yielded
fruitful results. ^{197}Au and ^{193}Ir should also be useful
isotopes for catalytic work in spite of the fact that, as
with ^{99}Ru, data must be collected at liquid He temperatures.
Use of ^{57}Fe to probe non-iron catalysts (7) promises to extend
the application range still further. Limitations of the technique due to the fact that it is not surface-specific have not
been severe, partly because many catalytic materials have high
surface to volume ratios. Applications to date have included
bulk and surface characterization of catalyst precursors,
pretreated catalysts, and catalysts in the presence of chemi-
sorbing or reacting gases. The present maturity of the field
is well represented by the elegance with which answers to
important catalytic questions have been extracted from
Mössbauer spectra in the following papers of this symposium.

ACKNOWLEDGEMENTS

It is a pleasure to thank R. L. Garten for critical
comments and E. H. Delgass for editorial assistance in preparation of this manuscript.

REFERENCES

1. P. A. Flinn, S. L. Ruby, and W. L. Kehl, Science, 143, 1434 (1964).
2. H. M. Gager and M. C. Hobson, Jr., Catal. Rev-Sci. Eng., 11, 117 (1975).
3. J. A. Dumesic and H. Topsøe, Adv. in Catal., to be published.
4. O. M. Poltorok and V. S. Boronin, Intern. Chem. Engin., 7, 452 (1976).
5. J. Müller, Rev. Pure and Appl. Chem., 19, 151 (1969).
6. C. H. Bartholomew and M. Boudart, J. Catal., 29, 278 (1973).
7. R. L. Garten and D. F. Ollis, J. Catal., 35, 232 (1974).
8. R. L. Garten and M. A. Vannice, J. Molecular Catal., in press.
9. R. L. Garten, J. Catal., in press.
10. J. M. Blakely and J. C. Shelton, in Surface Physics of Materials, Vol. I, J. M. Blakely,Ed., Academic Press, New York, 1975, p. 189.
11. R. Bauman, G. J. M. Lippits and W. M. H. Sachtler, J. Catal., 25, 350 (1972).
12. F. L. Williams and M. Boudart, J. Catal., 30, 438 (1973).
13. L. D. Schmidt, Cat. Rev.-Sci. Eng., 9, 115 (1974).
14. E. K. Rideal, Concepts in Catalysis, Academic Press, New York, 1968, p. 41.
15. R. Gomer, Solid State Physics, 30, 93 (1975).
16. J. W. Gadzuk, in Surface Physics of Materials, Vol. II, J. M. Blakely, Ed., Academic Press, New York, 1975, p. 339.
17. R. Van Hardeveld and F. Hartog, Surf. Sci., 15, 189 (1969).
18. M Boudart, Adv. in Catal., 20, 153 (1969).
19. R. C. Baetzold and R. E. Mack, J. Chem. Phys,62,1513(1975).
20. K. Johnson and R. P. Messmer, J. Vac. Sci. Technol., 11, 236 (1974). J.G. Fripiat, K.T. Chow, M. Boudart, J.G. Diamond, K.H. Johnson, J. Molecular Catal., 1, 59 (1975).
21. G. A. Somorjai, Catal. Rev., 7, 87 (1972).
22. A. N. Karasev, Yu. A. Kolbanovskii, L. S. Polak, and E. Sh. Shlikhter, Kinet. i Katal., 8, 232 (1967).
23. C. F. Cook, P. R. Gray and H. M. Barton, Jr., Proceedings of The Second Symposium on Low Energy X- and Gamma Sources and Applications, ORNL-11C-10, p. 130 (1967).
24. I. P. Suzdalev, A. M. Afanas'ev, A. S. Plachinda, V. I. Gol'danskii, and E. F. Makarov, Soviet Phys. JETP, 28, 923 (1969).

25. I. W. Burton, R. P. Godwin, and H. Frauenfelder,
    Applications of the Mössbauer Effect in Chemistry and
    Solid State Physics (Vienna 1966), IAEA Tech. Rep.
    Ser. 50, p. 73, and I. W. Burton and R. P. Godwin,
    Phys. Rev., 158, 218 (1967).
26. G. W. Simmons, E. Kellerman, and H. Leidheiser, Jr.,
    Corrosion, 29, 227 (1973).
27. W. N. Delgass, L.-Y. Chen and G. Vogel, Rev. Sci.
    Instrum., 47, 136 (1976).
28. C. A. Clausen, III, and M. L. Good, J. Catal., 38, 92
    (1975).
29. P. N. Ross, Jr., and W. N. Delgass, in Catalysis, Vol.1,
    J. W. Hightower, Ed., North Holland Publishing Co.,
    Amsterdam, (1973) p. 597.
30. H. M. Gager, Ph.D. Thesis, Virginia Commonwealth
    University (1972).
31. L. V. Skalkina, I. P. Suzdalev, I. K. Kolchin and
    L. Ya. Margolis; Kinet. i Katal., 10, 456 (1969).
32. M. C. Hobson, Jr., Nature, 214, 79 (1967).
33. I. P. Suzdalev and E. F. Makarov, Proceedings of the
    Conference on the Application of the Mössbauer Effect
    (Tihany, 1969), Akademiai Kiado, Budapest, 1971, p. 201.
34. I. P. Suzdalev, A. S. Plachinda, and E. F. Makarov,
    Soviet Phys. JETP, 26, 897 (1968).
35. H. Topsøe and M. Boudart, J. Catal., 31, 346 (1973).
36. E. A. Samuel and W. N. Delgass, J. Chem. Phys., 62,
    1590 (1975).
37. P. N. Ross, Jr. and W. N. Delgass, J. Catal., 33, 219
    (1974).
38. G. A. Somorjai, Principles of Surface Chemistry, Prentice
    Hall, Englewood Cliffs, N.J., 1972, p. 99.
39. K. S. Singwi and S. A. Sjölander, Phys. Rev., 120,
    1093 (1960).
40. V. I. Gol'danskii and I. P. Suzdalev, Russian Chem. Rev.,
    39, 609 (1970).
41. W. Kundig, H. Bömmel, G. Constabaris, and R. H. Lindquist,
    Phys. Rev., 142, 327 (1966).
42. I. P. Suzdalev, Proceedings of the Conference on the
    Application of the Mössbauer Effect (Tihany, 1969),
    Akademiai Kiado, Budapest, 1971, p. 193.
43. H. Hobert and D. Arnold, ibid, p. 325.
44. A. M. Rubashov, P. B. Fabrichnyi, B. V. Shakhov and A.
    M. Babeshkin, Zh. Fiz. Khim., 46, 1327 (1972).
45. Y.-Y. Huang and J. R. Anderson, J. Catal., 40, 143 (1975).
46. F. A. Fortunato and W. N. Delgass, unpublished results.

47. J. A. Dumesic, H. Topsøe, S. Khammouma, and M. Boudart, J. Catal., 37, 503 (1975).
48. R. L. Garten, W. N. Delgass, and M. Boudart, J. Catal., 18, 90 (1970).
49. M. Boudart, R. L. Garten and W. N. Delgass, Memoires de la Soc. Roy. Sc. Liège, 6ᵉ Série, I(4), 135 (1971).
50. Y. Y. Huang, J. E. Benson, and M. Boudart, Ind. Eng. Chem. Fundam., 8, 346 (1969).
51. R. L. Garten, J. Gallard-Nechtschein and M. Boudart, Ind. Eng. Chem. Fundam., 12, 299 (1973).
52. E. A. Samuel, Ph.D. Thesis, Yale University (1973).
53. H. Topsøe, J. A. Dumesic and M. Boudart, J. Catal., 28, 477 (1973).
54. P. H. Emmett and S. J. Brunauer, J. Amer. Chem. Soc., 59, 1553 (1937) and 62, 1732 (1940).
55. R. Hosemann, A. Preisinger, and W. Vogel, Ber. der Bunseng, 70, 796 (1966).
56. R. Hosemann, K. Lemm, A. Schonfeld, and W. Wilke, Kolloid-Z. Z. Polym. 216-217, 103 (1967).
57. R. Hosemann, Chem. Ing. Tech., 42, 1252 and 1325 (1970).
58. A. A. Firsova, N. N. Khovanskaya, A. D. Tsyganov, I. P. Suzdalev, and L. Ya. Margolis, Kinet. i Katal., 12, 792 (1971).
59. L. Ya. Margolis, J. Catal., 21, 93 (1971).
60. I. P. Suzdalev, A. A. Firsova, A. U. Aleksandrov, L. Ya. Margolis, and D. A. Baltrunas, Dok. Akad. Nauk. SSSR., 204, 408 (1972).
61. Yu. V. Maksimov, I. P. Suzdalev, V. I. Gol'danskii, O. V. Krylov, L. Ya. Margolis, and A. E. Nechitailo, Chem. Phys. Letters, 34, 172 (1975).
62. G. K. Boreskov, Kinet. i Katal., 11, 374 (1970).
63. D. S. Shihabi and W. N. Delgass, unpublished results.
64. L.-Y. Chen, Ph.D. Thesis, Purdue University (1975).
65. K. Tamaru, Adv. in Catal., 15, 65 (1964).
66. R. J. Kokes, Cat. Rev., 6, 1 (1972).

MÖSSBAUER SPECTROSCOPY STUDIES OF THE STATE OF HETEROGENEOUS CATALYSTS DURING CATALYSIS[†]

J. A. Dumesic[*], Yu. V. Maksimov[**] and
I. P. Suzdalev[**]
Stanford University, Stanford, CA, USA

Institute of Chemical Physics, Moscow, USSR

ABSTRACT

The magnetic state of a metallic catalyst for the ammonia synthesis and the electronic state of oxide catalysts for the propylene and methanol oxidations were investigated using Mössbauer spectroscopy during the respective catalytic reactions. The experiments were carried out in cells which served simultaneously as catalytic reactors and units for Mössbauer spectroscopy -- "Mössbauer catalytic reactors."

The metallic catalyst for the ammonia synthesis consisted of small iron particles (ca. 5 nm in size) supported on MgO. The superparamagnetic relaxation frequency for these particles was found to be sensitive

----

[*]Permanent address: University of Wisconsin
Department of Chemical Engineering
Madison, Wisconsin 53706 USA

[**]Permanent address: Institute of Chemical Physics
Academy of Sciences of the Soviet Union
Moscow, USSR

[†]Written while one of us (JAD) was a fellow in the USA-USSR exchange program in chemical catalysis; Stanford University (USA) - Institute of Chemical Physics (USSR)

to both hydrogen chemisorption on the particles and to treatments of the particles which were shown to change their catalytic properties. These results can be understood in terms of Néel's phenomenological theory of magnetic-surface anisotropy; in addition, surface iron atoms with seven nearest neighbors seem to be particularly active for the ammonia synthesis.

For the soft oxidation of propylene and methanol, cobalt molybdate containing minor amounts of iron and non-stoichiometric iron molybdates were used as catalysts. In both cases, the iron initially entered the catalyst structure mainly in the trivalent state. During both catalytic processes, however, additional, partially reduced forms of iron were detected, and these forms disappeared upon termination of the reaction. These electronic states of iron can be treated in terms of vacancy models.

INTRODUCTION

In recent years Mössbauer spectroscopy has proven to be an effective method for the investigation of various surface phenomena, such as the dynamics of atom vibrations on surfaces, the electronic and magnetic states of surface atoms, and the interaction of adsorbed species with solid surfaces. Detailed analyses of these results can be found in a number of surveys (1-5). In addition Mössbauer spectroscopy, along with other physical methods for investigating the structure of solids, has penetrated deeply into the problems of real catalyst systems, e.g. problems dealing with catalyst preparation, modification and genesis. Indeed, the above approaches have been very successful and significant advances will continue to be made.

These types of studies, however, are typically carried out (1) either before or after the sample is actually used as a catalyst in a given chemical process, or (2) with the sample exposed to a given adsorbing gas. In the first case, conclusions about the surface state (electronic and geometric) of the catalyst during catalysis carry mostly the character of more or less well-grounded extrapolations; and, in the second case, those forms of surface compounds observed during chemisorption experiments alone should be

identified with intermediate states of catalysis with
extreme caution. Therefore, in both cases the conclusions
reached must be confirmed by investigations involving
observation of the state of the catalyst directly during
the associated catalytic reaction.

The idea behind these experiments on observing the
state of catalysts during catalysis is very simple.
However, its realization meets with certain difficulties,
the most serious of which are: (1) the selection of an
active (possibly highly dispersed) catalyst that contains a
Mössbauer atom which either directly participates in or is
a "witness" of the catalytic process; (2) designing a unit
that is simultaneously a catalytic reactor and a cell for
Mössbauer spectroscopy; and (3) selection of a reaction
with a sufficiently high rate, the products of which do
not accumulate on and poison the surface during prolonged
contact.

In the present paper we discuss results from our two
laboratories of investigations during catalysis of several
catalyst systems. Specifically, the magnetic state of
metallic catalysts during the ammonia synthesis (studied
at Stanford University) and the electronic states of oxide
catalysts during oxidation reactions (studied at the
Institute of Chemical Physics) will be discussed. It is
hoped that these examples will then illustrate most of the
concepts outlined above.

MEASUREMENT OF THE STATE OF CATALYSTS DURING CATALYSIS

The problem statement for the design of Mössbauer
spectroscopy cells for the study of surface and chemisorption phenomena is the following: to design an in situ cell
which has "windows" transparent to the γ-radiation, and
which can operate at temperatures of interest for surface
studies (often this means high temperatures, e.g. 700 K)
and of interest for observation of the various hyperfine
interactions (this may mean low temperatures, e.g. 77 K).
The construction of cells of this type has recently been
discussed elsewhere (5). For the study of catalytic
phenomena during catalysis, however, the additional
constraint is imposed that the "dynamic state" of the
catalyst be well defined. In general, this dynamic state

is specified by (1) the concentrations of the reactants and products of the catalytic process, (2) the mode of operation of the catalytic reactor (e.g. integral or differential), and (3) the flow rate of the reactant gas mixture through the reactor.

For illustration, a simple system for the study of catalytic processes at temperatures higher than 300 K is shown in Fig. 1. The "Mössbauer catalytic reactor" is cylindrically shaped, and the two ends of this cylinder serve as the γ-ray windows. The entire cell is made of pyrex glass allowing studies to be made at temperatures up to ca. 700 K, and the windows for γ-ray transmission are ca. 150 μm thick. During operation the catalyst, in powder form, lies in the cell as a thin layer over the cylinder base with the cell placed in a vertical γ-ray beam (see Fig. 1); in this position the reactant gas mixture flows horizontally over the thin catalyst layer.

Connected to the reactor is the gas handling system which supplies the reactant gas mixture. In one mode of operation (simple flow operation) the reactant gas mixture passes through the reactor only once, and after this single contact with the catalyst the gas mixture is analyzed to obtain the product concentrations. During this single pass, a measurable conversion of reactants to products must take place; therefore, the overall reaction rate and the catalyst surface coverages by reaction intermediates at given temperature are dependent on the gas phase concentration gradients within the reactor, e.g. integral flow reactor. In the study of the methanol oxidation reported later in this paper, the Mössbauer catalytic reactor was that illustrated in Fig. 1, and it was operated in the simple flow mode.

In a second mode of operation (as for the propylene oxidation study reported in this paper) the reactant gas mixture is brought into contact with the catalyst many times using a circulation pump (Fig. 1), but during each contact the extent of conversion of reactants to products is very small. After each pass of the reactants over the catalyst the products formed are frozen out of the gas phase using appropriate cryogenic traps, and the total pressure in the circulation loop is measured using a manometer (Fig. 1). To begin operation, the circulation loop is filled

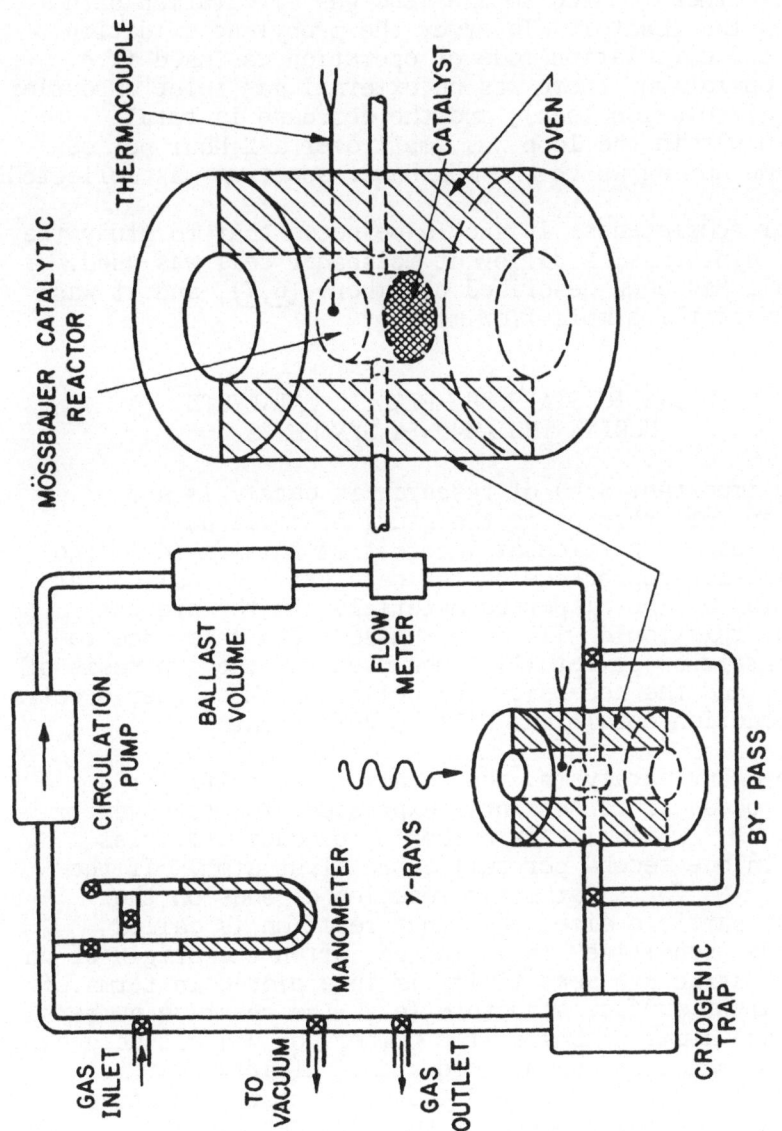

Figure 1   "Mössbauer Catalytic Reactor" and Circulation System

with the reactant gas mixture and the gas circulation is
started with the catalytic reactor by-passed. The stop-
cocks are then rotated so that the gas circulation path
includes the reactor. To study the propylene oxidation,
the closed-circulation mode of operation was used (i.e.
during operation, there was no external gas inlet or outlet
to the circulation loop), and the decrease in total
pressure within the loop was small over a 2 hour period
(the time during which the Mössbauer spectrum was collected).

For completeness it should be noted that to study the
ammonia synthesis, a different Mössbauer cell was used.
This cell has been described elsewhere (6,7), and it was
operated in the simple flow mode.

MAGNETIC STATE OF METALLIC CATALYSTS
DURING THE AMMONIA SYNTHESIS

An important area of research in catalysis and
Mössbauer spectroscopy is the study of small particle
systems, and of particular interest in this field is the
particle size dependence of the catalytic and solid state
properties of the dispersed material. In the present
section, this topic will be discussed with reference to
small particles of metallic iron used for the synthesis of
ammonia, and the necessity for collecting Mössbauer spectra
under reaction conditions will be demonstrated.

For many catalysts, the catalytic activity for a
given process is conveniently expressed as a turnover
number, N, which is the number of product molecules
formed in one second per surface catalyst atom. If the
value of N for a catalytic reaction depends on the
catalyst particle size, then that reaction is called
"structure sensitive" (8). Indeed, often the origin of an
observed structure sensitivity is interpreted in terms of
the catalyst surface structure (e.g. the relative numbers
of surface atoms with specific numbers of nearest neighbors),
and its dependence on particle size. Clearly what is
required to fully understand these structure sensitive
reactions is a measurement of the catalyst surface
structure under reaction conditions.

Toward the realization of this goal, relations are
sought between the surface structure and the Mössbauer

parameters. Indeed, the Mössbauer parameters for "surface" and "bulk" atoms may be different (5); unfortunately, the quantitative prediction of these differences and the effects of surface structure thereon is not yet possible in most cases. However, those Mössbauer parameters which are sensitive to the symmetry of the surface may be quite useful for the measurement of surface structure changes. For example, changes in the anisotropy of the recoil-free fraction (as determined in the Goldanskii-Karyagin effect) and the quadrupole splitting for the "surface" atoms may be related to surface symmetry changes if other variables, such as the chemisorption coverage and particle size, do not change appreciably. In addition, the superparamagnetic relaxation time for small particles may be sensitive to the surface symmetry, as will be seen below.

The above concepts of structure sensitivity and surface structure measurement using Mössbauer spectroscopy are illustrated by the ammonia synthesis over metallic iron. This reaction was carried out at atmospheric pressure in the temperature range between 570 and 700 K over metallic iron particles supported on MgO. For these particles after H_2 reduction, the ammonia synthesis turnover number, N, was found to decrease with decreasing particle size, d, i.e. the reaction was structure sensitive (9). However, by pretreating the small particles (less than 5 nm in size) with ammonia, so as to form iron nitride, followed by decomposition of this nitride in a $H_2:N_2$ gas mixture, the value of N was increased compared to that value before the treatment. Subsequently, treatment of the particles in pure H_2 completely erased the effect of the ammonia treatment. Finally, for the large iron particles (greater than ca. 10 nm in size) the ammonia and the H_2 treatments had no effect on N. Thus, the origin of the structure sensitivity (change of N with d) appeared to be related to the understanding of the ammonia and H_2 treatments. Since these treatments led to surface structure changes at constant d, they were well suited for study using Mössbauer spectrocopy.

It was shown in separate experiments that the superparamagnetic relaxation time, τ_{sp}, for small metallic iron particles (less than ca. 5 nm in size) was sensitive to the surface (10). Specifically, the fraction of the metallic iron that appeared magnetically hyperfine split

in the Mössbauer spectrum at 670 K depended whether the
sample was in flowing H_2 or He; the magnetic susceptibility
of the particles was found to be sensitive at low temperatures (77 K) to hydrogen chemisorption on the particles;
and, at higher temperatures (greater than ca. 300 K) the
magnetic susceptibility was the same in H_2 and He. At
670 K for Mössbauer spectroscopy and at 77 K for magnetic
susceptibility, τ_{sp} was of the order of the time scale
for the experimental measurement, τ_{exp} (ca. 10^{-8}s for
Mössbauer spectroscopy and 1 s for magnetic susceptibility);
however, at 300 K for magnetic susceptibility this condition
was not met. Thus, for $\tau_{sp} \sim \tau_{exp}$, an effect of hydrogen
chemisorption was observed while for $\tau_{sp} < \tau_{exp}$ no such
effect was found. These results point to a surface
sensitive value of τ_{sp}.

Mössbauer spectra were then collected for the small
iron particles after the ammonia and H_2 treatments, with
the catalyst under reaction conditions. This latter point
must be stressed, because the catalytic properties of these
particles are sensitive to the history and gaseous environment of the sample. It was thereby seen that the ammonia
treatment, which increases N, decreases the metallic iron
hyperfine split spectral area, A_{hfs}, at a given temperature during the reaction, as seen in Fig. 2. This result,
due to superparamagnetic relaxation (11), rules out
sintering of the particles during the ammonia treatment
(since an increase in d would produce an increase in τ_{sp}
and thus also an increase in A_{hfs}), and instead points
to a decrease in τ_{sp} after the ammonia treatment.
Subsequent H_2 treatment of the particles, which decreased
N to its original value, was then found to increase A_{hfs}
to its original value, as determined by again collecting
the Mössbauer spectra under reaction conditions. Thus, a
correspondence was found between N and τ_{sp}, i.e. as N
increased, τ_{sp} decreased for constant d.

Since it was previously shown that τ_{sp} was surface
sensitive, Néel's phenomenological theory of magneto-surface
anisotropy (12,13) was used to interpret these results.
The surface structure was first expressed by the relative
concentrations of C_i surface atoms, where a C_i atom is
one with i number of nearest neighbors. The correspondence
between N and τ_{sp} then led to a correlation between N
and C_i, with a C_7 surface atom being particularly active

for the ammonia synthesis (11). In addition, because C_7 atoms are not normally characteristic of small particle surfaces (14), the correlation between N and C_7 also explains the observed structure sensitivity (i.e. the correlation between N and d).

The possible application of Mössbauer spectroscopy for surface structure measurement can now be seen. The interpretation of catalytic properties in terms of the measured surface structure, however, is only meaningful when the latter is determined under reaction conditions.

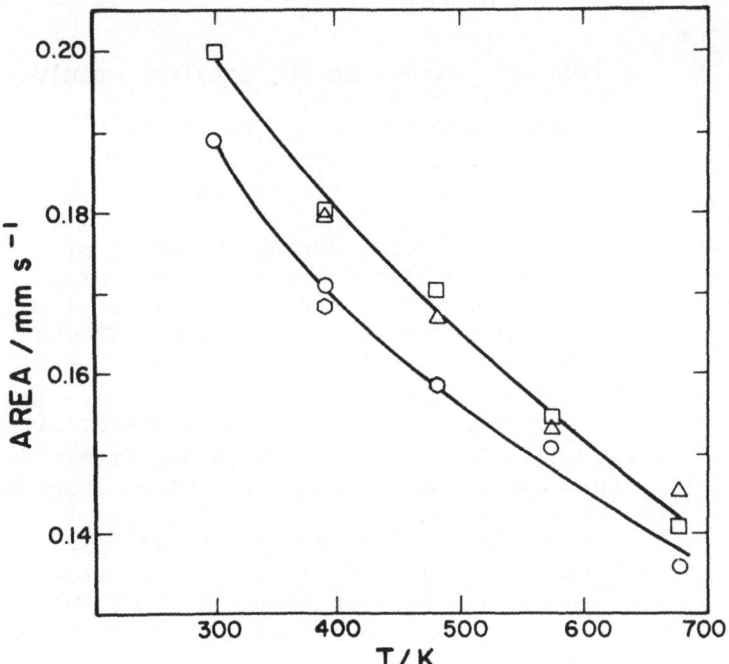

Figure 2   Hyperfine Split Spectral Area Versus Temperature
            for Ammonia Synthesis Catalyst
Pretreatment Sequence: ◻ H_2 reduction, ○ NH_3, △ H_2, ◯ NH_3

ELECTRONIC STATES OF OXIDE CATALYSTS
DURING OXIDATION REACTIONS

Our first studies of the catalytic state during catalysis for oxide systems were carried out with non-supported catalysts for partial propylene and methanol oxidations (15,16).

In the reaction of the soft propylene oxidation to acrolein, the catalyst was $CoMoO_4$ doped with 3 at. % ^{57}Fe. It is important to note that the addition of minor amounts of iron to $CoMoO_4$ not only increased the catalytic activity, but also improved the selectivity to acrolein of the reaction. Consequently, the iron ions played an active role in the formation of the catalytically active solid. The experimental results for this system have been published elsewhere (15), and the main results of this work are the following:

(i)   The iron was present in the catalyst mainly in the form of high spin, octahedrally coordinated Fe^{3+} ions substituted for cobalt in the $CoMoO_4$ lattice (δ (300 K) = 0.74 ± 0.07 mm s^{-1} relative to SNP).

(ii)   Along with this form, during the study of the catalyst in static air at temperatures between 80 and 820 K, a reduced form of iron with relative content, R, of about 4% was observed in the sample (δ (580 K) = 1.2 ± 0.2 mm s^{-1}, ΔE_Q (580 K) = 2.2 ± 0.2 mm s^{-1}). This state probably arose due to the transfer of an electron to the Fe^{3+} ion from a negatively charged imperfection of the structure (e.g. a vacancy, V) formed by removal of oxygen ions from the lattice during the catalyst preparation. This reduced complex of iron [Fe^{2+}---V] may thus be termed a "biographical" donor state (i.e. a state with an electron localized on it), and this state may contribute to the conductivity of the system.

(iii)   After the propylene oxidation reaction was started (580 K, C_3H_6/O_2 = 1/10, 1 atm, circulation velocity = 10 ℓ h^{-1}) the relative content of [Fe^{2+}---V] increased

more than two times; termination of the
reaction restored the initial state of the
catalyst (R = 4%).

In accord with ESR data, propylene adsorption on Mo^{6+}
is accompanied by charge transfer, leading to the
propylene activation and a $Mo^{6+} \rightarrow Mo^{5+}$ reduction (17).
This means the generation of a new donor state localized
on molybdenum. Simultaneously, near the Fe^{3+} ion the
formation of a neutral vacancy takes place connected
with the process of water removal. The thermally
activated jump of an electron from Mo^{5+} to the Fe^{3+}
leads to the formation of new "catalytic" states
$[Fe^{2+}---V]$. These states persist only during the
catalytic reaction, and consequently they appear to act
as agents of charge transfer to the adsorbing oxygen
molecules.

For the methanol oxidation to formaldehyde, the
catalyst studied was non-stoichiometric iron molybdate
with the ratio of Mo/Fe = 2 (18). X-ray studies showed
that some lattice spacings of the pseudorhombic crystal
were appreciably different from those calculated for the
stoichiometric compound (19). In accord with Pernicone
(20), this may be explained by the presence of an excess
of O^{2-} ions in the bulk iron molybdate. The oxygen
packing in $Fe_2(MoO_4)_3$ is indeed quite "loose", and it is,
therefore, possible to find interstitial positions large
enough to accommodate the excess O^{2-} ions.

Mössbauer spectra of the catalyst were collected at
temperatures between 530 and 590 K under the following
conditions: (1) in static air, (2) in flowing air, (3) in
vacuum after vacuum treatment of the sample at 670 K for
2 h, and (4) in a flowing methanol-air mixture (i.e. under
reaction conditions).

Figure 3A shows the spectrum of the catalyst in static
air at 560 K. The broad single line with isomer shift
equal to 0.60 ± 0.06 mm s^{-1} is due to high spin Fe^{3+} ions
in the octahedral sites of the oxygen lattice (19). An
additional, low intensity line at ca. 1.5 mm s^{-1} seems to
correspond to the right component of a spectral doublet
arising from a new iron complex (δ = 0.8 ± 0.2 mm s^{-1} and
ΔE_Q = 1.4 ± 0.2 mm s^{-1}). At temperatures between 530 and
590 K the fraction of the iron present in this state, R',

is about 6%; at 490 K this fraction is less than 5%; and, at 300 K this new complex is not observable in the Mössbauer spectrum.

The spectrum of the catalyst after vacuum treatment and also in flowing air is shown in Fig. 3B. Compared to the case of static air, R' has now decreased more than two times. Under reaction conditions for the methanol oxidation, the spectrum of the catalyst is qualitatively similar to that in Fig. 3A; however, for this case the value of R' increases with increasing temperature from 6% at 530 K up to 11% at 590 K. Termination of the reaction restored the initial state of the catalyst.

The changes in the local symmetry of the Fe^{3+} ion upon formation of the new complex before the methanol

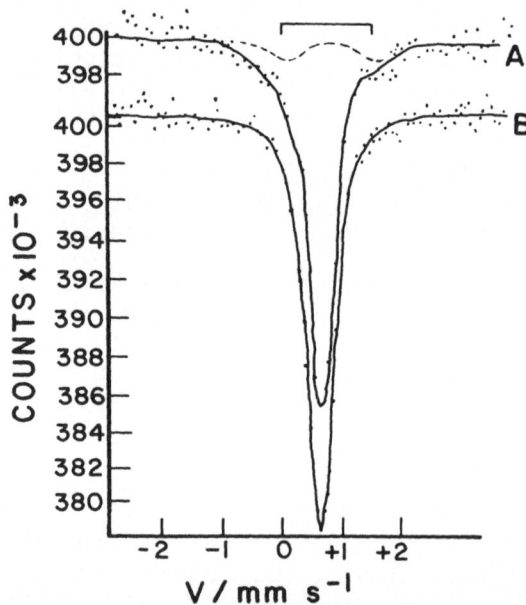

Figure 3   Mössbauer Spectra at 560 K of Methanol Oxidation
        Catalyst in (A) static air and (B) in vacuum

reaction may be again associated with the interaction of iron atoms with neighboring vacancies. However, for this case, in view of the Mössbauer parameters, the vacancy may be neutral, as formed for example by the process of dehydroxylation of the catalyst pre-surface layer (20):

$$2 \ OH^- \underset{-1}{\overset{1}{\rightleftharpoons}} O^{2-} + H_2O \uparrow + V$$

The temperature dependence for the formation of the complex $[Fe^{3+}---V]$ in static air then seems to reflect the temperature dependencies of the rates of reactions 1 and -1. The processes for the destruction of vacancies may be connected with (i) the hydroxylation of the surface during the flowing air treatment (reaction -1), and (ii) the diffusion of non-stoichiometric oxygen ions from the bulk of the material to the surface (followed by vacancy annihilation) under the high temperature vacuum treatment of the sample.

In accord with the vacancy model for methanol oxidation (20), methanol adsorption on neutral vacancies localized near Mo^{6+} is accompanied by the generation of new neutral vacancies near the iron and negatively charged vacancies near the molybdenum. Therefore, after the reaction is started the steady state concentration of neutral vacancies near the Fe^{3+} ions may increase, thereby producing an increase in the spectral signal from $[Fe^{3+}---V]$. Thus, for the partial methanol oxidation over iron molybdate, one may again distinguish two sorts of surface defects: (1) "biographical" vacancies formed probably during dehydroxylation of the catalyst at elevated temperatures, and (2) "catalytic" defects which again appear only during the catalytic process.

CONCLUDING REMARK

We would like to close by noting that these first results of the investigation of the magnetic and electronic states of catalysts during catalysis carry mainly the character of qualitative or semi-quantitative observations. The next step is to carefully correlate the behavior of these states with the velocity of reaction (the turnover number, N) measured during the collection of the Mössbauer spectrum.

REFERENCES

1. Goldanskii, V. I., and Suzdalev, I. P., Proc. Conf.
    Appl. Mössbauer Effect, Tihany (Hungary), p. 269,
    1969.

2. Hobson, M. C., Jr., Surface Membrane Sci. 5, 1 (1972).

3. Gager, H. M., and Hobson, M. C., Jr., Catal. Rev. 11,
    117 (1975).

4. Goldanskii, V. I., Maksimov, Yu. V., and Suzdalev,
    I. P., Proc. Fifth Inter. Conf. Mössbauer
    Spectroscopy, Cracow (Poland), 1975.

5. Dumesic, J. A., and Topsøe, H., Adv. Catal., in press.

6. Khammouma, S., Ph.D. Dissertation, Stanford University,
    1972.

7. Boudart, M., Delbouille, A., Dumesic, J. A.,
    Khammouma, S., and Topsøe, H., J. Catal. 37,
    486 (1975).

8. Boudart, M., Robert A. Welch Foundation
    Conferences on Chemical Research. XIV.
    Solid State Chemistry, Milligan, W. O. (Ed.),
    Houston, Texas, p. 299, 1971.

9. Dumesic, J. A., Topsøe, H., Khammouma, S., and
    Boudart, M., J. Catal. 37, 503 (1975).

10. Boudart, M., Dumesic, J. A., and Topsøe, H.,
    in The Physical Basis of Heterogeneous
    Catalysis, Jaffee, R. (Ed.), Gstaad
    (Switzerland), 1975.

11. Dumesic, J. A., Topsøe, H., and Boudart, M.,
    J. Catal. 37, 513 (1975).

12. Néel, L., Compt. Rend. Acad. Sci. 237, 1468 (1953).

13. Néel, L., J. Phys. Radium 15, 225 (1954).

14. Van Hardeveld, R., and Hartog, F., Surface Sci.
    15, 189 (1969).

15. Maksimov, Yu. V., Suzdalev, I. P., Goldanskii, V. I.,
    Krylov, O. V., Margolis, L. Ya., and Nechitailo,
    A. E., Chem. Phys. Letters 34, N1,172 (1975).

16. Maksimov, Yu. V., Suzdalev, I. P., Goldanskii, V. I.,
    Matveev, A. I., Makarov, E. F., and Margolis, L. Ya.,
    Dokl. Akad. Nauk USSR, to be published.

17. Sancier, K. M., Aoshime, A., and Wise, H., J. Catal.
    34, 257 (1974).

18. Bibin, V. N., and Popov, B. I., Kinet. Katal.
    (Russian) 10, 1326 (1969).

19. Plasova, L. M., Klefsova, R. F., Borisov, S. V.,
    and Kefeli, L. M., Crystallography (Russian)
    12, 939 (1967).

20. Pernicone, N., J. Less Common Metals 36, 289
    (1974).

A MOSSBAUER INVESTIGATION OF A PLATINUM-TIN PARAFFIN DEHYDROGENATION CATALYST

Peter R. Gray and Floyd E. Farha

Phillips Petroleum Company

Bartlesville, Oklahoma

During the development of a steam active paraffin dehydrogenation catalyst, it was observed that the addition of tin to platinum supported on zinc aluminate had a dramatic effect on the stability and selectivity of the catalyst (1,2). Without tin the catalyst activity declined rapidly with time requiring frequent reactivation with an oxygen-containing gas to burn off accumulated coke. The role of the tin was not known. Tin on zinc aluminate is inactive for paraffin dehydrogenation, whereas platinum on zinc aluminate is an active, albeit short-lived, paraffin dehydrogenation catalyst.

Mossbauer spectroscopy offered the potential for determining the mechanism by which tin stabilizes this catalyst. Tin is supported on the surface in low concentrations, and the surface reactions can be studied by Mossbauer spectroscopy without interference or dilution by large concentrations of tin in the bulk of the material. The catalyst is used in the absence of air; therefore, reduced tin species may be stable and observable under the conditions of the catalytic reaction. Tin is one of the few elements which can be studied by Mossbauer spectroscopy at elevated temperatures making in situ studies possible. Finally, tin systems have been extensively studied by Mossbauer spectroscopy and interpretation of experimental spectra is facilitated (3).

Although most Mossbauer studies of tin systems have been made cryogenically or at temperatures only moderately above room temperatures (3), previous studies in our

laboratory had indicated that many tin species had a
significant recoil-free fraction at temperatures as high
as 800-900°C, e.g., stannic oxide (Figure 1). Other tin
materials, such as tin metal, which melts at 231°C could
not be observed at the temperatures of interest in this
study. Preliminary studies of the platinum-tin catalyst
showed that the tin could be seen in spectra made at 500-
600°C and offered the possibility of determining the

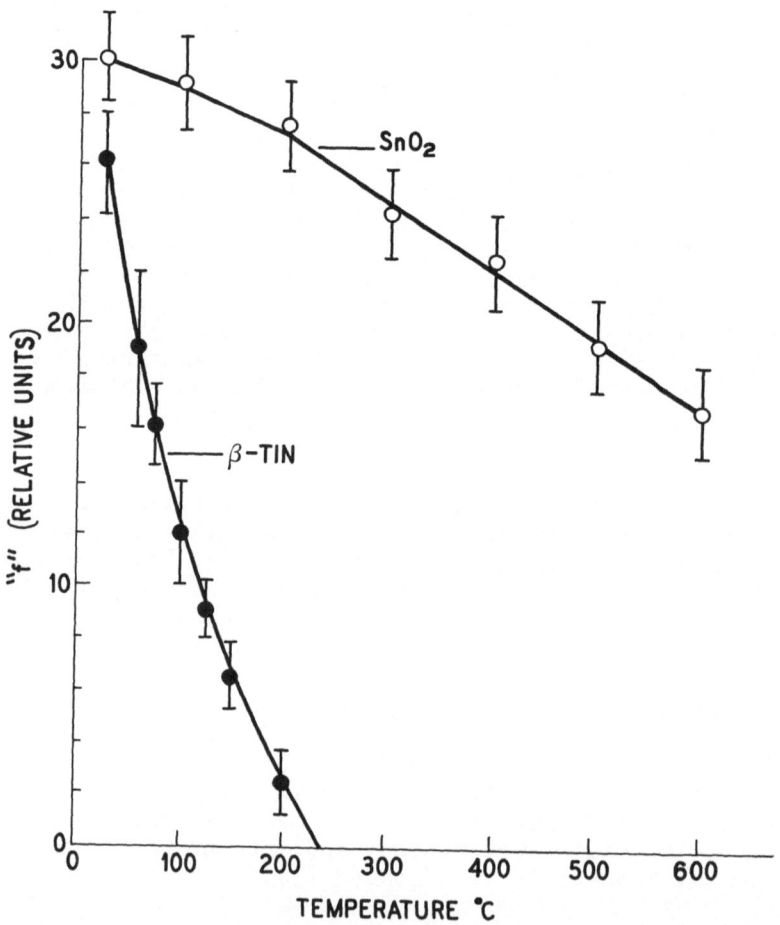

Figure 1. Relative Recoil-Free Fractions of SnO_2 and
β-Tin as a Function of Temperature.

environment of the tin under reaction conditions (butane and steam at 550°C).

Catalyst samples were prepared by impregnating zinc aluminate containing a small excess of zinc oxide with 0.4 weight per cent platinum and tin. Platinum was added by dissolving chloroplatinic acid in water. Metallic tin, enriched to 90 per cent tin-119, was dissolved in hydrochloric acid and diluted with water before mixing with the platinum solution. The characteristic platinum-tin complex was impregnated on 20-40 mesh zinc aluminate. After drying at low temperature, the catalyst was calcined in air at 600°C for several hours. No other activation of the catalyst was required before use.

In situ Mossbauer spectra of the catalyst were made in a hot cell shown in Figure 2. The stainless steel cell, with beryllium windows made leak tight with gold o-rings, holds a nominal 0.5 to 1 gram of catalyst. Reactant gases can be passed across and through the

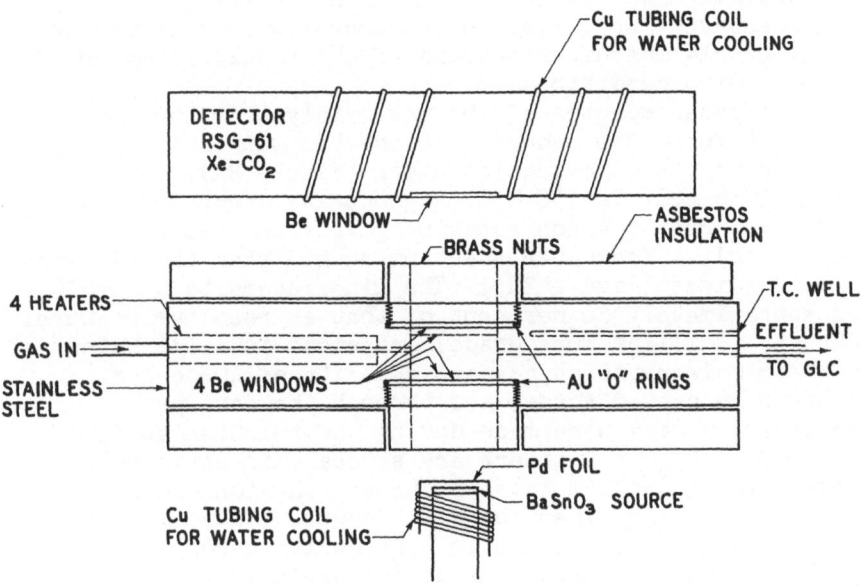

Figure 2. Hot Cell for Mossbauer Spectra of Tin Materials at Temperatures to 1000°C.

material and the exiting gases collected for analysis or
the cell can be directly coupled to a gas chromatographic
column. The cell is heated by four 400-watt cartridge
heaters which have the capacity of heating the cell to
temperatures in excess of 1000°C. A thermocouple is used
for temperature measurement and control. The cell is
mounted horizontally in the Mossbauer spectrometer and
7/16-inch sheets of pressed asbestos on the top and bottom
of the cell (with the exception of the beryllium windows)
shield the source and detector from heat radiation. The
20-mCi barium stannate source was further shielded from
the heat of the sample cell by a water-cooled 4-mil
palladium filter. This palladium filter was soldered to
a small cooling coil - the filter being kept cool by
conduction. Similarly, the proportional counter detector
was water cooled. These techniques were sufficient to
maintain both the source and the detector at temperatures
only nominally above ambient. Cooling was necessary
since relatively close geometry was required for making
spectra at these elevated temperatures where the recoil-
free fractions are small. Even with the closest geometry
practical with the hot cell, and with the 20-mCi source,
long Mossbauer runs, e.g., eight hours, were required to
obtain usable Mossbauer spectra of the catalytic material
under in situ conditions.

Mossbauer spectra of the new catalyst in the hot cell
at room temperature showed that the tin was entirely
oxidized to the +4 oxidation state (Figure 3A). After
heating the cell to 550°C in high purity argon (Matheson
Ultra High Purity Argon containing less than 10 ppm
impurities), a Mossbauer spectrum showed that the tin was
still oxidized (Figure 3B). The line intensity at 550°C
was approximately 60 per cent of that at room temperature.
N-butane (Phillips Pure Grade) and steam in a 1/6 butane
to steam mole ratio at a space velocity of 40 cc per
minute were passed across and through the catalyst.
Mossbauer spectra were made during the reaction at 550°C.
As shown in Figure 4A there are several tin environments
present in the active catalyst under reaction conditions.
This spectrum was taken for eight hours. The cell was
cooled to room temperature in high purity argon and
another Mossbauer spectrum made (Figure 4B). Except for
the slight shifts in the lines due to the difference in
temperature and some changes in the relative intensities
the spectra are very similar. The intensity differences
are expected since the spectrum made at 550°C is an

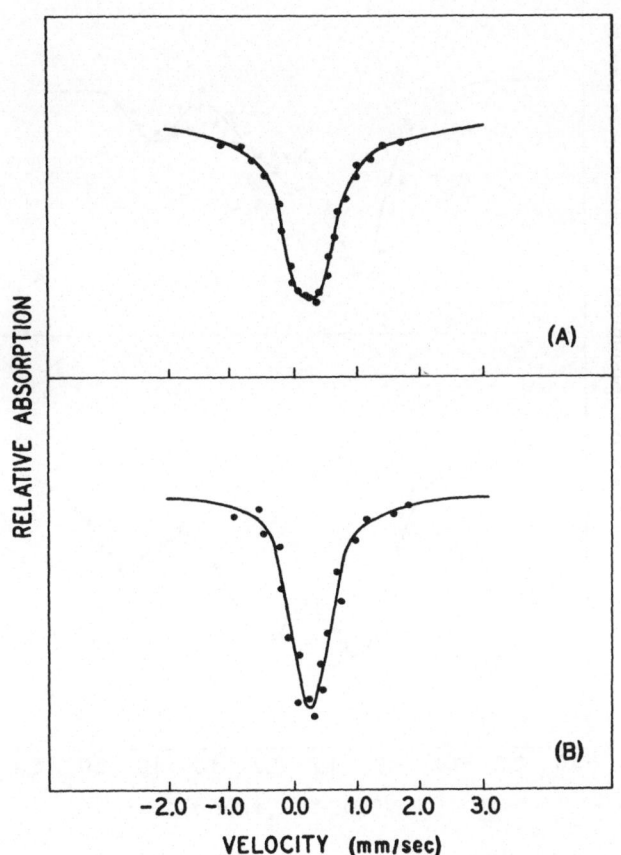

Figure 3. Mossbauer Spectra of New Catalyst in Hot Cell at (A) Room Temperature and (B) 550°C.

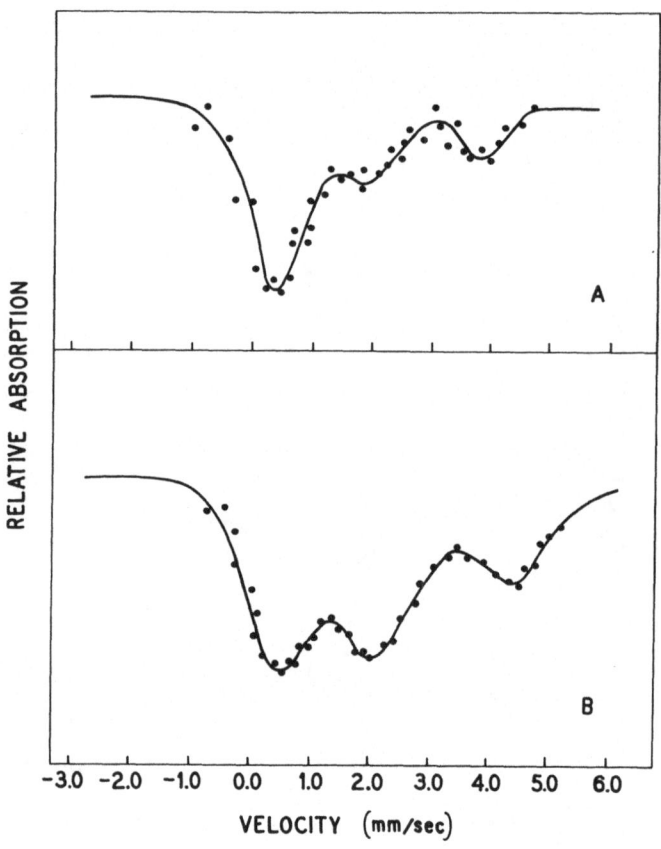

Figure 4. Mossbauer Spectra of Catalyst (A) During Dehydrogenation at 550°C and (B) After Cooling to Room Temperature in High Purity Argon.

integrated spectrum of the tin environments present during
the eight hours of reaction, whereas the spectrum made at
room temperature is of the tin environment present at the
end of the reaction.

Techniques were developed for "freezing" the environment of the active tin under reaction conditions and
cooling to room temperature where Mossbauer spectra can be
obtained faster with better statistics and resolution. It
was necessary to cool the materials in the absence of
oxygen - the reduced tin being very susceptible to reoxidation by molecular oxygen, particularly at elevated
temperatures but markedly so at room temperature. Subsequent experiments were made in small quartz reactors as
shown in Figure 5. Catalyst samples (normally 0.5 cc)
were heated to reaction temperatures in high purity argon
in a vertical tube furnace. A thermocouple was used for
controlling and measuring the temperature of the catalyst
material. When the temperature was stabilized, butane
and steam flow were started and the gases from the reactor
monitored by gas chromatography. At the end of the
reaction period the reactor was removed from the furnace
and cooled to room temperature in high purity argon. The
catalyst material was transferred to a thin-window glass
cell which could be evacuated or sealed in argon. The
transfer was made in a stream of high purity argon. This
technique of cooling and transfer was satisfactory for
maintaining the tin in the environment that was present
when the catalytic reaction was stopped. After sealing
the glass cell, the tin environments were, for all
practical purposes, permanently stable. Whenever new and
different reaction conditions were studied, spectra were
made in the hot cell under reaction conditions and compared with those made using the quartz reactor and glass
cell to insure that the tin environments seen were
representative of those present under reaction conditions.

Identifications of the tin seen in the Mossbauer
spectra were made by comparisons with parameters of known
tin compounds. The tin in the catalyst as prepared was
in the +4 oxidation state. Its chemical shift, 0.26
mm/sec, and quadrupole splitting of 0.6 mm/sec, suggest a
mixed oxide of tin, possibly stannic aluminate (see table
below). Cassiterite, SnO_2, normally has a slight smaller
chemical shift of 0.18 mm/sec in our system and an
indeterminate quadrupole splitting. Its Mossbauer spectrum

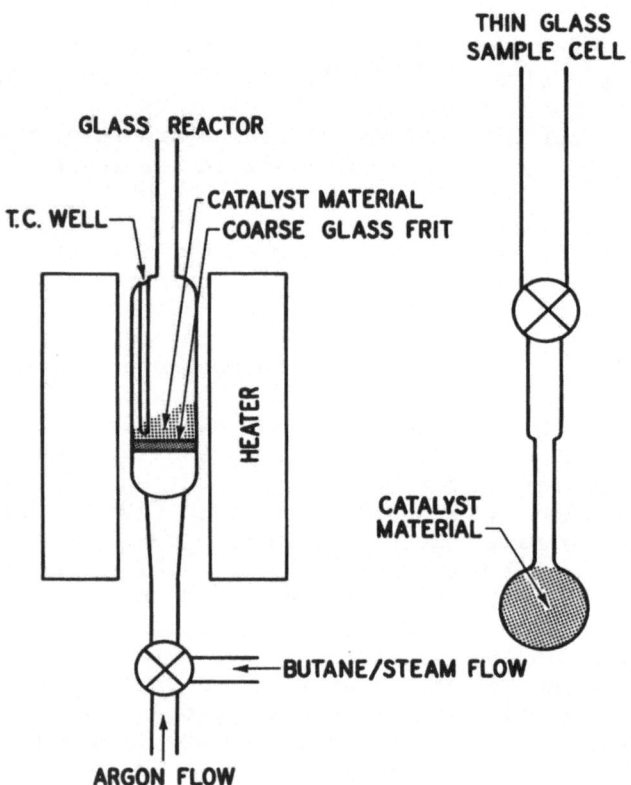

Figure 5. Quartz Reactor for Catalyst Dehydrogenation
Studies and Glass Cell for Recording Mossbauer Spectra.

MOSSBAUER PARAMETERS (ROOM TEMPERATURE)

Material	Chemical Shift mm/sec.	Quadrupole Splitting mm/sec.
Catalyst System:		
Stannic (New Catalyst)	0.26	0.6
Stannic (Reduced Catalyst)	0.4-0.7	-
Stannous	3.2	2.3
Alloy	2.0	-
Reference Materials:		
Stannic Oxide, SnO_2	0.17	0.5
Stannic Aluminate, $SnAl_2O_5$ (?)	0.12	0.65
Ammonium Chlorostannate, $(NH_4)_2SnCl_6$	0.57	-
Cobalt Chlorostannate, $CoSnCl_6$	0.62	-
Stannic Chloride Hydrate, $SnCl_4 \cdot H_2O$	0.46	-
Stannous Oxide (Blue Black), SnO	2.81	1.34
Stannous Oxide (Red), SnO	2.73	1.95
Tin Metal (White), Sn	2.70	-
Platinum-Tin Alloy, PtSn	2.00	-
Stannous Aluminate, $SnAl_2O_4$ (?)	3.3	2.2

appears usually as a wide single line (poorly resolved doublet), whereas the stannic material in the new catalyst is a definite doublet. The possibility that the stannic tin is present in two environments - one a stannic-oxygen bonded material such as stannic aluminate and the other a non-oxygenated material such as stannic chloroaluminate can not be discounted. The chemical shift of the stannic tin in the catalyst during the dehydrogenation reaction appears to be a function of the stannic concentration, becoming more positive as the intensity of the stannic tin decreases. When approximately 70-80 per cent of the tin has been reduced during the reaction, the chemical shift of the remaining stannic tin is 0.6 to 0.7 mm/sec. If the tin is present initially in two environments, the tin-oxygen material is apparently reduced preferentially.

There are at least three tin environments present during the catalytic reaction (Figure 6). Their parameters are given in the table together with parameters of related reference materials. In addition to the residual stannic tin, there are peaks at 2.0 and 4.4 mm/sec. The parameters of the peak at 2.0 mm/sec suggest an alloy of tin. The

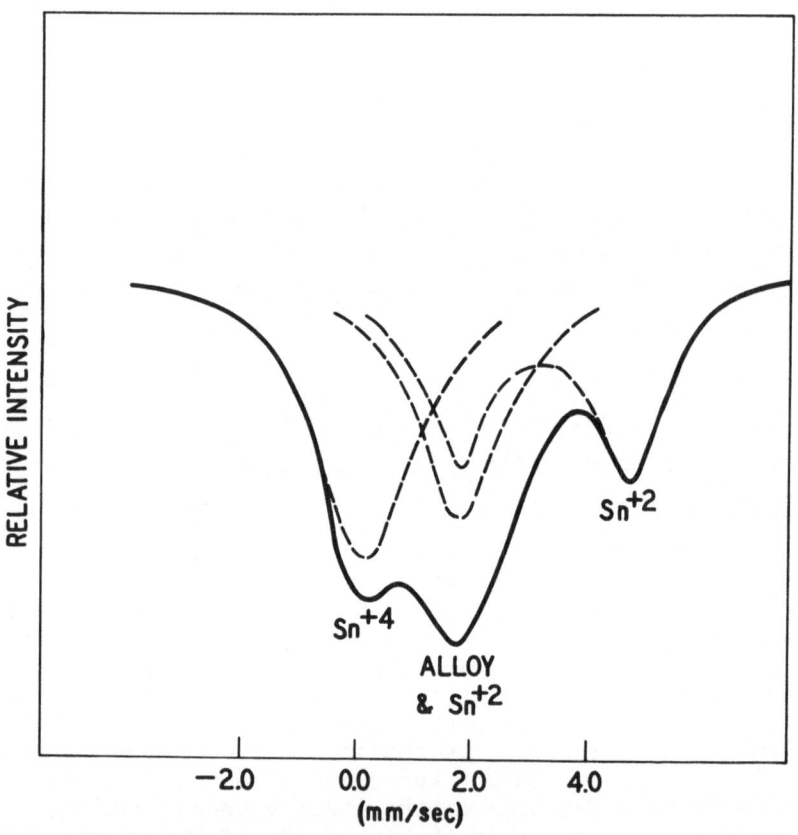

Figure 6. Mossbauer Spectrum of Platinum-Tin Catalyst Following Butane Dehydrogenation at 550°C.

peak at 4.4 mm/sec is indicative of a stannous compound.
Subsequent experiments in which tin alone was impregnated
on zinc aluminate and the material reduced with hydrogen
gave a spectrum similar to Figure 6 except the intensities
of the lines at 2.0 and 4.4 mm/sec were equal. These
spectra suggest the peak at 2.0 mm/sec in the Pt-Sn catalyst
is complex and representative of an alloy and one line of
a quadrupole split stannous compound. This stannous material has not been uniquely identified - it is not one of
the simple stannous oxides, but has parameters similar to
what would be expected for stannous aluminate. Attempts
were made to prepare both stannic and stannous aluminates.
Although their structures could not be confirmed by x-ray
diffraction, the measured Mossbauer parameters of these
tin aluminate preparations are similar to the stannic and
stannous compounds seen in the catalyst spectra. The
parameters of the stannous material in the reduced catalyst
are not constant but change with the length of time the
reaction is carried out on the catalyst without reactivation. The chemical shift and the quadrupole splitting
decrease very slowly, indicating a changing electronic
environment around the stannous tin (hydration?). After
reactivation, the reduced stannous tin environment is
again similar to what it was on the freshly reduced new
catalyst.

The alloy was identified as PtSn. A series of alloys
and solid solutions of tin and platinum were prepared and
as shown in Figure 7 there is a linear relationship
between the atom per cent tin in the material and the
chemical shift (4,5). Since the PtSn alloy is seen in the
reduced catalyst in spectra made at 550°C the melting
point of the alloy must be significantly higher than
550°C. An approximate determination of the melting point
showed it is between 800 and 1000°C.

The recoil-free fractions of the tin compounds on the
catalyst were measured relative to the stannic tin. The
factors were determined by slowly oxidizing the alloy
and stannous tin in the reduced catalyst by small injec-
tions of air at room temperature, and finally oxidizing
the alloy at 500°C with air. Mossbauer spectra were made
after each air injection and the changes in the concentrations of the three tin species were correlated with
their respective Mossbauer factors. The recoil-free
fractions for the PtSn alloy and stannous material were

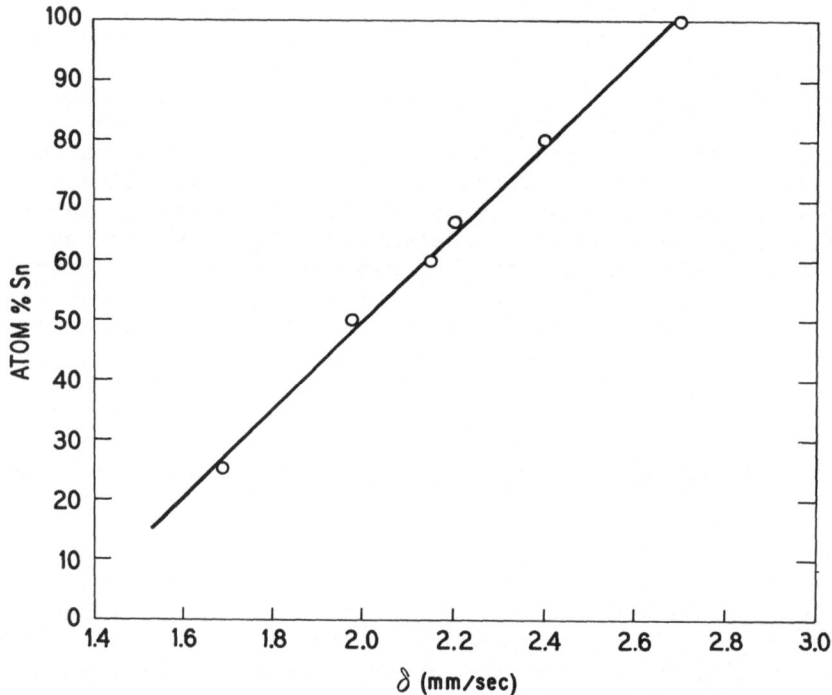

Figure 7. Chemical Shifts of Platinum-Tin Alloys and
Solid Solutions as a Function of Atom % Tin.

$0.7\overset{+}{-}0.1$ and $0.75\overset{+}{-}0.1$, respectively, relative to stannic
tin.   It was necessary to determine these factors on the
catalytic material under study (6).   A determination of
the factor for the PtSn alloy using 100 mesh alloy
particles gave a factor of 0.1 relative to the stannic
material on the catalyst.   The dispersed alloy on the
reduced catalyst gave a more realistic measure of the
recoil-free fraction.

       Variations in the concentrations of the tin compounds
during the dehydrogenation reaction were studied as a
function of reaction temperature and time of reaction.
Separate 0.5 cc batches of new catalyst material (calcined) were heated in the quartz reactor in high purity
argon to the desired temperature.   Butane and steam
(1/6 mole ratio) were passed through the material at 40

cc/minute. At the end of the dehydrogenation reaction time the catalyst was quickly cooled in high purity argon and transferred under argon to glass cells and sealed. Mossbauer spectra were made at room temperature. The catalytic reaction, the dehydrogenation of butane to butenes, was followed by periodic gas chromatographic measurements of the gas effluent, with the final chromatographic determination being made one minute before the termination of the run. The results of these temperature studies are shown in Figure 8 where the percentages of the tin phases present (corrected for their recoil-free fractions) are plotted as a function of the reaction temperature. Constant thirty minute reaction times were used for this temperature study. The concentration of the butenes as determined chromatographically are given in Figure 8 using an arbitrary concentration scale.

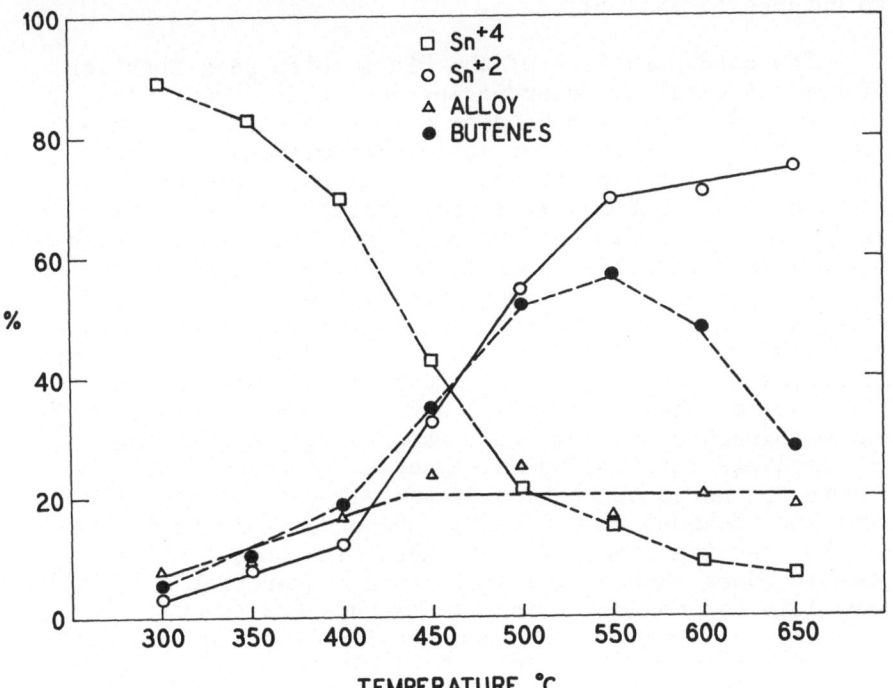

Figure 8. Relative Tin and Butene Concentrations After 30 Minute Butane Dehydrogenation as a Function of Reaction Temperature.

The catalytic reaction begins at about $300^{\circ}C$ and is characterized by an increasing conversion of the butane to butene as the temperature increases. The concentration of the butenes increased steadily up to $550^{\circ}C$, after which the selectivity to butenes decrease and the conversion to cracked products and oxides of carbon increase. Coincident with the onset of catalytic activity is a reduction in the concentration of stannic tin with the subsequent appearance of both the PtSn alloy and stannous tin. The concentration of the stannic tin decreases regularly as the temperature increases. The alloy concentration increases until its concentration is about 20 per cent at $400^{\circ}C$ and then remains essentially constant as the temperature is increased. The stannous tin concentration increases with temperature and parallels the increase in concentration of butenes - at least to the temperatures where cracking and carbon oxide formation begin to cause the conversion to butenes to fall off.

The concentrations of the tin species as a function of time of catalytic dehydrogenation at $550^{\circ}C$ are shown in Figure 9. Separate batches of new catalyst were used for each run, the catalyst being used without reactivation for the duration of that experiment. The concentration of the butenes from the catalytic reaction is again plotted in Figure 9 using an arbitrary scale. The butenes were determined one minute before the reaction was terminated. Since times from five minutes to 40 hours were used in this study, a logarithmic time coordinate was used in Figure 9 to facilitate data presentation. As far as was possible, the conditions of the dehydrogenation were maintained constant during the run. Frequent chromatographic analyses were made to insure that the catalyst was functioning as expected. As shown in Figure 9, the concentration of the alloy is essentially constant at about 20 per cent, independent of the length of the reaction time. The stannous tin concentration, however, does change, and again in a manner that closely parallels the concentration of the butene product of the catalytic reaction. The concentration of the stannous tin increases slightly for the first 30 to 60 minutes and then decreases logarithmically. The concentration of the butenes follows a similar pattern. Although not plotted in Figure 9, the concentration of the stannic tin increases with time as expected from the observed changes in the stannous tin concentrations.

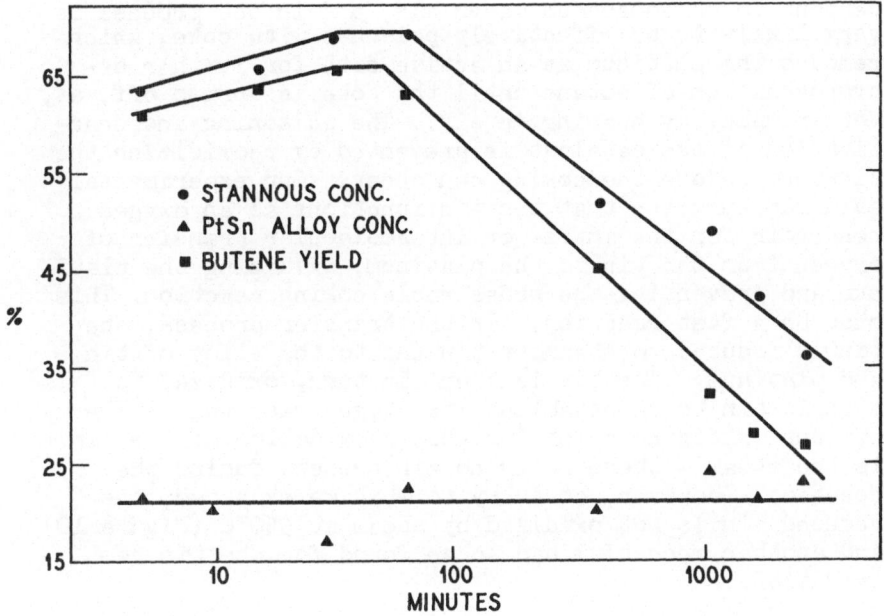

Figure 9. Relative Tin and Butene Concentrations After
Butane Dehydrogenation at 550°C as a Function of
Reaction Time.

     Although we have observed only the tin environments
on these materials the changes seen as well as other confirming experiments have suggested a mechanism for the
catalytic dehydrogenation reaction. The time and temperature studies suggest that the stannous tin is an important
factor in the mechanism since it closely parallels the
conversion of the butane feed to butene with both time
and temperature. The tin is not the active catalyst in
the system as it has been shown that without platinum,
tin is not active for paraffin dehydrogenation under
these conditions.

     Oxidized platinum on the surface, probably surface
layers of PtO_2 (7), is the active species abstracting
hydrogen from the butane, giving butene, water, and a
reduced platinum site. Reduced or elemental platinum is
wellknown to be extremely efficient for reducing hydro-

carbons to carbonaceous materials, and in the process is
very likely to be effectively poisoned with coke, which
removes the platinum as an active site for further dehydrogenation of butane until the coke is burned off, as,
for example, by heating in air. The poisoning and deac-
tivation of the catalyst is prevented by reoxidizing the
platinum before the coking can occur. Our experimental
evidence suggests that the tin functions as an oxygen
reservoir for the intra- or intermolecular transfer of
oxygen from the tin to the platinum, oxidizing the platinum and preventing the undesirable coking reaction. This
must be a fast reaction. In the transfer process, the
tin is reduced to stannous tin and to the alloy of tin
and platinum. The tin is then, in turn, oxidized to
stannic tin to re-establish its oxygen reservoir. The
obvious oxidizing agent for this reoxidation of the tin
is the steam - there being no air present during the
reaction. However, contrary to what we expected, the
reduced tin is not oxidized by steam at 550°C (Figure 10)
and another mechanism had to be found for the tin reoxidation.

A characteristic of this catalyst is that platinum
and tin are supported on a zinc aluminate base which contains a slight excess, a few per cent, of zinc oxide. If
stoichiometric zinc aluminate is used as the base material,
the catalyst deactivates, by coking, relatively quickly,
even with tin present. This suggests that the excess
zinc oxide reoxidizes the reduced tin, either by an intra-
or intermolecular transfer of oxygen from the zinc to the
tin, or less likely at the temperatures used in the
reaction, by oxygen which has diffused through the material and which has its origin in the thermal decomposition
of zinc oxide (8,9). Under the conditions of the catalytic
reaction, zinc is readily oxidized to zinc oxide by steam.

The overall mechanism for this reaction is suggested
to be:

1. $PtO_x + C\text{-}C \longrightarrow Pt + C\text{=}C + H_2O$

2. $(SnO_2)_{oxidized} + Pt \longrightarrow PtO_x + (Sn)_{reduced}$

3. $(Sn)_{reduced} + (ZnO)_{oxidized} \longrightarrow (SnO_2)_{oxidized} + Zn$

4. $Zn + H_2O \longrightarrow (ZnO)_{oxidized} + H_2$

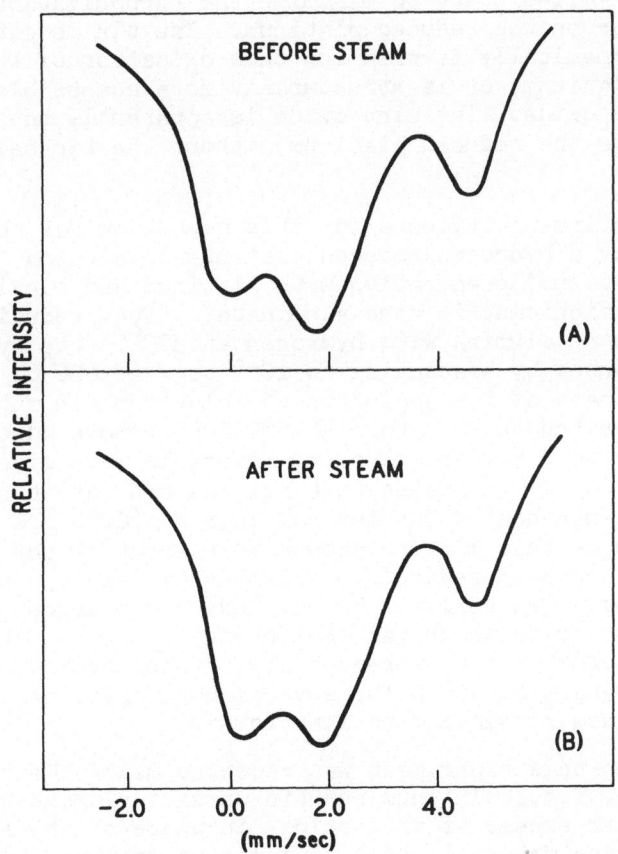

Figure 10. Mossbauer Spectra of Reduced Platinum-Tin Catalyst (A) Before Steam and (B) After Steam.

Overall: $C-C \rightleftharpoons C=C + H_2$

This mechanism is consistent with our experimental data for this platinum-tin paraffin dehydrogenation catalyst. The function of the tin appears to be as an oxygen reservoir to maintain the surface of the platinum in an oxidized state to minimize the carbonization which can occur on the reduced platinum. The tin is either thermodynamically favored for this oxidation of the reduced platinum, or is structurally more accessible than the zinc oxide. The zinc oxide is apparently not able to reoxidize the reduced platinum without the tin being present.

Confirming evidence for this mechanism was obtained by mixing a hydrogen reduced platinum on alumina (10 per cent platinum) preparation with platinum and tin impregnated on stoichiometric zinc aluminate. After reducing the platinum on alumina with hydrogen at 500°C, the hydrogen was desorbed by evacuating to 10^{-5} torr at 600°C. In another part of the apparatus the platinum-tin material was evacuated to 10^{-5} torr at 600°C to remove adsorbed and entrapped oxygen. The two materials were mixed in the vacuum. After verifying that all the tin was stannic tin, the cell was heated for several days at 600°C. A Mossbauer spectrum of this mixture showed that about 50 per cent of the tin had been reduced to stannous tin and the platinum tin alloy. The cell was opened under pure argon, mixed with zinc oxide which had been evacuated to 10^{-5} torr to remove adsorbed and entrapped air, evacuated and resealed. After heating at 600°C for several more days, all of the tin was now reoxidized to stannic tin.

When this experiment was repeated using the normal catalyst, i.e., platinum and tin on zinc aluminate containing an excess of zinc oxide, in place of the stoichiometric base material, only 14 per cent (rather than 50 per cent) of the stannic tin was reduced after heating for several days. When stannic oxide, SnO_2, was used in place of the catalyst material, none of the tin was reduced by the reduced platinum.

Additional confirming data for this dehydrogenation mechanism were obtained during a study of platinum and tin impregnated on other base materials. A preparation using an alumina as the base material showed very similar

activity and Mossbauer spectra as the zinc aluminate catalyst material. Other preparations using different aluminas
and silicas were poor catalysts, i.e., low activity and
excessive coking. The Mossbauer spectra of these latter
materials were also different. Little or no stannous tin
was seen, and the alloy peaks were abnormally large.
Different alloys or solid solutions of tin and platinum
were observed which usually contained a lower percentage
of tin than in the PtSn alloy. An analysis of the alumina
that gave good activity showed it contained 1.44 per cent
zinc oxide which probably accounted for the observed
activity; the zinc oxide maintaining the tin in an oxidized
state.

The above experiments support the suggested mechanism
for this catalyst. The stannic tin is capable of oxidizing the reduced platinum, and without excess zinc oxide in
the base material, remains reduced until mixed with added
zinc oxide. The failure of the stannic oxide, SnO_2, to
oxidize the reduced platinum is consistent with other
experimental evidence on this platinum-tin catalyst system.
It has been found that as the catalyst is used for several
months, reactivation fails to restore the activity of the
catalyst to the conversion levels of the new catalyst.
Mossbauer spectra of these used catalysts indicate that the
tin is primarily present as stannic oxide, with a chemical
shift of 0.17 mm/sec, rather than the stannic oxide material (stannic aluminate?) with a chemical shift of 0.26 and
a quadrupole splitting of 0.6 mm/sec. Apparently during
long usage, the tin is gradually converted to stannic
oxide, a form which is incapable of oxidizing the reduced
platinum, and activity of the catalyst decreases.

Although the addition of tin to the catalytic material
stabilizes the catalyst and significantly increases the
time between reactivations, the catalyst does eventually
coke up and become inactive. This is probably caused by
the failure of the tin to reoxidize all the reduced
platinum sites sufficiently fast, possibly because the
tin had not as yet been reoxidized by the zinc oxide
following a previous oxidation of the platinum by the tin.

The Mossbauer spectra of the reduced Pt-Sn catalysts
are similar to those observed by Firsova, et al., (10) in
their investigation of surface compounds of propylene and
acrolein on a tin-molybdenum catalyst. They have inter-

preted their spectra as being due to a surface compound of
divalent tin in which the bonding of the organic molecule
with the tin ion in Sn-Mo-O is realized through oxygen
(Sn-O-C-) after treatment of the catalyst surface with
oxygen. Although we can not rule out such a surface compound in our system, the experimental evidence appears to
favor the formation of stannous aluminate or some other
similar material. The stannous compound is produced by
reduction of the catalyst with hydrogen as well as by
butene and this stannous species is "stable" at temperatures as high as 550°C. Firsova, et al., found it necessary
to pretreat their surface with oxygen before forming the
complex. Although the normal Pt-Sn catalyst is calcined
in air prior to use, we have found that evacuation of the
catalyst to 10^{-5} torr prior to the catalytic reaction did
not change the nature of the stannous compound formed.
Our mechanism is similar to that proposed by Firsova, et
al., in that they also propose an electron transfer between the tin and the molybdenum.

Our experimental data as well as that of Firsova's
can not "a priori" rule out the mechanism reported by
Haber (11) for the oxidation of olefins over bismuth-
molybdenum oxidation catalysts. Haber has proposed that
the activation of the olefin by abstraction of the first
hydrogen atom occurs on active bismuth (III) sites and
abstraction of the second hydrogen atom at the active
molybdenum (VI) sites. This is followed by desorption of
the product with removal of an oxygen ion from the crystal
lattice of the catalyst. Haber's mechanism does not require the transfer of electrons between metal oxides in
the lattice.

The stabilization function of the tin on the platinum-
tin catalyst (tin does not appear to be primarily active
in the dehydrogenation process) and the inability of steam
to reoxidize the stannous tin to give the overall non-
oxidative dehydrogenation reaction led us to postulate the
electron transfer mechanism between metal oxides in the
lattice. We believe this mechanism is more consistent
with our experimental data than are the mechanisms proposed by Firsova, et al., and Haber. The proposed
mechanism accounts for lattice oxygen participation (12,13,
14) and oxygen migration (12,15,16,17) and provides a
reasonable explanation for the stabilization properties
shown by tin in these tin mixed oxide systems.

REFERENCES

1. E. O. Box and D. Uhrick, U.S. Patent 3,641,182 to
    Phillips Petroleum Co.

2. E. O. Box and D. Uhrick, U.S. Patent 3,880,776 to
    Phillips Petroleum Co.

3. J. G. Stevens and V. E. Stevens, Mossbauer Effect Data
    Index, IFI/Plenum Data Corp., New York, 1971.

4. C. R. Kanekar, K.R.P. Mallikarjuna Rao and V. Udaya
    Shankar Rao, Physics Letters, 19, 95 (1965).

5. N. S. Ibraimov and R. N. Kuz'min, Soviet Physics-
    Doklady, 10, 1071 (1966).

6. H. M. Gager and M. C. Hobson, Jr., Catal. Rev.-Sci.
    Eng., 11(1), 117 (1975).

7. T. P. Pignet, L. D. Schmidt and N. L. Jarvis, J.
    Catal., 31, 145 (1973).

8. E. A. Secco, Can. J. Chem., 38, 596 (1960).

9. K. Kodera, I. Kusunaki and S. Shimizu, Bull. Chem.
    Soc. Japan, 41, 1039 (1968).

10. A. A. Firsova, N. N. Khovanskaya, A. D. Tsyganov,
    I. P. Suzdalev and L. Ya Margolis, Kinetika i Kataliz,
    12, 792 (1971).

11. J. Haber, Intern. Chem. Eng., 15(1), 21 (1975).

12. G. W. Keulks, J. Catal., 19, 232 (1970).

13. Ph. A. Batist, C. J. Kapteijns, B. C. Lippens and
    G.C.A. Schuit, J. Catal., 7, 33 (1967).

14. M. Blanchard and G. Louguet, Kinetika i Kataliz,
    14(1), 30 (1973).

15. R. D. Wragg, P. G. Ashmore and J. A. Hockey, J. Catal.,
    22, 49 (1971).

16. R. D. Wragg, P. G. Ashmore and J. A. Hockey, J. Catal.,
     28, 337 (1973).

17. Yu A. Mishchenko, N. D. Gol'dshtein and A. I.
     Gel'bshtein, Zh. fiz. khimi, 47(3), 511 (1973).

MÖSSBAUER SPECTROSCOPY OF SUPPORTED BIMETALLIC CATALYSTS

R. L. Garten

Corporate Research Laboratories
Exxon Research and Engineering Company
Linden, New Jersey 07036

INTRODUCTION

In recent years there has been renewed interest in
catalysis by alloys from both the fundamental and practical points of view. Much of this interest has been stimu-
lated by the advent of bimetallic catalysts in catalytic
reforming. Recent studies of Sinfelt and co-workers (1)
have demonstrated that alloying can have a marked effect
on catalytic specificity. For example, the specific
activity of bulk CuNi alloys for carbon-hydrogen bond
breaking (dehydrogenation) is nearly independent of composition up to ∿80 atomic percent Cu whereas the specific
activity for carbon-carbon bond breaking (hydrogenolysis)
is decreased by about four orders of magnitude when ∿30
atomic percent Cu is added to Ni. Similar results obtained by Sinfelt (2) for SiO_2-supported CuRu and CuOs
catalysts strongly point to the formation of bimetallic
clusters in these well-dispersed metal catalysts. The
term bimetallic cluster is preferred to alloy for well-
dispersed particles where surface effects dominate since
catalytic evidence indicates that bimetallic clusters may

be formed even for cases where no corresponding bulk alloy is known (2).

As is frequently the case, the bulk metals and alloys which are amenable to surface and bulk characterization by a number of techniques do not have great practical value whereas the more useful supported metal catalysts are much more difficult to characterize by analytical tools. This is due to the small size and low concentration of the metal particles which are deposited within the pores of the catalyst support.

At present, Mössbauer spectroscopy appears to be the only technique which can provide <u>direct</u> information on the chemical nature of metal clusters in practical supported metal catalysts. It is the purpose of this paper to demonstrate this unique applicability of the Mössbauer technique to the study of supported metal catalysts, placing emphasis primarily on the qualitative features of the spectra. Part of the work described here has been discussed in more detail elsewhere (3). In this paper, the use of Mössbauer spectroscopy to investigate the question of bimetallic cluster formation in Al_2O_3-supported PdFe and PtFe and SiO_2-supported PtFe bimetallic catalysts is presented. We find that the Mössbauer effect provides direct evidence for bimetallic clusters as well as information on the chemical nature, adsorption interactions and factors affecting the surface composition of the clusters.

EXPERIMENTAL

The Pd catalysts were prepared by impregnation of η-Al_2O_3 (245 m^2g^{-1}) with acidic $PdCl_2$ solution, while the

Pt on γ-Al_2O_3 (214 m^2g^{-1}) and SiO_2 (Cabosil HS-5,
300 m^2g^{-1}) catalysts were prepared with H_2PtCl_6 solution.
For the alumina and silica 0.5 cc g^{-1} and 2.2 cc g^{-1},
respectively, of metal solution was used for the impregnation. The metal concentrations of the solutions were ad-
justed to give the desired metal loading in the final
catalyst. The Pd/η-Al_2O_3 catalyst was dried 16 hrs. at
393°K and calcined 3 hrs. at 773°K in air. The Pt/γ-Al_2O_3
was dried 16 hrs. at 393°K and calcined 4 hrs. at 530°K in
air. The Pt/SiO_2 was dried 16 hr. in air at 393°K. Each
sample was then impregnated as described above with a ferric nitrate solution, 93% isotopically enriched in Fe[57],
followed by a repetition of the respective drying and cal-
cining procedures described above. Catalysts prepared in
this manner are termed co-impregnated in this paper.

Dispersion measurements using hydrogen chemisorption
were carried out in a conventional volumetric adsorption
apparatus (4). For PdFe/η-Al_2O_3 the method of Aben (5)
was used whereas monolayer coverage of the metal for SiO_2-
and γ-Al_2O_3 supported PtFe catalysts was determined from
the zero pressure intercept of the hydrogen isotherm (6).
Dispersion measurements were made on samples pretreated by
a procedure identical to that used in the Mössbauer
studies.

The Mössbauer apparatus, controlled atmosphere and
temperature sample cell, procedure for the Mössbauer experiments and computer analysis of the data have all been
described previously (7). The "goodness of fit" parameter
(x^2/df) for each spectrum was determined from the quotient
of x^2 and the number of degrees of freedom. Mössbauer

data were obtained using a Co^{57}-Cr source and all isomer
shifts are reported with respect to that source. The
minimum linewidth obtainable with the spectrometer was
0.27 mm sec^{-1} for a sodium nitroprusside absorber containing 8.7 x 10^{17} Fe atoms cm^{-2}. The isomer shift and quadrupole splitting for the sodium nitroprusside absorber was
-0.11 mm sec^{-1} and 1.69 mm sec^{-1}, respectively, in good
agreement with literature values.

Hydrogen for the reductions and dispersion measurements was obtained from Matheson Co. and was 99.95% pure.
It was purified further by passage through a Deoxo unit
followed by a zeolite drying trap. Oxygen of 99.5% purity
and helium of 99.9999% purity were used without additional
treatment.

All spectra reported here were recorded at 298 $\pm$ 2°K
in the gas ($\sim$760 Torr) used in a particular pretreatment.

RESULTS AND DISCUSSION

Supported Iron Catalysts

The Mössbauer spectra of iron on η-Al_2O_3, γ-Al_2O_3 and
SiO_2 at the same loading and for the same reduction conditions to be considered later for the bimetallic catalysts
are shown in Fig. 1. For all three samples, the iron
could only be reduced to the ferrous state. This behavior
has been reported and discussed in detail in many
studies (8) of low concentrations of iron on SiO_2 and
Al_2O_3. In general, the spectra can be decomposed into at
least two ferrous doublets. For SiO_2-supported iron,
doublet A, with smaller isomer shift and quadrupole
splitting than doublet B, is altered by adsorbates (NH_3,

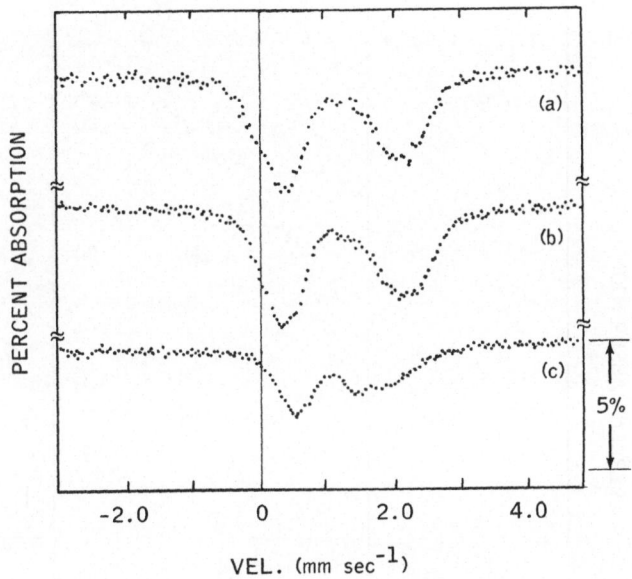

Fig. 1. Mössbauer spectra of hydrogen reduced iron catalysts. (a) 0.1% Fe/η-Al₂O₃, reduced 1 hr., 673°K; (b) 0.1%
Fe/γ-Al₂O₃, reduced 1 hr., 773°K; (c) 0.1% Fe/SiO₂, reduced
2 hr., 973°K.

H_2O, H_2S), whereas doublet B is not affected. It has been
suggested (8a) that doublet A is due to coordinately unsaturated ferrous ions at the surface and doublet B to
ferrous ions in the interior of ferrous oxide micro-
crystals. The important result for this investigation is
that the iron can only be reduced to the ferrous state.

$PdFe/\eta\text{-}Al_2O_3$ Catalysts

The effect of Pd on the chemical nature of Fe in
$PdFe/\eta\text{-}Al_2O_3$ is shown in Fig. 2, while pertinent Mössbauer
parameters and dispersions are tabulated in Table 1. When
0.2% $Fe/\eta\text{-}Al_2O_3$ was mixed by grinding with 9.5% $Pd/\eta\text{-}Al_2O_3$

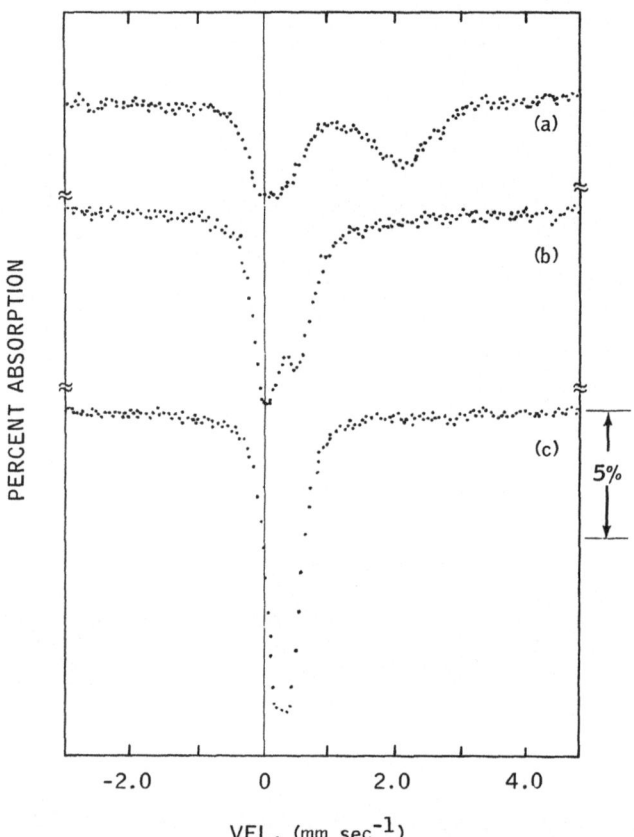

Fig. 2. Mössbauer spectra of hydrogen reduced 0.1% Fe, 4.75% Pd/η-Al$_2$O$_3$ catalysts. (a) Physical mixture, reduced 1 hr., 673°K; (b) Co-impregnated, reduced 1 hr., 673°K; (c) Co-impregnated reduced 2 hr., 973°K.

to give a sample of nominal composition of 0.1% Fe/4.75% Pd/η-Al$_2$O$_3$ and then reduced, Fig. 2a was obtained. Fig. 2a is essentially the same as Fig. 1a, indicating little or no effect of the Pd in the physical mixture. In a co-impregnated catalyst of the same composition, however, ferrous ions were not present in the reduced catalyst and

TABLE 1

Mössbauer Parameters for PdFe/η-Al$_2$O$_3$ for Various Treatments

Treatment[a]	Peaks	δ (mm sec^{-1})	Δ (mm sec^{-1})	χ^2/df	D
Fig. 2b	1-2	0.28	0.47	1.22	54 (2.5)[b]
Fig. 2c	1-2	0.33	0.24	1.95	11 (14.6)[b]
Fig. 3a	1-2	0.48	1.45	3.68	54
Fig. 3b	1-3	0.29	0.55	1.02	54
	2-4[c]	1.26	1.70		
Fig. 3c	1-4[d]	0.51	1.37	1.05	11
	2-3	0.30	0.26		

(a) Treatment described under corresponding figure caption.

(b) Calculated average particle size in nm given as the length of an edge of an fcc octahedron assuming adsorption on all eight faces.

(c) ε and Γ for peaks 1 and 4 constrained to be equal.

(d) ε and Γ for peaks 2 and 4 constrained to be equal.

a quadrupole-split spectrum (Fig. 2b) with isomer shift (δ) close to the value of 0.34 mm sec^{-1} observed for dilute Fe in bulk PdFe alloys (9) was obtained. Figs. 2a and b show clearly that the Pd and Fe must be present on the same support particle for Pd to affect the chemical nature of the Fe.

Reduction of the co-impregnated PdFe/η-Al$_2$O$_3$ catalyst at 973°K caused a collapse of the quadrupole splitting (Δ) (Fig. 2c) and gave a δ nearly identical to that for the bulk alloy. Concurrent with the collapse of the quadrupole splitting was a decrease in dispersion of the metals by a factor of $\sim$5 (Table 1). These results provide evidence for the formation of PdFe bimetallic clusters in the reduced samples of the co-impregnated catalyst. The quadrupole splitting observed in Fig. 2b indicates that a

large fraction of the iron atoms occupied sites of non-
cubic symmetry in agreement with the dispersion measurements which showed that ∿50% of the metal atoms in the
catalyst were surface atoms. Agglomeration of the PdFe
clusters by reduction at 973°K decreased the fraction of
surface atoms and increased the fraction of iron atoms
with a more symmetrical bulk-like environment. This
accounts for the collapse of the quadrupole splitting and
the agreement between the isomer shifts for the iron in
the catalyst and dilute iron in bulk PdFe alloys.

Further support for PdFe bimetallic clusters was
obtained from the chemical behavior of the iron in PdFe/
η-Al_2O_3 as shown in Fig. 3. When the sample giving Fig.
2b was evacuated and exposed to O_2 at 298°K, Fig. 3a was
obtained. The Mössbauer parameters for Fig. 3a given in
Table 1 indicate the formation of high spin Fe^{3+} ions on
oxidation. The fact that all the iron is oxidized confirms that the iron is well-dispersed as discussed previously. Exposing the oxidized sample to H_2 at 298°K gave
Fig. 3b. Part of the iron was reduced to the ferrous
state, while the remainder gave a doublet with parameters
nearly identical to that obtained for Fig. 2b. Figs. 3a-b
demonstrate that the iron in reduced PdFe/η-Al_2O_3 can be
reversibly oxidized and reduced at 298°K and strongly
supports the case for PdFe bimetallic clusters. Small
iron particles oxidize at 298°K but do not reduce in H_2 at
this temperature (10). A chemisorbed oxygen layer on Pd,
however, is readily removed by H_2 titration at 298°K (10).
When Fe is bonded to Pd atoms, either the Fe-oxygen bond
is sufficiently weakened that it titrates at 298°K or Pd

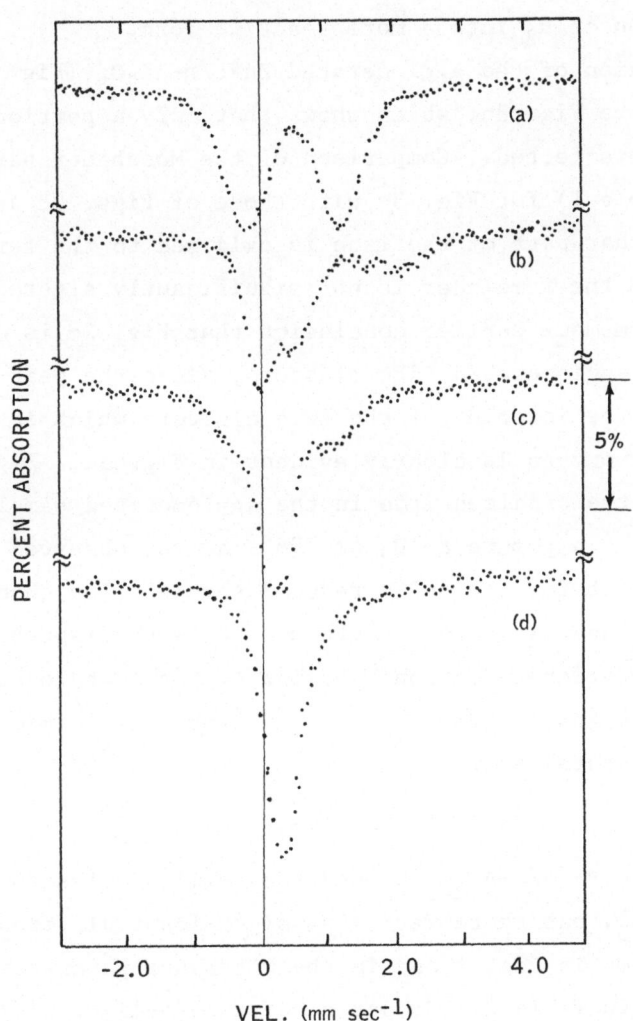

Fig. 3. Oxidation-reduction behavior of co-impregnated 0.1% Fe, 4.75% Pd/η-Al$_2$O$_3$. (a) Reduced 1 hr., 673°K, evacuated and exposed to O$_2$ at 298°K; (b) Following treatment (a), evacuated and exposed to H$_2$ at 298°K; (c) Reduced 2 hr., 973°K, evacuated and exposed to O$_2$ at 298°K; (d) Following treatment (c), evacuated and exposed to H$_2$ at 298°K.

catalyzes the reduction by providing a route for the
dissociation of H_2 into a more reactive form.

Oxidation of the agglomerated PdFe/η-Al$_2$O$_3$ (Fig. 2c)
at 298°K gave Fig. 3c, which shows that only a portion of
the iron is affected. Comparison of the Mössbauer parameters (Table 1) for Fig. 3c with those of Figs. 2c and 3a
indicates that part of the iron is oxidized to the ferric
state while the remainder is not significantly altered.
This confirms our earlier conclusion that Fig. 2c is due
to iron in agglomerated PdFe clusters, since the peak due
to iron in the interior of the PdFe clusters which is not
affected by oxygen is clearly evident in Fig. 3c. Fig. 3d
shows that the oxidized iron in the agglomerated sample is
re-reduced on exposure to H_2 at 298°K as was observed in
Figs. 3a and b for the 663°K reduced sample. The changes
in the Mössbauer spectra of PdFe/η-Al$_2$O$_3$ with dispersion
and the oxidation-reduction behavior of the iron leave
little doubt that PdFe bimetallic clusters are formed in
the reduced catalysts.

PtFe/γ-Al$_2$O$_3$ Catalysts

Iron in PtFe/γ-Al$_2$O$_3$ showed some similarities to
PdFe/η-Al$_2$O$_3$, but there were also significant differences.
This is shown in Fig. 4 and in the Mössbauer parameters
summarized in Table 2. Reduction of PtFe/γ-Al$_2$O$_3$ at 773°K
gave the asymmetric quadrupole-split spectrum shown in
Fig. 4a. The δ (Table 1) determined from the peak positions is much different from that observed for PdFe/
η-Al$_2$O$_3$ and is close to the value of 0.50 mm sec^{-1}
observed for dilute iron in bulk PtFe alloys (9) and to
the values of 0.46-0.50 mm sec^{-1} reported by Bartholomew

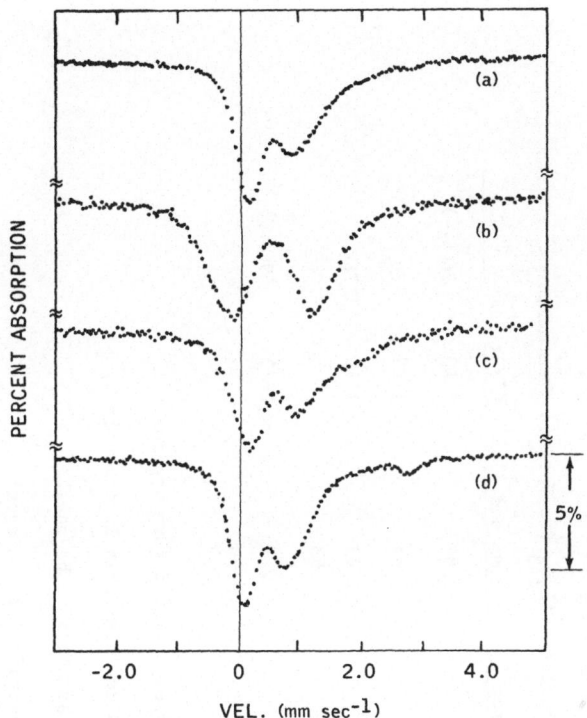

Fig. 4. Mössbauer spectra of co-impregnated 0.1% Fe,
1.75% Pt/γ-Al$_2$O$_3$. Sequential experiments on one sample.
(a) H$_2$ reduced 1 hr., 773°K; (b) Evacuated and exposed to
O$_2$ at 298°K; (c) Evacuated and exposed to H$_2$ at 298°K;
(d) H$_2$ reduced 1 hr., 973°K.

and Boudart (11) for carbon-supported PtFe bimetallic
clusters. The quadrupole splitting observed in Fig. 4a is
consistent with well-dispersed PtFe bimetallic clusters
and this is substantiated by the high dispersion determined for this catalyst (Table 2). Figs. 4b-c and Table 1
show the effect of oxygen and hydrogen at 298°K in sequential experiments. As with PdFe/η-Al$_2$O$_3$, it was observed

TABLE 2

Mössbauer Parameters for PtFe/γ-Al$_2$O$_3$ for Various Treatments

Treatment[a]	Peaks	δ(mm sec^{-1})	Δ(mm sec^{-1})	χ^2/df	D
Fig. 4a	1-2	0.53	0.82	4.63	76(1.4)[b]
Fig. 4b	1-2	0.52	1.46	1.42	76
Fig. 4c	1-4[c]	0.90	1.93	1.10	76
	2-3	0.59	0.88		
Fig. 4d	2-4[d]	1.52	2.57	2.73	59(2.0)
	1-3	0.45	0.79		

(a) Treatment described under corresponding figure caption

(b) Calculated average particle size (see Table 1, footnote b)

(c) ϵ and Γ for peaks 1 and 4 constrained to be equal.

(d) ϵ and Γ for peaks 2 and 4 constrained to be equal.

that the iron reversibly oxidized and reduced at 298°K, a
result which we take as convincing evidence that PtFe
clusters were formed.

Reduction of PtFe/γ-Al$_2$O$_3$ at 973°K gave Fig. 4d
which contrasts to the behavior observed for PdFe/η-Al$_2$O$_3$.
Determination of the dispersion of PtFe/γ-Al$_2$O$_3$ reduced at
973°K (Table 2), however, showed that severe agglomeration
did not occur for PtFe/γ-Al$_2$O$_3$ and the collapse of the
quadrupole splitting associated with the large decrease in
surface/volume ratios observed for PdFe/η-Al$_2$O$_3$ was not
expected for PtFe/γ-Al$_2$O$_3$. The sample giving Fig. 4d also
reversibly oxidized and reduced at 298°K giving spectra
similar to Fig. 4b and c. Some Fe^{2+} was still present in
the PtFe/γ-Al$_2$O$_3$ sample after reduction at 773 or 973°K,
as indicated by the small peak at $\sim$2.8 mm sec^{-1}, which we
assign to the right-hand peak of a Fe^{2+} doublet. This,
along with a distribution of chemical environments for the
Fe associated with Pt, as discussed later, probably
accounts for the poor χ^2/df in Table 1 for Fig. 4a.

PtFe/SiO$_2$ Catalyst

In this section, the ability of the Mössbauer effect
to detect changes in the surface concentration of PtFe
bimetallic clusters will be presented. Since the heat of
adsorption of oxygen on Fe is 569 kJ/mole compared to
268 kJ/mole for Pt, we expect, based on current theories
of surface enrichment in binary alloys (12), that oxidation of PtFe clusters will enrich the surface in Fe. This
was confirmed by the present work. The results of this
study are shown in Fig. 5 and the Mössbauer, parameters are
summarized in Table 3. The fresh catalyst was pretreated

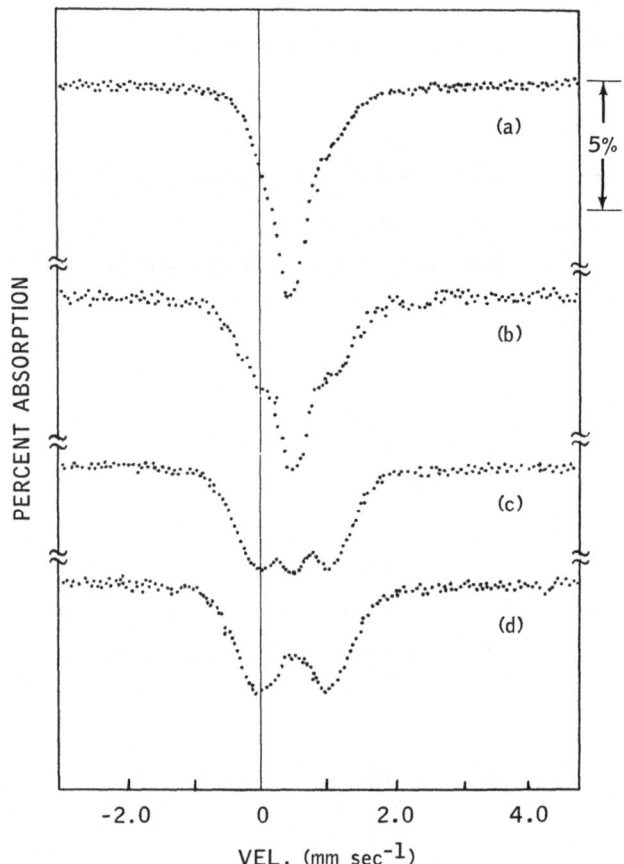

Fig. 5. Mössbauer spectra of 0.1% Fe, 2% Pt/SiO₂. Sequential experiments on one sample. (a) Helium treated 1 hr., 573°K, 100% H₂ introduced at 573°K, reduced 2 hr., 973°K; (b) Evacuated and exposed to O₂ at 298°K; (c) O₂ 2 hr., 773°K, reduced 1 hr., 573°K evacuated and exposed to O₂, 298°K; (d) O₂ 2 hr., 873°K, reduced 1 hr., 573°K, evacuated and exposed to O₂, 298°K.

in He at 300°C and then reduced at 973°K to promote ag-

glomeration of the metals. The low dispersion given in

Table 3 and intense Pt lines in the X-ray diffraction

pattern of the sample confirmed that agglomeration occur-

TABLE 3

Mössbauer Parameters for PtFe/SiO$_2$ for Various Treatments

Treatment	Peak 2 δ (mm sec^{-1})	Peaks 1-3 δ (mm sec^{-1})	Δ	χ^2/df	D
Fig. 5a	0.47	0.55	0.94	1.63	15 (10) [a]
Fig. 5b	0.47	0.51	1.27	0.98	
Fig. 5c	0.47	0.49	1.19	2.21	
Fig. 5d	0.41	0.49	1.09	1.46	

(a) Calculated particle size (nm) (See Table 1, footnote b).

ed. The Mössbauer spectra of the reduced sample (Fig. 5a) and the oxidized sample (Fig. 5b) also indicate PtFe clusters with poor dispersion. The central peak in Fig. 5b is assigned to iron inside the PtFe clusters which is not oxidized, while the peaks on either side of the central peak are assigned to iron at the surface which was oxidized to Fe^{3+} on exposure to oxygen. This is in accordance with the previous interpretation of Fig. 3c for agglomerated PdFe clusters exposed to oxygen at 298°K. Surface and bulk peaks were also discernible in the reduced sample (Fig. 5a) and computer analysis gave the results shown in Table 3, assuming three peaks. The δ for the surface peaks, based on the three peak analysis, is positive with respect to the bulk peak by 0.08 mm sec^{-1}, a result in agreement with that of Bartholomew and Boudart (11) for carbon-supported PtFe clusters. Oxidation had no effect on the δ for the bulk iron peak but substantially increased the quadrupole splitting of the surface peaks when Fe^{3+} was formed.

The effect of oxidation on the surface concentration

of iron in the PtFe clusters is shown in Figs. 5c and d.
In each case the sample was oxidized at the temperature
noted, reduced at 573°K and exposed to O_2 at 298°K.
Mössbauer spectra following the reduction at 573°K showed
that all of the iron was reduced. The 573°K reduction was
designed to remove oxygen from the metals while preserving
the surface composition induced by the prior oxidation.
That the surface composition was preserved by the 573°K
reduction is supported by calculating the expected diffusion distance for Fe in Pt during 1 hr. at 573°K. From
the Einstein diffusion equation and the reported (13)
diffusion coefficient for Fe in Pt this distance is negligible (10^{-6} nm). The exposure to O_2 at 298°K following
the 573°K reduction was designed to show changes in the
concentration of iron in the surface layers of the clusters
as indicated by the fraction of iron converted to the
ferric state. Figs. 5c and d show that with increasing
oxidation temperature the concentration of iron in the
surface layers of the PtFe clusters increases. The diffusion distance of the iron calculated as described above are
∿.1 nm (2 hr., 773°K) and ∿1.7 nm (2 hr., 873°K). Even
though this diffusion distance is calculated on the basis
of measurements on bulk PtFe alloys, the values are of the
right order of magnitude and not in contradiction to the
results in Fig. 5.

Table 3 shows that the isomer shifts for the iron inside the PtFe clusters and oxidized iron at the surface
are essentially constant as the surface concentration of
iron increases. The low value for peak 2 in Fig. 5d
probably reflects its poor resolution in the spectrum.

Re-reduction of the sample giving Fig. 5d at 973°K gave a spectrum nearly identical to the spectrum of the initial sample (Fig. 5a). The dispersion of the sample after this series of experiments was unchanged from that of the original sample. The treatment at 973°K in H_2 we conclude allows the PtFe clusters to equilibrate once again to the original surface composition of Fig. 5a. This is consistent with a calculated diffusion distance during 1 hr. at 973°K of 18 nm, a value larger than the average particle size calculated from the dispersion (Table 3).

Details of the Spectra of Bimetallic Catalysts

When fitted to two peaks the Mössbauer spectrum of PdFe/η-Al$_2$O$_3$ reduced at 673°K (Fig. 2b) gave a doublet with peaks of unequal intensities ($\varepsilon_2/\varepsilon_1$ = .77), line-widths (Γ_2/Γ_1 = 1.36) and areas (A_2/A_1 = 1.06). The area ratio increased with increasing sample temperature. These results are consistent with a combination of a distribution of chemical environments for the iron and a Goldanskii-Karyagin (G-K) effect. A multiplicity of iron environments giving a distribution of isomer shifts and quadrupole splittings would give peaks with unequal intensities and line-widths but equal areas. The G-K effect can account for the unequal areas and the increasing area ratio with increasing sample temperature.

Consideration of a model of a small metal cluster on a support (Fig. 6) shows why a multiplicity of environments and a G-K are to be expected. Regardless of the crystallite model chosen, the iron can occupy a number of

Fig. 6 – Model of octahedral fcc crystallite showing multiplicity of chemical environments for surface atoms.

sites with different symmetries and coordination numbers
which are expected to produce a distribution of isomer
shifts and quadrupole splittings. Points of contact
between the crystallite and the support add to the number
of possible environments. Furthermore, the atoms at the
surface are expected to exhibit vibrational anistropy and
a G-K effect.

For Fig. 2b the isomer shift based on the peak minima
was 0.28 mm sec^{-1}. When a distribution of environments is
present, the average isomer shift is determined by the
centroid of the spectrum. For Fig. 2b this value is
0.34 mm sec^{-1}, in good agreement with the value for the

bulk alloy. When a distribution of environments and a G-K effect are present, however, the centroid shift will be biased in direction of the peak with larger area (right-hand peak) in Fig. 2b. The average isomer shift for Fig. 2b, therefore, lies between 0.28 and 0.34 mm sec^{-1}.

Qualitatively the features of the spectrum of reduced PtFe/γ-Al$_2$O$_3$ (Fig. 4a) were the same as those for PdFe/ η-Al$_2$O$_3$ (Fig. 2b). Computer analysis of Fig. 4a for two peaks gave unequal intensities (ϵ_2/ϵ_1 = .70), linewidths (Γ_2/Γ_1 = 2.43) and areas (A_2/A_1 = 1.70). Fitting the spectra for four peaks including a ferrous doublet with peak intensities and widths constrained to be equal to that of the small peak at $\sim$2.8 mm sec^{-1} had little effect on these values due to the very small intensity of the ferrous peaks. The area ratio was found to increase with increasing sample temperature. The same interpretation as discussed in the previous paragraphs for PdFe/η-Al$_2$O$_3$ (Fig. 2b), therefore, may be applied to Fig. 4a for PtFe/γ-Al$_2$O$_3$. Determination of the centroid shift for Fig. 4a is complicated by the presence of Fe^{2+} peaks. When treated as two peaks, a centroid shift of 0.58 mm sec^{-1} is obtained for Fig. 4a compared to the value of 0.53 mm sec^{-1} determined from the peak minima. The centroid value, however, is biased too high by both the G-K asymmetry and the ferrous peak at $\sim$2.8 mm sec^{-1} and probably lies closer to 0.53 than 0.58 mm sec^{-1}.

The dispersions of the samples which gave Fig. 2b (PdFe/η-Al$_2$O$_3$) and Fig. 4d (PtFe/γ-Al$_2$O$_3$) were 50-60% as determined by hydrogen chemisorption. Why then were "bulk" peaks not resolved in the Mössbauer spectra of the

reduced samples? Several possible explanations must be
considered for this result. First, this result could be
explained by surface enrichment of Fe in the reduced
clusters but, as discussed by Williams and Nason (12),
such enrichment is not expected for small PdFe or PtFe
clusters. Second, a distribution of particle sizes is
frequently observed for supported metal catalysts (4) and
sequential addition of iron to the noble metal catalyst as
was done in this work may lead to preferential association
of the iron with the smallest particles of the distribution. Mössbauer spectra of the reduced samples, for this
case, would then reflect a higher dispersion than chemisorption which is averaged over all the particles. Third,
a cluster with ∿50% dispersion contains ∿85% of the atoms
in the first two atomic layers. Since the Thomas-Fermi
screening length in metals is of the order of the interatomic spacing (14), electric fields produced by chemisorbed hydrogen, by the ions of the support and by defects
in the cluster surface may also be felt by at least the
second layer atoms in the crystallite. The quadrupole
interactions for the second layer atoms may be of sufficient magnitude that a clear distinction between "surface"
and "bulk" peaks in the Mössbauer spectra of reduced
samples with ∿50% dispersion is not possible. On this
basis the absence of "bulk" peaks in the PdFe/η-Al$_2$O$_3$
(Fig. 2b) and PtFe/γ-Al$_2$O$_3$ (Fig. 4d), for which dispersion
was 50-60%, does not appear to be unreasonable. As the
particles increase in size, however, a point will be
reached where the "bulk" atoms do not sense the surface
effects and both "surface" and "bulk" peaks may be re-

solved if the quadrupole splitting of the "surface" peaks
is sufficiently large. Such resolution was possible for
agglomerated PtFe/SiO$_2$ (Fig. 5a) but not for agglomerated
PdFe/η-Al$_2$O$_3$ (Fig. 2c) even though the dispersions of
these samples were nearly the same. This may be attributed to the smaller quadrupole splitting for PdFe
"surface" peaks (Table 1, Fig. 2b) compared to PtFe
"surface" peaks (Table 2, Fig. 4a).

The spectra of samples oxidized at 298°K following
reduction allowed much better resolution of "surface"
peaks (as Fe^{3+}) and "bulk" peaks in agglomerated samples
(Figs. 3c and 5b). If the recoil-free fractions for the
respective species are known and the areas, corrected for
absorber thickness, can be accurately determined, the
dispersions of the samples can be calculated from the
Mössbauer spectra. Values determined in this manner are
usually much larger than the dispersions determined by
chemisorption (3). This may be explained by concentration
of the iron in the smaller PtFe or PdFe clusters as discussed previously. In addition, the exothermicity of the
oxidation at 298°K may cause local heating of the
particles (15) and disruption of several surface layers
into which oxygen is incorporated as indicated by the work
of Ratnasamy et al. (16). It thus appears that the
determination of dispersion from Mössbauer spectra of
oxidized samples will not be reliable.

CONCLUSIONS

The results of this investigation provide firm
evidence for the formation of "bimetallic clusters" in

catalysts prepared by conventional techniques and using
common catalyst supports. The fact that bimetallic
clusters of iron with a number of Group VIII metals can be
formed on a number of supports allows the extension of the
Mössbauer technique to a variety of questions important to
catalytic chemists. These include the factors which
affect multi-metallic cluster formation, adsorption interactions, metal-support interactions, effects of promoters,
surface composition and in situ catalytic studies. The
iron incorporated into a bimetallic cluster may act as a
sensitive probe for interactions not previously accessible
by other techniques. In addition, iron may also be incorporated in small concentrations as the third metal in a
bimetallic catalyst and act as a probe for cluster formation between two other Group VIII metals. This approach
has recently been utilized by us to establish composition
of matter in a bimetallic reforming catalyst. It appears
that the Mössbauer technique has great potential for increasing our knowledge of the science of small particles,
an important branch of the science of catalysis.

ACKNOWLEDGMENTS

I thank H. W. Dougherty and L. Turaew for experimental assistance in this work. The suggestions of G. Perlow
and helpful discussions with W. N. Delgass are gratefully
acknowledged.

REFERENCES

1. J. H. Sinfelt, J. L. Carter and D. J. C. Yates, J.
   Catal., 24, 283 (1972).
2. J. H. Sinfelt, J. Catal., 29, 308 (1973).
3. R. L. Garten, J. Catal., in press.
4. For example see J. R. Anderson, "Structure of Metallic
   Catalysts", Academic Press, New York, 1975.
5. P. C. Aben, J. Catal., 10, 224 (1968).
6. G. R. Wilson and W. K. Hall, J. Catal., 24, 306
   (1972).
7. R. L. Garten and D. F. Ollis, J. Catal., 35, 232
   (1974).
8. (a) M. C. Hobson, Jr. and H. M. Cager, J. Colloid and
   Interface Sci., 34, 357 (1970). (b) H. M. Cager,
   J. F. Lefelhoez and M. C. Hobson, Jr., Chem. Phys.
   Lett., 23 386 (1973). (c) T. Tachibana, T. Ohya,
   T. Yoshioka, J. Koezuka and H. Ikoma, Bull. Chem. Soc.
   Japan, 42, 2180 (1969).
9. S. M. Qaim. Proc. Phys. Soc., 90, 1065 (1967).
10. R. L. Garten, unpublished results.
11. C. H. Bartholomew and M. Boudart, J. Catal., 29, 278
    (1973).
12. F. L. Williams and D. Nason, Surf. Sci., 45, 377
    (1974).
13. M. I. Dekhtyar, V. N. Kolesmk, V. I. Patoka,
    V. I. Silantev and I. Ya. Dekhtyar, Phys. Stat. Sol.,
    24, 699 (1974).
14. J. M. Ziman, "Principals of the Theory of Solids",
    2nd Edition, Cambridge University Press, Cambridge,
    Mass., 1972.
15. (a) J. A. Cusumano and M. J. D. Low, J. Catal., 17,
    98 (1970). (b) H. Mark and M. J. D. Low, J. Catal.,
    30, 40 (1973).
16. P. Ratnasamy, A. J. Leonard, L. Rodrique and
    J. J. Fripiat, J. Catal., 24, 374 (1973).

THE APPLICATION OF MÖSSBAUER SPECTROSCOPY TO STUDIES OF SUPPORTED RUTHENIUM CATALYST SYSTEMS

C. A. Clausen, III[*] and M. L. Good[**]

[*]Department of Chemistry
Florida Technological University
Orlando, Florida   32816

[**]Department of Chemistry
University of New Orleans
New Orleans, Louisiana   70122

INTRODUCTION

Heterogeneous catalysis is inherently a complex sub-ject, and progress toward making it a science rather than an art has required knowledge and techniques in many fields. Consequently, as they have been more fully developed, physical tools such as electron microscopy and optical and magnetic resonance spectroscopy have seen increasing use in attempts to obtain detailed information on the surface structure of catalyst systems. A relatively recent addition to the collection of physical tools is the spectroscopic technique based upon the Mössbauer effect.

Mossbauer spectroscopy has been used to study the oxidation and reduction processes in the pretreatment of supported iron catalysts,[1] to measure the crystallite size of ferric oxide on high surface area supports,[2-5] and to study the chemisorption of ammonia, water and hydrogen sulfide on supported iron catalysts.[6,7] Unfortunately, most Mössbauer studies have been concerned almost exclusively with iron and tin catalyst systems, while only two studies have been reported on the much more important noble metal systems. In the noble metal group, only supported gold and platinum catalysts have been studied by Mössbauer spectroscopy. Delgass and co-workers[8] performed a Mössbauer study of gold supported on MgO and η-Al_2O_3. The Mössbauer data taken

after heat treatments showed that thermal decomposition be-
gan at temperatures greater than 140°C and that decomposition to metallic gold was not always complete. In addition,
an unidentified gold species with an absorption peak at a
more negative velocity than metallic gold was observed in
the alumina sample. Similar studies on supported platinum[9]
suggest that the strength of binding in the supported
crystallites is similar to that of platinum in a foil. Unfortunately, the isomer shifts for platinum are small relative to the observed linewidths, therefore direct chemical
information about oxidation states of the platinum present
in the catalyst has not been obtained. Results of these
studies on platinum systems suggest that this element is
not ideally suited for Mössbauer studies. The initial
results for gold systems were promising although no follow-
up work has been done. Consequently, the total contribution
that Mössbauer spectroscopy may be able to make to studies
of supported noble metal catalysts has not been completely
determined. Thus, the search for a noble metal that is more
ideally suited to Mössbauer studies of supported metal
systems is of significant practical importance. A very promising candidate for such studies appears to be ruthenium.
The Mössbauer effect in this metal has been successfully
applied to the study of ruthenium coordination compounds as
well as to mixed oxidation state compounds.[10-16] These
studies have made significant contributions to the characterization of complex multicentered ruthenium systems and
in the generalization of structure and bonding effects in
ruthenium compounds.

This paper is a review of the results of our attempts
to apply Mössbauer spectroscopy to the solution of structure
and bonding problems in several supported ruthenium systems
which may be considered models for heterogeneous catalysts.
To provide the reader with an overview of the scope of the
studies carried out, all of the various systems studied so
far are described, although some of the work has been previously reported[17] and other segments have been submitted
for publication.[18,19]

EXPERIMENTAL METHODS

Mössbauer Spectrometer. The Mössbauer spectra were obtained with the apparatus previously described.[14,17] The use

of a germanium–lithium drifted detector (Elscint Ltd., Model
GP/GC) resulted in improved resolution over that previously
reported. All spectra were obtained at 4.2°K by use of a
Kontes/Martin glass Dewar system where both the source and
absorber were immersed directly in the liquid helium well
(see Figure 1). The source used to study the alumina and
silica supported catalyst samples consisted of approximately
7 mCi of 16 day ^{99}Rh contained in a host lattice of ruthenium
metal. This source exhibited linewidths of 0.28–0.32 mm/sec
for a natural ruthenium metal absorber. The source used to
study the zeolite and automotive emission control catalysts
consisted of approximately 7 mCi of ^{99}Rh contained in a host
lattice of rhodium metal. This source exhibited linewidths
of 0.45 $\pm$ 0.30 mm/sec for a natural ruthenium metal absorber.
Both sources were prepared by New England Nuclear Corp.,
Boston, MA.

In general, the base line for each spectrum contains
between 1 and 2 million counts per channel and the relative
percent absorption of the Mössbauer peaks are in the range
of 0.1–0.5 percent. Data reduction was carried out on a
PDP-10 computer system. The spectra were subjected to a
least-squares fit to a Lorentzian line shape with both the
experimental points and the calculated least-squares curve
plotted out directly by a Calcomp Model 563 plotter. The
Mössbauer hyperfine parameters were calculated from the
least-squares fit. Error analyses for the isomer shift,
quadrupole splitting, and peak full width at half maximum
values are given along with the data.

MATERIALS

Davison silica gel Grade 923 (100–200 mesh, surface
area approximately 285 m^2/g) and Davison η–alumina Grade
992-F (100–200 mesh, surface area approximately 210 m^2/g)
were used as support materials. The model catalysts were
prepared by impregnating the support materials with aqueous
solutions of ruthenium trichloride (RuCl$_3$·1–3H$_2$O, A. D.
Mackay, Inc.) by the incipient wetness method, followed by
oven drying at 100°C. Catalyst samples were impregnated
with 10 wt. % ruthenium. After drying, each sample was
placed in a quartz cell and all further treatments were
carried out on the sample in the cell shown in Fig. 1.

The stabilized automotive emission control catalysts
were prepared by the incipient wetness impregnation of

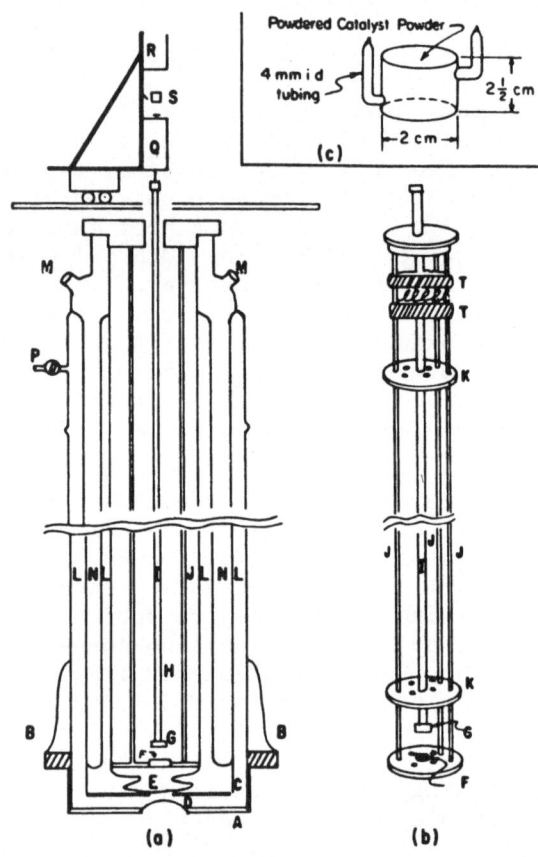

Fig. 1. Sketch of glass Dewar and quartz sample cell used
for the Mössbauer measurements reported in this work. (a)
Glass Dewar with drive assembly, linear motor and laser
calibrator in position as during a spectral run; (b) drive
assembly; (c) quartz sample cell: (A) aluminum base plate;
(B) O-ring seal; (C) Kovar radiation shield; (D) 0.001 in.
Al foil; (E) Mylar windows (0.005 in. thick); (F) absorber
filled quartz sample cell (c); (G) source; (H) helium well;
(I) drive rod (1/4 in. tube: SS, 4 mil thick); (J) support
rods (1/4 in. tubing: SS, 10 mil thick); (K) Teflon disks
to strengthen drive assembly; (L) vacuum insulation space,
silvered on both sides with a narrow unsilvered portion to
read liquid level inside; (M) liquid nitrogen fill ports;
(N) liquid nitrogen well; (P) evacuation port; (Q) motor;
(R) laser generator (He-Ne); (S) interferometer; (T)
Styrofoam insulation with glass wool in between.

the η-alumina support, first by a solution of barium nitrate followed by calcination at 900° C for 8 hours to convert the nitrate to the oxide, and secondly by a solution of ruthenium trichloride. The sample was then dried for 24 hours at 100° C. The dried samples were reduced in flowing hydrogen for 2 hours at 150° C, 2 hours at 300° C and finally 2 hours at 400° C. The very small ruthenium metal particles were then "fixed" by rapid heating in flowing air at 900° C for 1 hour, according to the procedure of Shelef and Gandhi.[20]

Linde Na-Y zeolite (63.5% SiO_2, 23.5% Al_2O_3 and 13.0% Na_2O) was used to prepare the zeolite catalyst samples. Ruthenium was exchanged into the zeolite support by use of the $[Ru(NH_3)_5N_2]Cl_2$ complex which was prepared according to the method of Allen, et al [21].   Cation exchange of this complex was performed in the following manner. The complex (1.8 - 2.2 g) was added to deoxygenated water (100 ml) and an appropriate amount of Na-Y zeolite (6 - 7 g) was then added and the exchange allowed to proceed for 12 - 16 hours with shaking under a nitrogen atmosphere. The zeolite was filtered, washed several times with water and then dried over P_2O_5 under vacuum in a desiccator for 48 hours. Based on the percent of sodium ions displaced, the exchange of $[Ru(NH_3)_5N_2]^{+2}$ was in the range of 60 - 70% for all samples prepared by this method.

Purified tank air was used in the calcination of catalyst samples. Hydrogen for the reduction steps was purified by passing it successively through a heated palladium catalyst, a 13X molecular sieve, a liquid nitrogen trap and finally through the cell. Anhydrous ammonia was purified by refluxing over sodium before distilling into storage bulbs. Matheson carbon monoxide, 99.5% pure, was passed through a trap at 195° K before use.

The average crystallite size of the supported ruthenium metal was determined both by x-ray line broadening using Warren's correction as described in Klug and Alexander [22]   and by hydrogen absorption measurements. The hydrogen adsorption isotherms were obtained with a conventional Pyrex glass, constant volume adsorption system using the method of Dalla-Betta.[23]

RESULTS AND DISCUSSION

Silica and Alumina Support. The samples listed in Table 1 may be classified into two categories according to treatment following impregnation. Samples 1-B and 3, after impregnation and drying at 110° C for 24 hrs, were reduced in flowing hydrogen for 2 hrs. at 150° C, 2 hrs. at 300° C and finally 2 hrs at 400° C. Samples 2-C and 5-C, after impregnation and drying at 110° C for 24 hrs., were calcined in flowing air for 2 hrs. at 150° C, 2 hrs. at 300° C and finally 3 hrs. at 400° C. These samples were then reduced in flowing hydrogen for 2 hrs. at 150° C, 2 hrs. at 300° C and 3 hrs. at 400° C. Sample 4, after impregnation and drying, was calcined for 2 hrs. at 150° C and 2 hrs. at 300° C. This sample was then reduced in flowing hydrogen for 2 hrs. at 150° C, 2 hrs. at 300° C and 2 hrs at 400° C.

The data in Table 1 show that the average particle size increases for these samples that are calcined before being reduced. The average particle size also increases as the temperature and length of the calcination step increases. Similar results have been observed by Dalla-Betta.[23]

Silica Support. Mössbauer spectral data obtained for ruthenium on a silica support during various stages of treatment are given in Table 2 and Fig. 2. Mössbauer data for a variety of known ruthenium compounds are given in Table 3 for comparison.

The Mössbauer spectrum for sample 1-A ($RuCl_3 \cdot 3H_2O$ impregnated on silica and then dried for 24 hr at 110°C shows that the impregnated ruthenium complex is absorbed on the surface of the silica support without undergoing a chemical change. The Mössbauer parameters for this sample are the same within experimental error as that observed for unsupported $RuCl_3 \cdot 3H_2O$.

After obtaining the Mössbauer spectrum for sample 1-A, it was reduced according to the previously described procedure. This reduced sample is called 1-B. After the accumulation of approximately 2 million counts in each channel, no absorption peaks could be detected in the Mössbauer spectrum for this sample. This was somewhat surprising since this sample gave a well-resolved spectrum

TABLE 1

PARTICLE SIZE OF RUTHENIUM METAL SUPPORTED ON SILICA AND ALUMINA

Sample No.	Support	Treatment	Wt % Ru	Ru surface area (m^2/g)[a]	Av diam (Å)	
					X-Ray	Ads
1-B	SiO_2	H_2 reduction	10	57	–	85
2-C	SiO^2	Calcined @ 400°C then reduced in H_2	10	22	240	230
3	$\eta-Al_2O_3$	H_2 reduction	10	45	95	108
4	$\eta-Al_2O_3$	Calcined @ 300°C then reduced in H	10	33	160	151
5-C	$\eta-Al_2O_3$	Calcined @ 400°C then reduced in H	10	18	295	275

[a]Calculated from hydrogen adsorption data.

C.A. CLAUSEN, III AND M.L. GOOD

TABLE 2

MÖSSBAUER PARAMETERS FOR RUTHENIUM SUPPORTED ON SILICA

Sample No.	Treatment	Absorber thickness (mg Ru/cm^2)	Isomer[a] shift (mm/sec)	Quadrupole splitting (mm/sec)	Peak width (Γ) @ half-height (mm/sec)
1-A	Before reduction	175	-0.34 ± 0.02	0	0.53 ± 0.04
1-B	After reduction	175	(No spectrum observed)		
2-A	Before reduction	165	-0.35 ± 0.02	0	0.54 ± 0.04
2-B	After calcination	165	-0.27 ± 0.02	0.46 ± 0.02	$\Gamma_1 = 0.37 \pm 0.04$
					$\Gamma_2 = 0.36 \pm 0.04$
2-C	After reduction	165	$+0.02 \pm 0.02$	0	0.34 ± 0.03

[a]Zero velocity is taken to be the center of the spectrum of a standard ruthenium metal sample.

TABLE 3

MOSSBAUER PARAMETERS OF SEVERAL WELL CHARACTERIZED RUTHENIUM COMPOUNDS

Ruthenium Species	Absorber thickness (mg Ru/cm)	Isomer shift[a] (mm/sec)	Quadrupole splitting (mm/sec)	Peak width (Γ) @ half height (mm/sec)
$RuCl_3 \cdot 1-3H_2O$	525	-0.34 ± 0.02	0	0.52 ± 0.04
Ru Metal powder	185	0.00 ± 0.02	0	0.32 ± 0.03
RuO_2	380	0.23 ± 0.03	0.51 ± 0.05	0.57 ± 0.03
RuO_4	340	$+1.06 \pm 0.01$	0	0.28 ± 0.02
$KRuO_4$	520	$+0.82 \pm 0.02$	0.37 ± 0.02	0.40 ± 0.04
$BaRuO_4 \cdot H_2O$	320	$+0.38 \pm 0.01$	0.44 ± 0.02	0.30 ± 0.02
$[Ru(NH_3)_6]Cl_2$	367	-0.72 ± 0.02	0	0.33 ± 0.05
$[Ru(NH_3)_5CO]Br$	151	-0.54 ± 0.02	0	0.39 ± 0.05
$[Ru(NH_3)_5NO]Cl_3 \cdot H_2O$	142	-0.16 ± 0.02	0.34 ± 0.02	0.31 ± 0.05
$[Ru(CO)_3Cl_2]_2$	181	-0.31 ± 0.02	0	0.42 ± 0.04
$Ru_3(CO)_{12}$	735	-0.24 ± 0.02	0	0.51 ± 0.05

[a]Zero velocity is taken to be the center of the spectrum of a standard ruthenium metal sample.

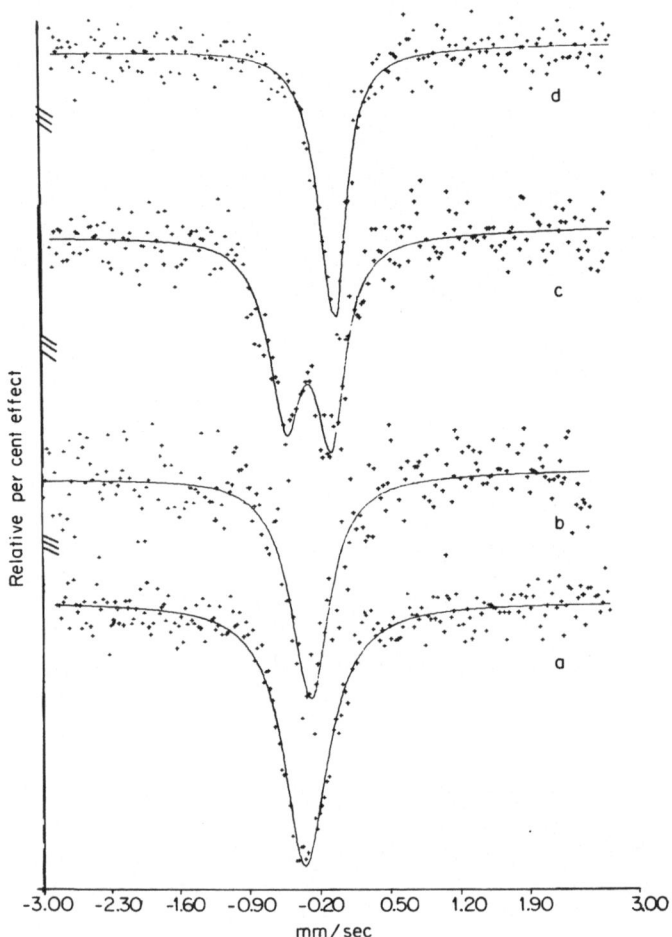

Fig. 2. Mössbauer spectra of: (a) $RuCl_3 \cdot 1-3H_2O$; (b) Sample 2-A ($RuCl_3 \cdot 1-3H_2O$ impregnated on a silica support); (c) Sample 2-B (ruthenium on a silica support after calcination); (d) Sample 2-C (ruthenium on a silica support after reduction).

prior to the reduction step. Chemical analysis showed that there was no loss of ruthenium from the catalyst sample during the reduction procedure. The absence of an observable spectrum for this sample must be the result of a decrease in the nuclear recoil-free fraction following the reduction of the complex to the metallic state.

It has been observed by Suzdalev, et al[24] that in highly dispersed tin the probability of the Mössbauer effect diminishes as the particle diameter decreases. It has also been shown by Van Wilringen[25] that in metal powders the particles may be so small that a single particle is unable to give a 'recoilless' Mössbauer transition. Van Wilringen defined the critical size of a particle as being that mass which is just sufficiently large to absorb the recoil of the gamma quantum without observable exchange of energy. The critical size can be calculated if it is assumed that the recoil energy is unobservable when it gives rise to a line displacement less than the natural line width.

Using a value for the ruthenium-99 gamma recoil energy given by Stevens and Stevens[26] it follows that the mass absorbing the recoil energy should be at least 19.4×10^5 times the mass of a single ^{99}Ru nucleus. For spherical particles of ruthenium (density 12.3 g/ cm^3) this leads to a critical particle diameter of 368 Å. Data in Table 1 show that the ruthenium particles (85 Å) in sample 1-B are much smaller than the critical size. Since no Mössbauer spectrum was observed for this sample, it must follow that either no recoil energy, or possible only an insignificant amount of the recoil energy, is transferred to the support. This suggests that the strength of the binding of the ruthenium to the silica support is very weak and that the binding forces between the atoms in the small catalyst particles are similar to those between ruthenium atoms in the powdered metal.

In order to increase the particle size of the supported metal we decided to calcine the sample before the reduction step. The new impregnated sample (2-A) before treatment exhibited a Mössbauer spectrum identical to that observed for sample 1-A. After the calcination step, a Mössbauer spectrum was obtained (sample 2-B). The data as given in Figure 2 and Table 2 show a well-resolved doublet corresponding to an isomer shift of -0.27 mm/sec and a quadrupole splitting of 0.46 mm/sec. These parameters agree very well with the isomer shift (-0.22 mm/sec) and quadrupole splitting (0.51 mm/sec) which have been observed for RuO_2.[11] The absence of any unidentified peaks in the spectrum indicates that essentially all of the ruthenium is present as small crystallites of RuO_2.

Sample 2-B was reduced according to the previously described procedure. Even though the average particle size of the metal (240 Å) in this sample is still less than the critical particle size, a Mössbauer spectrum was observed. This spectrum exhibited a single absorption peak with an isomer shift that agrees exactly within experimental error to that observed for powdered ruthenium metal. The absence of any other lines in the spectrum indicates that all of the ruthenium has been reduced to the zero valence, metal state. The fact that a Mössbauer spectrum was observed for this sample in spite of the subcritical particle size, indicates that the 'effective' Mössbauer mass of the particles must be greater than the critical mass. This suggests that weak binding forces exist between the small metal particles and the silica support. Another possible explanation is that the observed Mössbauer effect may be due to a small fraction of metal particles that are larger than 368 Å. In either case, additional work is necessary in order to establish the absolute minimum ruthenium particle size on silica for which a Mössbauer effect can be observed.

Attempts were made to obtain Mössbauer spectra of chemisorbed CO, NH_3 and H_2S on the reduced ruthenium catalyst. In each case, the chemisorbed species were introduced to a total pressure of 50 Torr at 25°C in the sample cell. The Mössbauer spectra obtained for each of these samples exhibited a single line that was identical within experimental error to that observed for the reduced catalyst. This suggests that either the ratio of surface ruthenium atoms to bulk ruthenium atoms is not great enough to observe surface effects, or that the chemisorption of these molecules on a ruthenium atom does not perturb its electronic structure enough to bring about an observable change in the Mössbauer spectrum.

Alumina Support. To investigate the nature of ruthenium supported on alumina, several samples of η-alumina were impregnated with ruthenium trichloride and treated in a manner as previously described. Mössbauer spectra of catalyst samples are given in Table 4 and Figure 3.

TABLE 4

MÖSSBAUER PARAMETERS FOR RUTHENIUM SUPPORTED ON ALUMINA

Sample No.	Treatment	Absorber Thickness (mg Ru/cm^2)	Isomer Shift (mm/sec)	Quadrupole Splitting (mm/sec)	Peak width (Γ) @ Half-Height (mm/sec)	% Abs
3	After Reduction	185	+0.01 ± 0.02	0	0.41 ± 0.04	0.2
4	Reduced After Low temp calcination	180	+0.01 ± 0.02	0	0.43 ± 0.04	0.3
5-A	Before Reduction	190	-0.41 ± 0.03	0.45 ± 0.02	Γ_1 = 0.35 ± 0.04 Γ_2 = 0.49 ± 0.04	--
5-B	After Calcination	190	-0.27 ± 0.03	0.49 ± 0.02	Γ_1 = 0.33 ± 0.04 Γ_2 = 0.35 ± 0.04	--
5-C	After Reduction	190	-0.02 ± 0.02	0	0.38 ± 0.04	0.6

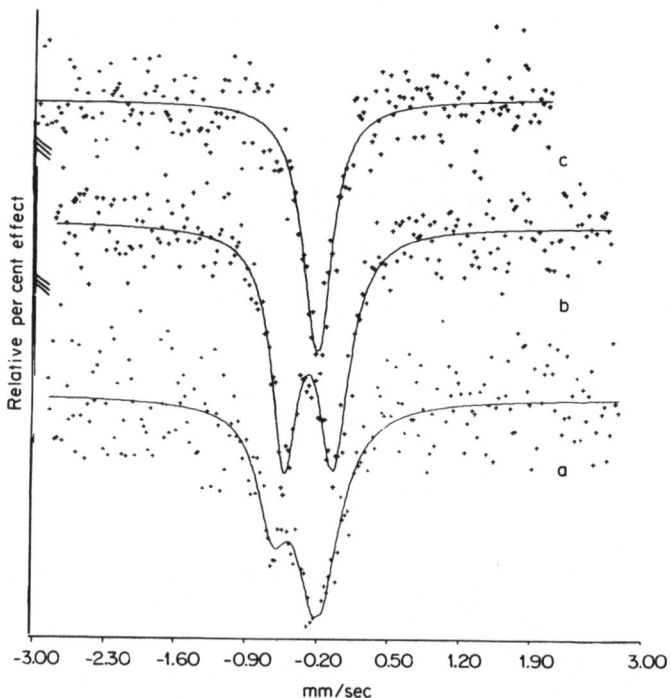

Fig. 3. Mössbauer Spectra of: (a) Sample 5-A (RuCl$_3$·1-3H$_2$O impregnated on alumina); (b) Sample 5-B (ruthenium on an alumina support after calcination); (c) Sample 5-C (ruthenium on an alumina support after reduction).

A single narrow absorption peak was observed in the Mössbauer spectra of catalyst samples 3 and 4. It should be noted that each of these samples was evacuated to a pressure of 10^{-6} Torr at 400° C, therefore their spectra represent a surface free of chemisorbed hydrogen. Within experimental error their Mössbauer parameters correspond exactly to those observed for ruthenium metal. As was observed in the case of the reduced silica catalysts, there is no evidence for the existence of any ruthenium species other than the reduced metal. Even though sample 3 was reduced directly, whereas sample 4 was calcined before reduction, the Mössbauer data show that other than for average particle size the state of the ruthenium is the same in both samples.

Both catalyst sample 3, with an average particle size
of 95 Å, and sample 4, with a particle size of 160 Å, contain supported ruthenium crystallites that are much smaller
than the critical particle size. The occurrence of a
Mössbauer effect in these samples indicates that fairly
strong binding forces exist between the metal particles and
the alumina support. The data also show that the Mössbauer
effect increases as the average particle size increases.

A Mössbauer spectrum was obtained for untreated impregnated ruthenium trichloride on an alumina support (5-A).
The spectrum as shown in Figure 3 exhibits an asymmetric
doublet with an isomer shift of -0.41 mm/sec and a quadrupole splitting of 0.45 mm/sec. This spectrum is significantly different from the spectrum obtained for unsupported
$RuCl_3 \cdot 1-3H_2O$ and $RuCl_3 \cdot 1-3H_2O$ supported on silica. The
isomer shift is slightly more negative than that observed
for the unsupported ruthenium trichloride and falls in a
region that borders on the upper end of isomer shifts observed for Ru(III) complexes and the lower end of isomer
shifts for Ru(IV) complexes. Therefore, it is difficult
to determine whether the ruthenium has undergone a change
in oxidation state or has been coordinated to the support.

Sample 5-A was calcined to form catalyst sample 5-B.
The Mössbauer spectrum for this sample shows that all of
the ruthenium was converted to RuO_2. This sample was then
reduced and evacuated to a pressure of 10^{-6} Torr at a
temperature of 400°C. The spectrum for this sample (5-C)
exhibited a single line with spectral parameters that agree
with those observed for the other reduced alumina samples.
The Mössbauer effect (0.6%) was greater than that observed
for the other reduced alumina samples. This is expected,
since the average particle size of sample 5-C (295 Å) is
twice the average particle size of the other samples.

Again, attempts were made to obtain Mössbauer spectra
for CO, NH_3, H_2O, O_2 and H_2S chemisorbed (at 25°C) on the
reduced catalyst samples. The Mössbauer spectra obtained
for each of these samples exhibited a single line that was
identical within experimental error to that observed for
reduced ruthenium on an alumina support. It is somewhat
surprising that no chemisorption effects were observed with
catalyst sample 3. The average metal particle size in this
sample is only 95 Å, which should give a favorable, surface
to bulk metal atom ratio.

It is possible that the application of Mössbauer spectroscopy to the study of chemisorption in ruthenium catalyst systems may be more effectively exploited by using zeolite catalysts, such as has been done for iron systems.[27,28] An initial attempt at such a study is described in the following section.

Zeolite Support. Listed in Table 5 are the Mössbauer parameters obtained for the $[Ru(NH_3)_5N_2]^{+2}$-Y zeolite samples. Representative Mössbauer spectra for two of the zeolite samples are shown in Figure 4. The synthetic faujasite Y type zeolite was choosen for this study because a great deal of information about its structure, catalytic activity and the chemical nature of the cation exchange sites has been published.[29] For the purpose of introducing ruthenium atoms into the zeolite framework, we chose to use the dinitrogen complex cation $[Ru(NH_3)_5N_2]^{+2}$ because we felt that it offered the possibility for stripping the NH_3's and N_2 from the co-ordination sphere, leaving the bare ruthenium ion in the zeolite. Secondly, we felt that the N_2 group might serve as a pathway for reversibly introducing such groups as CO, NO etc. into the ruthenium coordination sphere.

The sample referred to as 10-A in Table 5 corresponds to a portion of the $[Ru(NH_3)_5N_2]^{+2}$-Y zeolite after drying for 48 hours under vacuum. The sample was held under a vacuum of 10^{-5} Torr while the spectrum shown in Figure 4-(a) was obtained. The isomer shift and relative line intensities observed for this sample agree with those obtained for a crystalline sample of $[Ru(NH_3)_5N_2]Cl_2$. However, the quadrupole splitting for the zeolite sample (0.56 mm/sec) was greater than that observed for the $[Ru(NH_3)_5N_2]Cl_2$ sample (0.22 mm/sec). These data suggest that the $[Ru(NH_3)_5N_2]^{+2}$ group is exchanged without undergoing oxidation or ligand loss. However, the increase in quadrupole splitting indicates that some distortion in the coordination sphere has occurred upon exchange. The distortion may be produced by the rigid aluminosilicate backbone structure of the zeolite. For example, cations exchanged in a zeolite have been found to be capable of occupying several different sites within the zeolite framework.[30,31] Since these sites are located on the sides and at the distances of different

TABLE 5

MOSSBAUER PARAMETERS FOR RUTHENIUM EXCHANGED ZEOLITES

Sample No.	Treatment	Absorber Thickness (mg Ru/cm^2)	Isomer Shift (mm/sec)	Quadrupole Splitting (mm/sec)	Peak width (Γ) @ Half-Height (mm/sec)
10-A	$[Ru(NH_3)_5N_2]^{+2}$-Y dried and evacuated to 10^- Torr @ 25°C	125	-0.80 ± 0.04	0.56 ± 0.04	0.61 ± 0.05
10-B	Sample 10-A exposed to air for 2 days @ 25°C	125	-0.37 ± 0.03	0	0.79 ± 0.05
10-C	Sample 10-B reduced in H_2 at 400°C for 4 hours	125	$+0.02 \pm 0.03$	0	0.61 ± 0.04
10-D	A portion of Sample 10-A reduced in H_2 at 400°C for 4 hours	110	$+0.01 \pm 0.02$	0	0.52 ± 0.03
10-E	Sample 10-D exposed to air for 24 hours @ 25°C	110	-0.10 ± 0.03	0	0.67 ± 0.04
Ru Metal Powder -------		225	0.00 ± 0.02	0	0.45 ± 0.03
$[Ru(NH_3)_5N_2]Cl_2$ --------		175	-0.76 ± 0.04	0.22 ± 0.03	0.51 ± 0.03
$[Ru(NH_3)_5OH]Cl_2$ -------		190	-0.39 ± 0.03	0	0.49 ± 0.03

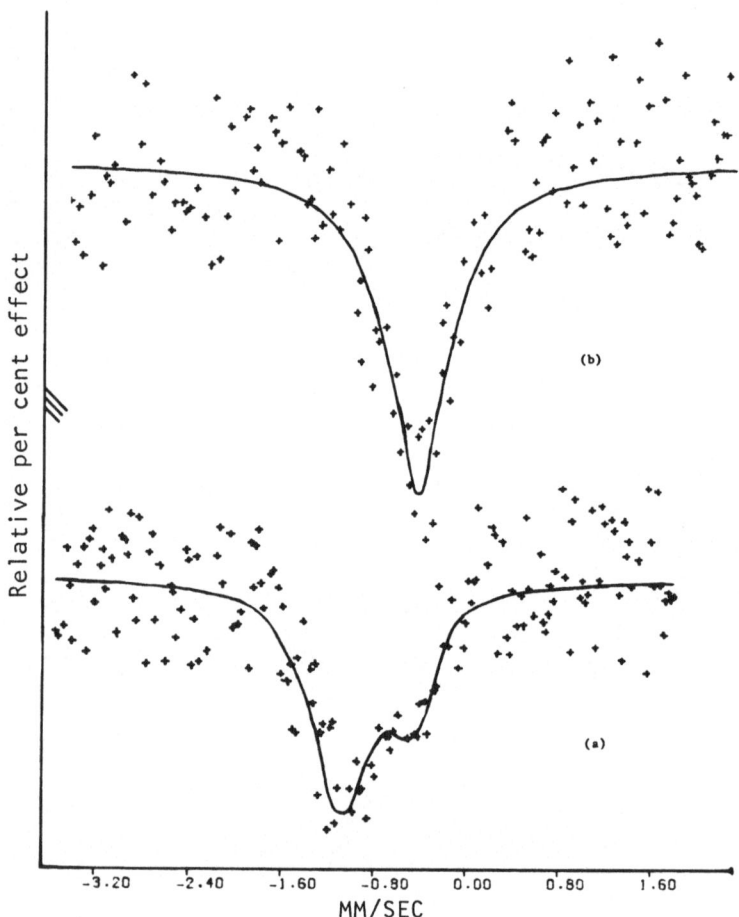

Fig. 4. Mössbauer spectra of: (a) Sample 10-A
($[Ru(NH_3)_5N_2]^{+2}$-Y zeolite after drying); (b) Sample 10-B
(this is sample 10-A after exposure to air).

size and shaped cavaties, each site would impose its own
characteristic structural and electronic requirements on
the cation. Unfortunately, because of the limited number
of studies dealing with this topic, it is not currently
possible to identify the position of the $[Ru(NH_3)_5N_2]^{+2}$
cation in the Y-zeolite from its Mössbauer spectral
parameters.

Upon exposure of sample 10-A to the atmosphere at 25°C, it slowly turned a wine color. After two days, the color appeared to stabilize and the spectrum shown in Figure 4-(b) was obtained. This sample is referred to as sample 10-B in Table 5. The Mössbauer parameters have changed significantly. The change in isomer shift from -0.80 to -0.37 mm/sec indicates that the ruthenium has undergone oxidation from the +2 state to the +3 state.[14] The only +3 wine colored ruthenium compound that could be found in the literature corresponds to the $[Ru(NH_3)_5OH]Cl_2$. The Mössbauer spectral parameters for this compound as shown in Table 5 agree with those observed for the wine colored compound in the zeolite. The broader linewidth for the zeolite sample may result from the presence of smaller concentrations of other ruthenium species or possibly $[Ru(NH_3)_5OH]^{+2}$ groups at different sites in the zeolite.

Laing et al.[32] have also observed that a $[Ru(NH_3)_5N_2]^{+2}$-Y zeolite sample decomposes in air to give a wine colored species. They proposed that the decomposition may occur by the following reaction:

$$[Ru^{II}(NH_3)_5N_2]^{+2}\text{-Y} + H_2O \rightarrow [Ru^{III}(NH_3)_5OH]^{+2}\text{-Y}$$

$$+ NH_4^{+} + \text{other products}$$

Sample 10-C in Table 5 corresponds to sample 10-B after treatment in a stream of hydrogen for 4 hours at 400°C. The Mössbauer parameters for this sample indicate that all of the ruthenium has been reduced to the metallic state. However, X-ray analysis of this sample indicated that a significant amount of crystallinity in the zeolite framework was lost upon reduction. Therefore, a new portion of sample 10-A was reduced in a hydrogen stream and this sample is referred to as 10-D. Again, the Mössbauer data show that all of the ruthenium has been reduced to the metallic state. X-ray analysis of this sample also indicated that most of the zeolite structure was maintained during the treatment and that all of the ruthenium metal particles were less than 80 Å in diameter. When this sample was exposed to the atmosphere, the Mössbauer data designated for sample 10-E in Table 5 was obtained. The observed change in isomer shift and linewidth upon exposure to the atmosphere indicates that some type of interaction has occurred between the small ruthenium metal particles and the gaseous components of air. The change in isomer

shift by −0.10 mm/sec is greater than the experimental error
in the measurement and indicates that the effective s−electron
density has been reduced in a majority of the ruthenium atoms.
This could be due to increased shielding brought about by
chemisorbed groups occupying p and d ruthenium orbitals, or
by direct s−electron withdrawal by chemisorbed groups. However, the important thing is that some form of interaction
was observed. This indicates that a favorable, surface atom
to bulk atom, ratio exists in the sample, since larger
ruthenium metal particles supported on silica and alumina
exhibited no interaction with air as observed by Mössbauer
spectroscopy. Therefore, it appears that zeolite supported
ruthenium can be used as a model system for studying chemisorption phenomena on ruthenium metal by Mössbauer spectroscopy.

Automotive Emission Control Catalysts. For the purpose
of illustrating the practical application of Mössbauer
spectroscopy to the solution of problems in the area of
ruthenium catalysis, we wish to report on a preliminary
study of emission control catalysts containing ruthenium.
Ruthenium−containing catalysts have been found to have a
pronounced selectivity for reduction of nitrogen oxides to
molecular nitrogen, and attention has recently been focused
on the development of these catalysts as a means of controlling nitrogen oxide emissions.[20,33,34] However, these
studies have shown that ruthenium catalysts exhibit poor
stability when the exhaust contains a net oxidizing composition. Analysis of spent catalysts revealed severe losses
of the active component, which was readily explained by the
formation and removal of the volatile ruthenium tetroxide.[20]

One method which has been prepared to minimize the
tendency of the ruthenium to volatilize is based on the
formation of the nonvolatile barium ruthenate.[20] The barium
ruthenate was prepared in situ on the alumina support by
impregnation first with a solution of barium nitrate followed by calcination to convert the nitrate to the oxide.
The support was then impregnated with a solution of ruthenium
trichloride. The catalyst was dried and reduced in hydrogen
and then 'fixed' by rapid heating in air at 900°C. Catalyst
samples prepared by this technique were found to exhibit
considerable improvement in the prevention of ruthenium
volatization while maintaining the desirable selective
catalytic reduction of nitric oxide to molecular nitrogen.[20]
However, under vehicle operating conditions the loss of

ruthenium from the stabilized catalyst was still found to
be higher than acceptable. The reason for this gradual loss
in ruthenium can be accounted for by referring to the
Mössbauer data in Table 6 and Figure 5.

Sample 20-A is a sample containing 12 weight % barium
and 4 weight % ruthenium prepared by the method described
above. The Mössbauer data for this sample was taken after
the 900°C 'fixation' step. The three lines in the spectrum
match those that would be found for a sample containing a
mixture of barium ruthenate and ruthenium dioxide. The peak
area ratio indicates that the barium ruthenate is present
in a greater concentration. This piece of data indicates
that in this sample, every ruthenium atom has not been deposited in the vicinity of a stabilizing oxide so as to
assure the formation of the ruthenate. Therefore, the loss
of the non-stabilized ruthenium during operation would
account for some of the ruthenium volatilization.

Sample 20-A was heated @ 700°C in a simulated auto
exhaust (SAE) having the following composition:

Component	Content, mole %
H_2	0.33
O_2	0.35
H_2O	10.00
CO	2.00
CO_2	13.00
C_3H_g	0.10
NO	0.10
N_2	74.12

The treated sample was called 20-B and its Mössbauer data
shown in Table 6 indicates that all of the ruthenium has
been reduced to the metallic state by this treatment. The
SAE was then made net oxidizing in composition by substituting 2% O_2 for the 2% CO. Sample 20-B was heated in the
oxidizing SAE mixture for 30 minutes at 700°C. This sample
after treatment is called 20-C. Its Mössbauer spectral
parameters show that the ruthenium has been oxidized back
to $BaRuO_3$ and RuO_2. However, the peak area ratio indicates
a smaller composition of $BaRuO_3$ than was present in the
initial sample. This sample was cycled at 700°C between
the net reducing SAE for 50 minutes and 10 minutes in the
net oxidizing SAE. The treatment was continued for 150

TABLE 6

MÖSSBAUER DATA FOR STABILIZED RUTHENIUM AUTOMOTIVE EMISSION CONTROL CATALYSTS

Sample No.	Treatment @ 700°C	No. of Lines in Spectrum	Isomer Shift (mm/sec)	Quadrupole Splitting (mm/sec)	Peak Area Ratio [a]
20-A	Initial Fixed Sample	3	-0.30 ± 0.04 -0.24 ± 0.04	0 0.53 ± 0.05	2.3:1.0
20-B	Sample 20-A heated for 10 hrs in SAE[b]	1	+0.02 ± 0.03	0	-------
20-C	Sample 20-B heated for 30 minutes in net oxidizing SAE atmosphere	3	-0.27 ± 0.04 -0.22 ± 0.03	0 0.53 ± 0.03	2.0:1.0
20-D	Sample 20-B cycled between net reducing SAE and net oxidizing SAE for 150 hours	3	-0.28 ± 0.05 -0.23 ± 0.03	0 0.52 ± 0.03	0.4:1.0
	Barium Ruthenate --------	1	-0.28 ± 0.03	0	-------
	Ruthenium Dioxide--------	2	-0.23 ± 0.03	0.51 ± 0.05	-------

[a] This is the ratio of the area of the single peak to the area of the pair of quadrupole split peaks.
[b] SAE = simulated auto exhaust

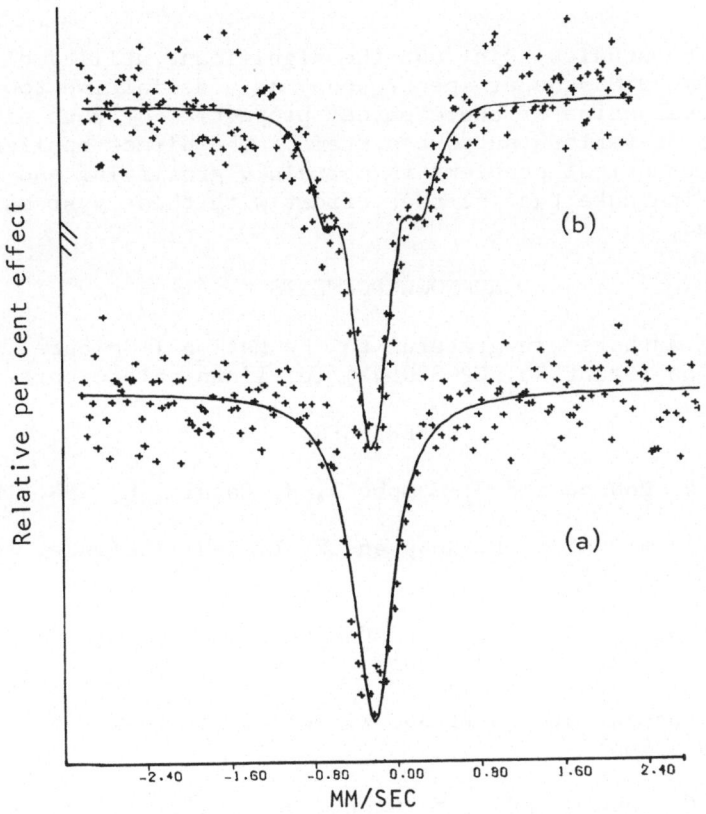

Fig. 5. Mössbauer spectra of: (a) Barium ruthenate; (b) Sample 4-A (12% barium and 4% ruthenium on an alumina support after initial 'fixation' step).

cycles over a period of 150 hours. The treated sample is referred to as sample 20-D. The Mössbauer spectrum for this sample indicates again that all of the ruthenium is in the form of barium ruthenate and ruthenium dioxide. However, the peak area ratio indicates that the BaRuO$_3$ concentration is now less than the RuO$_2$ concentration. This indicates that the cycling of these stabilized catalysts between a net reducing atmosphere and a net oxidizing atmosphere results in a significant separation between the ruthenium metal and the stabilizing agent. This explains why these catalysts do not have a satisfactory lifetime.

CONCLUSION

These studies point out the significant utility of ruthenium-99 Mössbauer spectroscopy as a definitive tool in the evaluation of the chemical properties of ruthenium moieties on various support systems. The direct application to practical problems is certainly gratifying and the results indicate that further effort with these systems is justified.

ACKNOWLEDGEMENTS

The authors are grateful to the National Science Foundation (Grant No. GP-38054X) for finanical support.

REFERENCES

1.   M. C. Hobson and J. Campbell, J. Catal., 8, 294 (1967).

2.   P. A. Flinn, S. L. Ruby and W. L. Kehl, Science, 143, 1434 (1964).

3.   G. Constabaris, R. H. Lindquist and W. Kündig, Appl. Phys. Lett., 7, 59 (1965).

4.   H. Dunken, H. Hobert and W. Meisel, Z. Chem., 6, 276 (1966).

5.   M. C. Hobson and H. M. Gager, J. Catal., 16, 254 (1970).

6.   M. C. Hobson, Nature (London), 214, 79 (1967).

7.   H. M. Gager, J. F. Lefelhocz and M. C. Hobson, Chem. Phys. Lett., 23, 386 (1973).

8.   W. N. Delgass, M. Boudart and G. Parravano, J. Phys. Chem., 72, 3563 (1968).

9.   W. N. Delgass and M. Boudart, Catal. Rev., 2, 145 (1968).

10.   O. C. Kistner, Phys. Rev., 144, 1022 (1966).

11. C. A. Clausen, R. A. Prados and M. L. Good, Chem. Commun., 1188 (1969).

12. G. Kaindl, W. Potzel, F. Wagner, U. Zahn and R. L. Mössbauer, Z. Phys., 226, 103 (1969).

13. C. A. Clausen, R. A. Prados and M. L. Good, J. Amer. Chem. Soc., 92, 7482 (1970).

14. C. A. Clausen, R. A. Prados and M. L. Good, in "Mössbauer Effect Methodology" (I.J. Gruverman, Ed.), Vol. 6, p. 31. Plenum, New York, 1971.

15. R. A. Prados, C. A. Clausen and M. L. Good, J. Coord. Chem., 2, 201 (1973).

16. M. L. Good, "A Review of the Mössbauer Spectroscopy of Ruthenium-99 and Ruthenium-101", Mössbauer Effect Data Index, (J.G. Stevens and V.E. Stevens, Eds.), pp. 51-69. Plenum, New York, 1972.

17. C. A. Clausen and M. L. Good, J. Catal., 38, 92 (1975). [A report of the alumina and silica supported systems].

18. C. A. Clausen and M. L. Good, J. Catal., submitted, [A report of the zeolite supported materials].

19. C. A. Clausen and M. L. Good, J. Catal., submitted, [A report on the automotive catalysts].

20. M. Shelef and H. S. Gandhi, Platinum Metal Rev., 18, No. 1, pp 2 (1974).

21. A. D. Allen, F. Bottomly, R. O. Hains, V. P. Reinsaln and C. V. Senoff, J. Amer. Chem. Soc., 89, 5595 (1967).

22. H. P. Klug and L. E. Alexander, "X-ray Diffraction Procedures", pp 504-509, Wiley, New York, (1954).

23. R. A. Dalla-Betta, J. Catal., 34, 57 (1974).

24. I. P. Suzdalev, M. Y. Gen, V. I. Goldanskii and E. F. Markarov, Sov. Phys., JETP, 24, 79 (1967).

25. J. S. Van Wilringen, Phys. Lett. A, 26, 370 (1968).

26. J. G. Stevens and V. E. Stevens, "Mössbauer Effect
     Data Index", P. 266, Plenum Data Corp., New York (1973).

27. W. N. Delgass, R. L. Garten and M. Boudart, J. Phys.
     Chem., 73, 2970 (1969).

28. R. L. Garten, W. N. Delgass and M. Boudart, J. Catal.,
     18, 90 (1970).

29. H. S. Sherry, Advan. Chem. Soc., 101, 350 (1971).

30. J. V. Smith, Advan. Chem. Soc., 101, 171 (1971).

31. D. H. Olson, J. Phys. Chem., 74, 2758 (1970).

32. K. R. Laing, R. L. Leubner and J. H. Lunsford, Inorg.
     Chem., 14, 1400 (1975).

33. M. Shelef and H. S. Gandhi, Ind. Eng. Chem., Prod.
     Res. Dev., 11, 393 (1972).

34. R. L. Klimisch and K. C. Taylor, Environmental Sci.
     Tech., 7, 127 (1973).

HEMES, IRON SULFUR CENTERS, AND SINGLE CRYSTALS: SOME ASPECTS OF RECENT BIOLOGICAL WORK

E. Münck and R. Zimmermann*

Freshwater Biological Institute/University of

Minnesota, Navarre, Minnesota 55392

I. INTRODUCTION

Nature abounds with a wide variety of iron containing biomolecules. These are intimately involved in practically all life sustaining processes and in that capacity they serve such functions as catalysis (nitrogenase and cytochrome P450, discussed below, are examples), electron transport (iron-sulfur proteins, many heme proteins, and rubredoxin), transport and storage of diatomic molecules (hemoglobin and myoglobin), and iron transport and storage. A typical biomolecule may have a molecular weight of 10^4 to 10^5, consisting for the most part of protein, a chain of amino acids folded in a specific way to stabilize a well-defined three-dimensional structure. Many biomolecules contain one or a few metal centers which are the focal point of action, i.e. the metal is at the "business center" of the molecule.

There are two classes of proteins which have attracted most of the attention of Mössbauer spectroscopists, the heme proteins and iron-sulfur proteins. Heme proteins perform many vital functions. Probably the best known example of this class is hemoglobin, the oxygen carrier in the blood. All heme proteins have essentially the same prosthetic group, namely Fe-protoporphyrin IX (heme). As shown in Figure 1, the iron atom is incorporated into an aromatic ring and coordinated to four nitrogen atoms. In general, a

---

*On leave from the University of Erlangen-Nürnberg, Germany

fifth (axial) position is occupied by an amino acid residue
(like histidine in hemoglobin, i.e. a nitrogenous ligand);
the sixth position is either vacant, or occupied by another
amino acid residue or by an extraneous ligand (like O_2, CO,
or a substrate, i.e. a compound that is to be modified by
the catalytic action of the heme protein). The multitude of
possible coordinations permits the diversity of biological
functions which characterize the heme proteins. It is truly
amazing that nature has invented a prosthetic group which is
capable of electron transport, oxygen transport and catalysis of a variety of chemical reactions.

The iron-sulfur proteins were discovered only a decade
ago (by EPR spectroscopy). Although these proteins have
been evolved by nature very early we had to wait until the
middle of this century before they were discovered. There
are principally two reasons for this late discovery. First
of all, iron-sulfur proteins are not very conspicuous with
respect to the principal tool of modern biochemistry--
spectrophotometry. Secondly, the classical approach of biochemistry, wet chemistry, had to fail because the iron
centers fall apart when the protein is treated in an un-
gentlemanly fashion. However, after the first inkling of
these new proteins was gleaned, a new field of biochemical
research was born which has grown and matured in the past
ten years. While it is meaningless to compare the relative
importance of biological molecules, it should be noted that
iron-sulfur proteins provide cornerstones for many key biochemical reactions, from the most primitive bacterium to
man. Their main function seems to be to transport electrons
at unusually low reduction potentials.

Iron-sulfur proteins derive their name from the fact
that they contain acid-labile or inorganic sulfur (sulfide)
which is always used to bridge iron atoms to yield spin-
coupled clusters. The simplest structure is found in the
so-called plant-type ferredoxins or 2Fe-2S* proteins. The
active center is made up of two iron atoms and two bridging
sulfur atoms. The whole structure is suspended into the
protein by four sulfur ligands which are furnished by
cysteine amino acid residues. The structure as depicted in
Figure 2 is the result of the application of a whole battery
of physical tools such as optical absorption spectroscopy,
circular dichroism, magnetic susceptibility, NMR, EPR, ENDOR,
and Mössbauer spectroscopy. The decisive clues as to the
nature of the active site structure, namely the inequivalence
of the iron atoms (one ferric and one ferrous in the reduced,

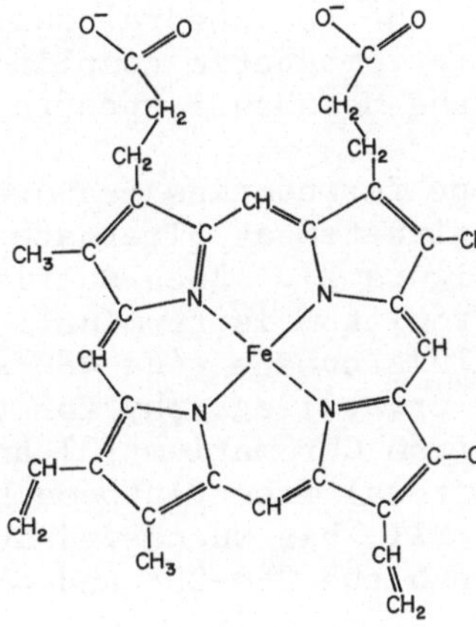

Figure 1. The heme group (Fe-protoporphyrin IX).

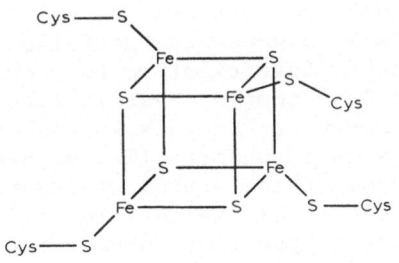

Figure 2. Structures of 4Fe-4S* and 2Fe-2S* clusters. The labile sulfur atoms bridge the iron atoms. In proteins the clusters are connected to the polypeptide chain by cysteine residues.

paramagnetic protein), the tetrahedral sulfur coordination, and the proof of antiferromagnetic coupling were provided by detailed analyses of the Mössbauer spectra [1,2].

The bacterial type ferredoxins feature 4Fe-4S* clusters, with four iron atoms situated at alternate corners of a distorted cube (see Figure 2). Each sulfide bridges three iron atoms and each iron atom is terminally coordinated to a cysteine residue. This cubane 4Fe-4S* structure has been determined from x-ray crystallography for the high-potential-iron-protein (HIPIP) from Chromatium [3] and for an eight-iron protein (two clusters) from Peptococcus aerogenes. R.H. Holm's group, at MIT, has succeeded in synthesizing model complexes for both the 2Fe-2S* and 4Fe-4S* clusters.

With so many exciting structures provided by nature the Mössbauer spectroscopist might ask at what level of investigation he can best apply his talents. In very complex and largely unexplored systems, like nitrogenase discussed below, we can use the Mössbauer effect to probe iron atoms which are not amenable to other spectroscopic tools. In the case of nitrogenase the Mössbauer effect revealed some structures which have yet not been seen in any other system. In Chapter II we will discuss the usefulness of combining an EPR and Mössbauer investigation to tackle the problem of nitrogenase. For less complex systems like rubredoxin (discussed by P. Debrunner in this Volume) and cytochrome P450 (discussed by us in Chapter III) we can make a detailed analysis to elucidate the electronic structure of the iron atoms, determine electronic zero-field splittings, and magnetic and electric hyperfine parameters. By comparing the results with information obtained from other proteins, or model complexes, we may be able to draw conclusions regarding the ligand structure and get some insight into the mechanism of catalysis.

In a few instances the proteins are available as single crystals. In that case we might want to determine the principal components and the orientation of the EFG tensor relative to, say, the crystallographic axes. Such work has been performed on a myoglobin single crystal which has two heme groups per unit cell which can be transformed into each other by a 180° rotation. Thus it seems that the problem is the same as for one site since the second site is connected to the first by a 180° rotation. Surprisingly, it is impossible to determine the principal axes values and the

orientation of the EFG tensor by measuring quadrupole spectra
for different orientations of the myoglobin crystal. We
will address this problem in Chapter IV.

Since this Methodology Symposium deals with catalysis
we think that it might be appropriate to describe briefly
the reactions which the proteins discussed catalyse. It is,
of course, not possible here to go into the details, but
we hope that the reader might appreciate a short, though
superficial, introduction into the biochemical problem.

II. NITROGEN FIXATION. THE MO-FE PROTEIN

A. The Biochemical Problem

Nitrogenase is an enzyme system capable of reducing
molecular nitrogen to ammonia. It has been estimated that
about 90 million tons of nitrogen are fixed annually by
biological systems, primarily by symbiotic N_2-fixing systems
and fixing systems of photosynthetic organisms. In contrast,
the industrial Haber process provides about 20 million tons
of fixed nitrogen as fertilizers for agriculture.

The reaction catalyzed can be written simply as

$$N_2 + 6H^+ + 6e^- \rightarrow 2 NH_3$$

This simple equation disguises the spectacular process
the biological system has to perform. An enzyme has to bind
the very unreactive N_2 molecule, provide 6 protons, and perform a reduction which requires 6 electrons, some of which
at very low redox potential. The biological system, called
nitrogenase, consists of two proteins, a molybdenum-iron
protein (MoFe protein) and an iron protein (Fe protein).
The reaction is energy-driven by coupling the hydrolysis of
ATP into ADP to the reduction of N_2.

Four basic ingredients are required to achieve biological N_2-fixation in vitro: MoFe protein (MW $\simeq$ 220,000, about
20 iron atoms, about 20 labile sulfur, and 1 or 2 Mo atoms),
Fe protein (MW $\simeq$ 60,000, 4 iron and 4 labile sulfurs;
Mössbauer studies implicate a 4Fe-4S* cluster), ATP (in
form of Mg · ATP), and a low potential reductant (usually
sodium dithionite).

Most of our knowledge about the mechanism of nitrogenase has come from EPR spectroscopy. The Mo-Fe protein
exhibits a unique EPR signal with principal g-values at
4.32, 3.67 and 2.01. The origin and the nature of the
structure giving rise to this signal has been a mystery for
a long time; both iron and molybdenum have been implicated
in the past. Under nitrogen fixing conditions, i.e. with
all components present, this EPR signal vanishes and it
returns when either the reductant or Mg·ATP is exhausted.
The Mössbauer investigation described below was specifically
addressed to the problem of the nature of this EPR signal.
(The MoFe protein certainly contains the catalytic, i.e.
N_2-binding site. The Fe protein serves as an electron
carrier which can transfer electrons to the MoFe protein.
Orme-Johnson and co-workers [4] have elucidated the process
of electron transfer in an elegant series of EPR experiments
and they have shown that Mg·ATP binding to the reduced Fe
protein lowers its reduction potential thereby allowing
electrons to be transferred to the MoFe protein).

Before going into the details we can make some state-
ments about the things we may expect. The MoFe protein has
about 20 Fe atoms and as many labile sulfur atoms; from past
experience we can therefore expect to find spin-coupled
structures. Since we are dealing with a six-electron
reduction process we can expect that the iron centers act
as intermolecular electron storage and transfer centers. In
the following we will describe how a coupled EPR and Mössbauer
investigation revealed the structure of the centers which
give rise to the observed EPR signal. We will show that
there are two EPR active centers per molecule, that the EPR
signal results from a $S = 3/2$ spin system and that eight
iron atoms are associated with these two centers, i.e. the
EPR signal originates from an iron complex consisting of
four iron atoms.

The discussion below follows closely the work by Münck
et. al. [5] who studied the protein from Azotobacter
vinelandii with EPR and Mössbauer spectroscopy. Smith and
Lang [6] have published a Mössbauer study of the proteins
from Klebsiella pneumoniae. The Mössbauer spectra of the
proteins from both species are almost identical.

B. Mössbauer and EPR Results

The rather unique g-values at 4.32, 3.65 and 2.01, observed for the MoFe protein, can be derived from an S=3/2
spin system. We will show that the Mössbauer and EPR results
can be described adequately in the framework of the spin
Hamiltonian

$$H_e(S=3/2) = D[(S_z^2 - 5/4) + \eta/3(S_x^2 - S_y^2)] + g_e\mu_B\vec{H}\cdot\vec{S} \quad (1)$$

The first term describes the fine structure of the spin
quartet, the zero field splitting. The second term expresses
the interaction of the electronic magnetic moment $g_e\mu_B\vec{S}$ with
an applied magnetic field $\vec{H}$. Note that we describe   the
Zeeman interaction by a simple g-value, the g-value of the
free electron, g_e=2.00. In general, one has to use a g-tensor;
the above assumption, however, is adequate to describe the
experimental results. The reader may note that the zero-
field splitting term in Eq (1) differs from the usual con-
vention used by EPR spectroscopists, D $[S_z^2-5/4 + \lambda(S_x^2-S_y^2)]$.
We have used the above notation to emphasize the
analogy of the S=3/2 zero-field splitting with the quadrupole term of the ^{57}Fe nuclear excited state. In Chapter III
we will equate g_e with the nuclear g-value of the 14.4 Kev
level. We then can use the EPR concept of effective g-values
to describe the response of the nuclear excited state to a
magnetic perturbation.

In zero applied magnetic field the spin quartet is split
in two <u>Kramers</u> doublets, separated by an energy
$\Delta = D\sqrt{1 + 1/3\eta^2}$. For small applied magnetic fields H such
that $g_e\mu_BH \ll |\Delta|$ it is convenient to describe the magnetic
properties of each doublet by a spin Hamiltonian with an
effective spin S'=1/2

$$H(S'=1/2) = \mu_B\vec{S}'\cdot\tilde{g}\cdot\vec{H} \quad (2)$$

The principal components of the g-tensor in Eq (2)
depend on η and g_e and can be computed from Eq (1) for each
doublet. In Figure (3) we have plotted the effective g-values
for each doublet as a function of η. For small values of η,
i.e. for nearly axial symmetry, the two doublets have drastically different EPR properties. One doublet gives rise to
an intense EPR signal, the other doublet is almost EPR
silent (the ±3/2 doublet, for η = 0). For the EPR active

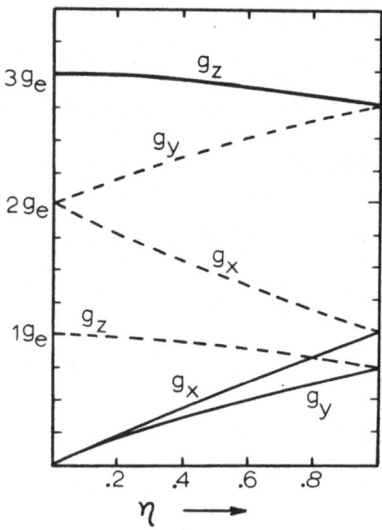

Fig. 3. The effective g-values of the two Kramers doublets
of an S=3/2 system are plotted as a function of η. The g-
values of the ±1/2 doublet are given by the dashed lines;
the solid curve refers to the ±3/2 doublet.

doublet we easily can compute the g-values; for g_e=2.00 and
η=0.165 we obtain g_x=3.66, g_y=4.32, and g_z=1.98, in good
agreement with the observed g-values for MoFe protein.

From the Mössbauer data we can infer (see below) that
the zero-field splitting is positive and that Δ/k ≃ 15°K.
Thus at 4.2°K only the EPR active doublet is populated. We
now come to a very important point regarding the biochemistry
of nitrogenase. How many S=3/2 centers do we have per
molecule? By carefully measuring the integrated EPR signal
intensity, we found 2.1 spins per molecule, i.e. the MoFe
protein seems to have two identical S=3/2 centers (Earlier
spin quantitations gave about 0.5 spins per molecule. The
EPR investigation here was substantially strengthened by
simultaneously investigating the sample with Mössbauer
spectroscopy. Even a minor amount of oxygen present can
oxidize the S=3/2 centers to an EPR-silent species which would
lead, of course, to an underestimation of the spin concentration. Since the EPR active species was distinguishable in
the Mössbauer spectra from all other iron atoms, it was

possible to verify that it was present in maximum concentration under the conditions of the EPR experiment.).

Our goal in Mössbauer spectroscopy was to decide whether the EPR signal originates from an iron or a molybdenum center. How can we establish such an association unambiguously? The answer is actually quite simple, although the details might be tricky for a protein which contains about 20 iron atoms. First, at sufficiently low temperatures, i.e. for long electronic relaxation times, we will find a Mössbauer spectrum showing paramagnetic hyperfine structure if the signal indeed originates from an iron center. This magnetic spectrum should be characteristic of a Kramers system, i.e. magnetic spectra should be observed in weak applied fields and in zero field. In addition the magnetic spectrum should reflect the magnetic properties of the EPR active Kramers doublet. Secondly, from the EPR studies we know that, at 20°K, the electronic spin relaxation is much faster than the nuclear precession frequency. Thus at 20°K our magnetic spectrum should have collapsed into a quadrupole doublet. Finally, the EPR signal vanishes under N_2-fixing conditions; hence the magnetic Mössbauer spectrum must vanish under these conditions also.

In Figures 4-6 we have displayed Mössbauer spectra of the native MoFe protein from Azotobacter vinelandii taken under various experimental conditions. The spectra in Figure 4, taken at 1.5°K, consist of a super-position of three quadrupole doublets and a spectral component showing paramagnetic hyperfine structure; the latter spectrum is the component we are interested in here.

The three quadrupole doublets, which reflect iron in sites of integer or zero electronic spin, are spectroscopic components which have not yet been seen by any other spectroscopic tool; the task of their elucidation falls squarely upon the shoulders of Mössbauer spectroscopy. Nevertheless, since we are interested here in the magnetic component, we discuss them only briefly. For a more detailed discussion we refer the reader to the original publications [5,6].

Two of the doublets are readily recognized. One appears in the center of the spectra (labeled D); it has a quadrupole splitting of $\Delta E_Q = 0.81$ mm/s (independent of temperature) and an isomeric shift $\delta = 0.64$ mm/s (relative to Fe metal).

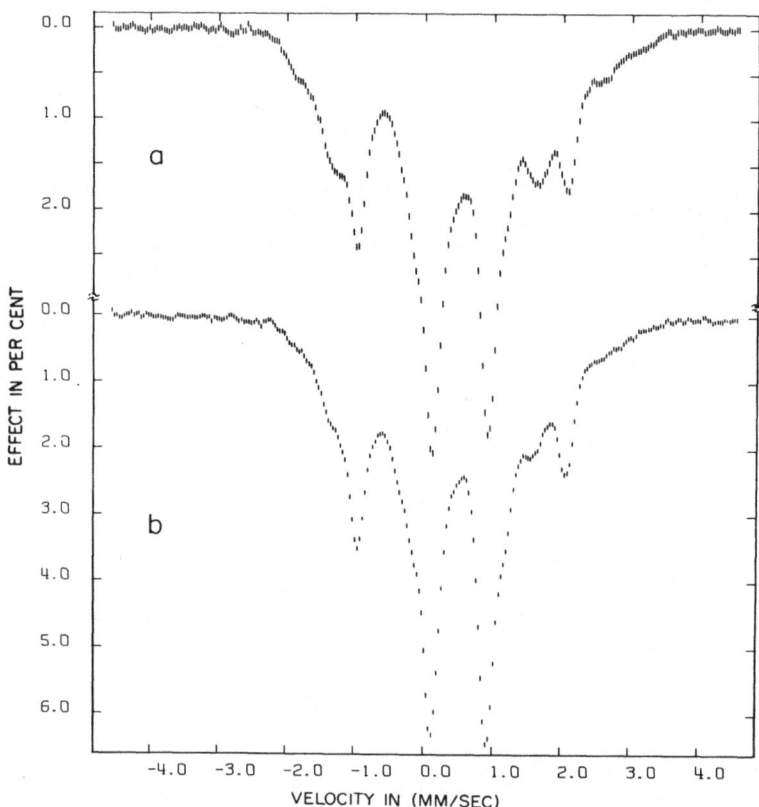

Figure 4. Mössbauer spectra of ^{57}Fe enriched MoFe protein from Azotobacter vinelandii. The data were taken at 1.5°K in a magnetic field of 360 gauss applied (a) parallel and (b) transverse the γ-radiation. The spectra in Figure 4-6 are plotted relative to a ^{57}Co (Rh) source, kept at room temperature. Consult this Figure together with Figure 5.

The other doublet (labeled Fe^{2+}) has absorption lines at −0.9 mm/s and +2.1 mm/s. The values ΔE_Q=3.02 mm/s and δ=0.69 mm/s suggest high-spin ferrous iron; indeed, this component is remarkably similar to reduced rubredoxin (see P. Debrunner's contribution in this volume). We are presently working on the elucidation of the physical nature of the sites which give rise to spectral components D and Fe^{2+} and we have good evidence that they reflect spin-coupled

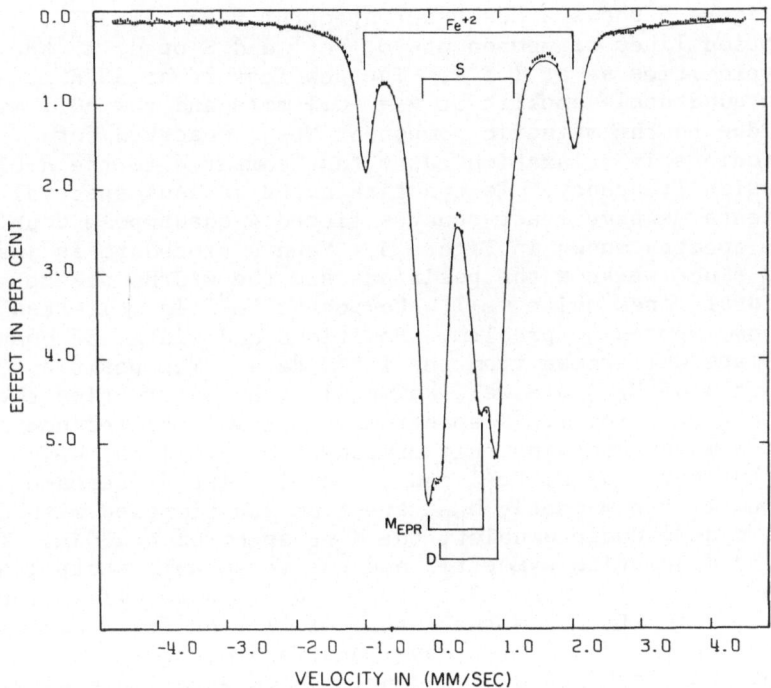

Figure 5. Mössbauer spectrum of the MoFe protein taken at
30°K. The solid line is the result of fitting four quadrupole doublets to the data (see text).

iron-sulfur centers. There is a third quadrupole doublet
(labeled S) in the 1.5°K spectra. It can only be recognized
with some imagination in the raw data but it is clearly
discernible when component D is subtracted from the data;
S gives rise to two symmetric absorption lines at -0.2 and
$+1.2$ mm/s, hence $\Delta E_Q \simeq 1.4$ mm/s and $\delta \simeq 0.6$ mm/s.

   The remainder of the spectra in Figure 4a and 4b is a
spectral component showing paramagnetic hyperfine structure.
The intensities of this component depend on the orientation
of the applied magnetic field relative to the γ-radiation;
this is quite apparent at velocities of -1.3 and $+1.8$ mm/s.
We will label it M_{EPR} and we will show that it represents
iron atoms associated with clusters which give rise to the
observed $S=3/2$ EPR signal.

Figure 5 shows a Mössbauer spectrum taken at 30°K. The absorption lines of components D, Fe^{2+} and S occur at the same velocities as at 1.5°K. The new feature at 30°K is a strong quadrupole doublet at v = -0.1 mm/s and v = +0.7 mm/s; it is due to the magnetic component M_{EPR}, observed for electronic spin-relaxation rates fast compared to the nuclear precession frequency. To quantitate the various spectral components we have least-squares fitted 4 quadrupole doublets to the spectra shown in Figure 5. Such a procedure is justified since we know the positions and the widths of the individual lines quite well. Component Fe^{2+} is well-resolved and thus imposes no problem. Positions and widths of component D are well-known from the 1.5°K data. The positions and widths of M_{EPR} are well-known also; by subtracting the 30°K data from the 1.5°K spectrum we obtain a difference spectrum which contains only component M_{EPR}, since the contributions from D, Fe^{2+} and S cancel. The difference spectrum is the magnetic M_{EPR} spectrum superimposed with an inverted quadrupole doublet, the M_{EPR} spectrum at 30°K. The inverted doublet is symmetric and has reasonably sharp lines (.3 mm/s); we find $\Delta E_Q = 0.76$ mm/s and $\delta = 0.40$ mm/s. The solid line in Fig.5 is the result of fitting four quadrupole doublets to the data. (We have constrained the fit by requiring that both lines of a doublet have equal intensity.) In terms of total iron absorption we find: 42.5% for D, 38.5% for M_{EPR}, 14% for Fe^{2+}, and 5% for component S. In translating these numbers into the numbers of iron atoms we assume that the Debye-Waller factor is the same for each component.

The story gets more interesting now. The 38.5% absorption found for spectral component M_{EPR} corresponds to approximately 8 iron atoms. Combining the EPR and Mössbauer results we can already draw some conclusions at this stage of the analysis: since there are two S=3/2 centers/molecule and since eight iron atoms are associated with these two centers, each center must contain four iron atoms. Moreover, the four iron atoms must be spin-coupled. Furthermore, the presence of a half-integer electronic spin implies an odd number of electrons in each cluster. Thus, if we have four iron atoms, they must be inequivalent or another paramagnetic atom (molybdenum?) must be involved. Inequivalent iron atoms are recognized clearly when we examine the magnetic spectra associated with M_{EPR}.

   To analyze the magnetic component M_{EPR} in the 1.5°K
spectra we have subtracted spectral components D, Fe^{2+}, and
S from the data in Figure 4 using the results of the least-
squares fitting procedure. The remainder, component M_{EPR}
in its magnetic form, is displayed in Figure 6.

   We first of all notice that the magnetic splitting of
M_{EPR} is rather small. This is typical for proteins which
contain clusters with a 4Fe-4S* tetramer core where extensive spin delocalization leads to a strongly reduced magnetic hyperfine interaction. The spectra in Figure 6 contain clearly two magnetic components, i.e. inequivalent
irons. One spectrum extends from -1.3 mm/s to +1.7 mm/s;
its shape is quite sensitive to the direction of the applied
magnetic field. The second component shows resolved structure
for velocities larger than +2 mm/s and a shoulder at -2 mm/s.
The relative intensities of both components do not change
in the temperature range from 1.5°K to 6°K; this excludes
the possibility that one spectrum results from the ground
state and the other from the excited state of the S=3/2
system. This observation then implies that the zero-field
splitting Δ > 8°K. (Values for Δ < 1.5°K can be excluded
from EPR work at 35 GHz and from Mössbauer investigations
in external fields H > 5kG.) Moreover, since the spectrum
in Figure 6 is sensitive to the direction of the applied
field it follows that it results from the EPR active Kramers
doublet (which is essentially a ±1/2 state). The ±3/2 state
should have g-values at 0.34, 0.32, and 6.0 for η = 0.165;
a Mössbauer spectrum associated with the ±3/2 doublet would
be insensitive to the direction of the applied field.

   To further substantiate the association of component
M_{EPR} with the observed EPR signal we have attempted a computer simulation for one of the magnetic components. The
relevant Hamiltonian may be written as

$$H(S=3/2) = H_e + A_o \vec{S} \cdot \vec{I} + H_Q \qquad (3)$$

with $H_Q = \dfrac{eQV_{zz}}{4I(2I-1)} \; [3 \; I_z^2 - I(I+1) + \eta(I_x^2 - I_y^2)]$

and H_e from Eq (1).

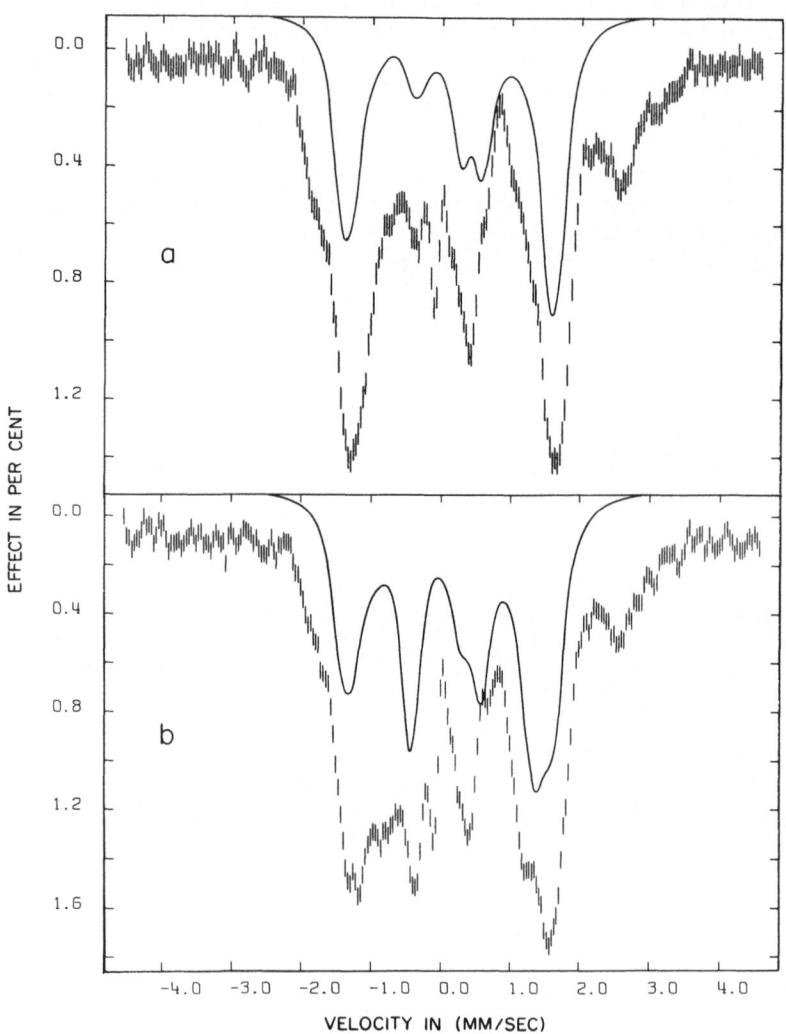

Figure 6. Low temperature spectrum of spectral component M_{EPR} in (a) parallel and (b) transverse field. Spectra were obtained from those shown in Figure 4 by subtracting components D, Fe^{2+}, and S using the fitting results. The displayed spectra contain at least two components. For one of them a computer simulation was attempted. (see text)

In Eq (3) we have assumed that the magnetic hyperfine interaction is isotropic (A_o). Since only the EPR active doublet
is populated at 1.5°K and since $g_e \mu_e << \Delta$, we may describe
the spectra in the effective spin formalism. This yields

$$H (S'=1/2) = \mu_e \vec{S}' \cdot \tilde{g} \cdot \vec{I} + \vec{S}' \cdot \tilde{A} \cdot \vec{I} + H_o \qquad (4)$$

From Eqs (4) and (1) it follows that the following relation
holds for each doublet

$$\frac{g_x}{A_x} = \frac{g_y}{A_y} = \frac{g_z}{A_z} = \frac{g_e}{A_o}$$

    Thus, since the g-values are known from EPR measure-
ments, the magnetic hyperfine tensor in Eq (4) is determined
except for a scaling factor. This factor is easily determined from the total magnetic splitting of the Mössbauer
spectrum. Since ΔE_o is known from the data taken at 30°K,
only the asymmetry parameter η is left unknown. The solid
curve in Figure 6 is the result of a computer simulation
using a program that computes Mössbauer spectra for paramagnetic S=1/2 systems for polycrystalline materials [7].
The following set of parameters was used for the computa-
tion: $g_x=3.65$, $g_y=4.32$, $g_z=2.01$, $A_o=4.0 \cdot 10^{-4}$ cm^{-1}(12.1 MHz),
$\Delta E_o=+0.78$ mm/s, and $\eta = 0.6$. The computed spectrum shown
in Figure 6 reproduces the features of one of the magnetic
components rather well. We have not attempted a computer
simulation for the second magnetic component, but we think
that its computation requires a fairly anisotropic A-tensor
in Eq (3).

    We may use the Mössbauer results to compute the hyper-
fine broadening of the EPR lines for ^{57}Fe enriched protein.
Using the quoted value for A_o of the computer simulated
spectrum (the one with the smaller magnetic splitting) we
expect a line broadening of about 5 gauss at g = 2.01, in
good agreement with the experimental results (7 gauss, see
[5]).

    We have mentioned above that the magnetic component
M_{EPR} has to vanish when the sample is studied under N_2-fix-
ing conditions. Indeed, M_{EPR} vanishes concommittantly with
the EPR signal and transforms into a quadrupole doublet
[5,6]. The Mössbauer data show that each S=3/2 cluster is

reduced by the addition of one electron. In contrast, components D, Fe^{2+}, and S remain in the same spectroscopic state as observed in the native MoFe protein. The elucidation of their role in the N_2 fixation process is the goal of Mössbauer studies which are presently being undertaken in our laboratory.

III.   OXYGEN FIXATION.   CYTOCHROME P450

A.   The Catalytic Reaction

In the last Chapter we dealt with an enzyme that has to cleave molecular nitrogen to produce two molecules of ammonia. In the following we will discuss an enzyme which cleaves molecular oxygen, incorporates one oxygen atom into the substrate in form of a hydroxyl group while the other oxygen atom ends up in H_2O. Such an enzyme is termed a monooxygenase, in contrast to dioxygenases which incorporate both oxygen atoms into a substrate. Perhaps the best understood monooxygenase system is the camphor hydroxylation system from Pseudomonas putida which has been studied extensively in I. C. Gunsalus' laboratory at the University of Illinois.

The heart of the camphor hydroxylation system in P. putida is the heme protein cytochrome P450, an enzyme with MW ≃ 40,000; it contains one heme group. The axial ligands coordinated to the heme are not known with certainty though a cysteine residue is implicated by many spectroscopic tools.

The proposed reaction mechanism is shown in Figure 7. We start at the lower left in Figure 7 and move clockwise around the reaction cycle. The native protein is in a low-spin ferric state (S=1/2). Upon binding of substrate, camphor, the electronic configuration of the heme iron changes, although the substrate does not bind directly to the iron atom. In the temperature range from 1.5°K to 200°K one observes a mixture of high-spin ferric and low-spin ferric material. The low-spin component is not due to uncomplexed P450; its spectroscopic signature is similar but clearly distinct from the native, substrate-free material. In the next step the P450-substrate complex is reduced in a one electron transfer reaction mediated by

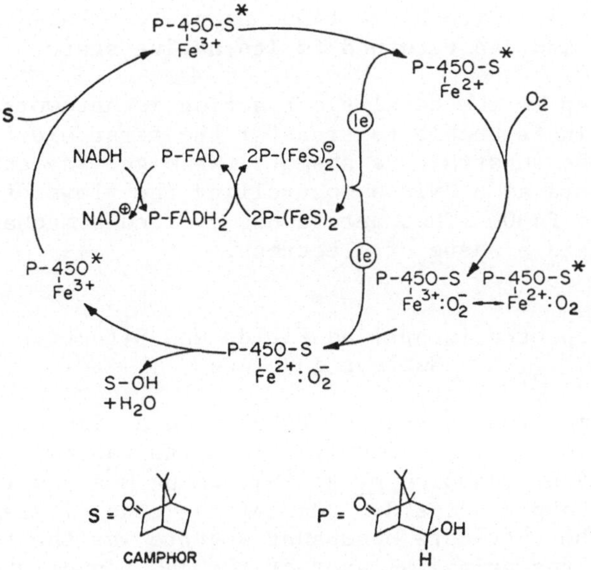

Figure 7. Proposed reaction mechanism for the hydroxylation of camphor in Pseudomonas putida. NADH (NAD+) is a biological reducing agent in its reduced (oxidized) state, P-FAD is a flavoprotein, P-(FeS)$_2$ is the 2Fe-2S* protein, putidaredoxin, S and S-OH stand for the substrate, D-camphor, and the product, hydroxylated camphor, respectively. Species marked with an asterisk have been studied by Mössbauer spectroscopy.

putidaredoxin, an electron transfer protein (putidaredoxin is an iron-sulfur protein with a 2Fe-2S* center as depicted in Figure 2; this protein has been studied extensively by Mössbauer spectroscopy [1]). After reduction the heme iron is in a high-spin ferrous (S=2) state; this intermediate will concern us in this chapter. The reduced P450-substrate complex binds to molecular oxygen to yield a diamagnetic complex. The Mössbauer data suggest that this complex is quite similar to oxygenated hemoglobin, i.e. the oxygen is bound to the heme iron [8]. Thus, at this stage, the P450 enzyme has assembled the reactants, camphor and O_2. The remaining steps, addition of a second electron and product formation are mediated by putidaredoxin and proceed through an unstable intermediate, or perhaps a

series of them. Cytochrome P450 has now completed a catalytic cycle and has returned to its native state.

One step in the catalytic reaction is noteworthy.
Putidaredoxin is unable to transfer the first electron to
P450 when the substrate is absent. We might say that the
substrate acts as a switch controlling the flows of
electrons to P450. Thus nature has evolved a mechanism
which prevents a waste of electrons.

B.   Spectra in Applied Fields and Effective Nuclear g-Values

The heme iron in reduced P450 is in a high-spin ferrous
configuration (S=2). Recently this state was extensively
investigated by Champion et al [9], using Mössbauer spectroscopy in applied fields. One of the crucial steps in
analysing the intricate Mössbauer spectra was the recognition that the principal axes of the EFG tensor are
tilted substantially relative to the frame describing the
electronic zero-field splitting. The decisive clues of a
rotated EFG came from a systematic investigation of spectra
taken at low temperatures (<10°K) in magnetic fields
H < 10kG. For P450, as for many other high-spin ferrous
compounds it is fruitful to discuss the magnetic behavior
of the excited nuclear state by effective g-values. (Details
of evaluating data with the aid of effective nuclear g-
values have been described by Champion [10]).

The hyperfine splittings of the nuclear excited state
are usually described by the Hamiltonian

$$H_n = -\mu_n g_e \vec{H}_{eff} \cdot \vec{I} + H_Q \qquad (5)$$

In the absence of an effective magnetic field the nuclear
excited state is split by the quadrupole interaction H_Q in
two Kramers doublets. (For simplicity we refer to them as
$\pm 3/2$ and $\pm 1/2$ states). If these doublets are perturbed by
an effective magnetic field such that $g_e \mu_n H_{eff} << |\Delta E_Q|$ we
can treat the magnetic Hamiltonian as a perturbation to the
quadrupole Hamiltonian, i.e. we can ignore matrix elements
connecting the two nuclear Kramer doublets. We then can
assign to each doublet an effective nuclear spin $I'=1/2$

and we can describe the magnetic splitting of each doublet
by an effective Hamiltonian H'

$$H' = -\mu_n \vec{H}_{eff} \cdot \tilde{g}_e \cdot \vec{I}' \tag{6}$$

where $\tilde{g}_e$ is the g-tensor of the pertinent nuclear doublet.
The tensor $\tilde{g}_e$ is diagonal within the principal axes system
of the EFG   and the g-values depend on η only. The
magnetic splitting of each Kramers doublet is given by a
formula well known in EPR work

$$\Delta E = \mu_n \sqrt{g_x^2 H_{eff,x}^2 + g_y^2 H_{eff,y}^2 + g_z^2 H_{eff,z}^2}$$

For the ^{57}Fe state with nuclear spin I=3/2 we can utilize
the formal analogy of the quadrupole Hamiltonian with the
zero field splitting term of an electronic S=3/2 system
(cf Eq (1)). This analogy allows us to employ the diagram
in Figure 3 to obtain the g-tensor defined in Eq (6).

     From the diagram in Figure 3 we can see the g-values
of the ±3/2 doublet are extremely anisotropic and that this
doublet does not split for η = 0 if the magnetic field acts
anywhere in the x-y plane of the EFG tensor. Moreover, the
diagram shows that the g-values of both doublets are the
same for η = 1; this reflects the well-known fact that a
spectrum perturbed by an isotropic magnetic field is
symmetric for η = 1.

     The effective g-values describe the response of the
nuclear excited state to a perturbing effective field. In
paramagnetic compounds H_{eff} is the vectorial sum of the
internal and the external magnetic field. The internal
magnetic field which arises from a polarisation of the
electronic shell can be computed from a variety of theoretical models (molecular orbitals, ligand field, spin
Hamiltonian). The appropriate model to use depends on the
problem at hand. For compounds with an orbital singlet
ground state which is well removed from excited orbital
states the spin Hamiltonian approximation has been proven
successful in explaining the Mössbauer data of some high-
spin ferrous compounds [11,9,12]. In this approximation
the effects of spin-orbit coupling and of the applied
field H on the spin quintet associated with the orbital

ground state can be expressed by the spin Hamiltonian (S=2)

$$H_e = D(S_z^2 - 2) + E(S_x^2 - S_y^2) + \mu_B \vec{H} \cdot \tilde{g} \cdot \vec{S} \qquad (7)$$

To describe the Mössbauer spectra Eq (7) can be augmented by terms describing the nuclear hyperfine interaction,

$$H_{hf} = \vec{S} \cdot \tilde{A} \cdot \vec{I} + H_Q - g_n \mu_n \vec{H} \cdot \vec{I} \qquad (8)$$

For an extensive discussion of Eqs (7) and (8) the reader is referred to the book by Abragam and Bleaney [13]. High-spin ferrous iron compounds usually show fast relaxation; just a few exceptions have been reported [11,14]. In the fast relaxation limit the Hamiltonian (8) can be replaced by the Hamiltonian (5). H_{eff} is now given by

$$H_{eff} = -\frac{1}{g_n \mu_n} \tilde{A} \cdot <\vec{S}>_T + \vec{H} \qquad (9)$$

where $<\vec{S}>_T$ is the expectation value of the electronic spin thermally averaged over the five spin states,

$$<\vec{S}>_T = \frac{\text{Tr } (S \exp(-H_e/kT))}{\text{Tr } (\exp(-H_e/kT))} \qquad (10)$$

By a judicious choice of experimental parameters the magnitude of $<\vec{S}>_T$ can be controlled in such a way that one can study high-spin ferrous compounds under conditions where the concept of effective nuclear g-values is applicable. Before applying the above considerations to P450 we first discuss a somewhat simpler compound, ferrous fluosilicate hexahydrate ($FeSiF_6 \cdot 6H_2O$)

Ferrous fluosilicate has been studied with a variety of physical tools. Spiering et al have published a detailed analysis of the high-field Mössbauer spectra [15]. For our discussion here it is sufficient to note that at low temperatures the compound has approximately trigonal symmetry, $E \simeq 0$, and that $D \simeq +15°K$. These parameters yield an energy level scheme with a spin singlet ground state, a doublet at 15°K, and another doublet at 45°K. Thus for low temperature ($T \simeq 4.2°K$) $<\vec{S}>_T$ is equal to the expectation value $<\vec{S}>$ for the singlet ground state. The latter can easily be computed

from Eq (7) by treating the external field in second order
perturbation theory,

$$<S_x> \simeq <S_y> \simeq -6 \frac{\mu_B g_\perp H_\perp}{D} , \quad <S_z> \simeq 0 \qquad (11)$$

From Eqs (9) and (11) we can see that the internal magnetic
field lies in the x-y plane. Since ferrous fluosilicate, at
4.2°K, has (approximately) trigonal symmetry [15], the principal EFG component of maximum absolute value is normal to the
x-y plane and $\eta \simeq 0$. The zero field splitting and EFG tensors are aligned. With this information we can use Figure 3.
The nuclear $\pm 3/2$ level is not split by the effective
($\simeq$ internal) field because of $g_x = g_y = 0$ for $\eta = 0$. On the
other hand, the $\pm 1/2$ level will split by an amount
$2 |g_e| \mu_n H_{eff}$. The observed splittings in the Mössbauer
spectrum are, of course, the sums of the ground state and
excited state splittings. For the splitting ratios we
obtain

$$\frac{\Delta_{1/2}}{\Delta_{3/2}} = \frac{g_g + 2|g_e|}{g_g + 0} = \frac{0.18 + 0.2}{0.18} \simeq 2$$

This ratio is in good agreement with the data on ferrous
fluosilicate (Figure 8). (In evaluating such data one has to

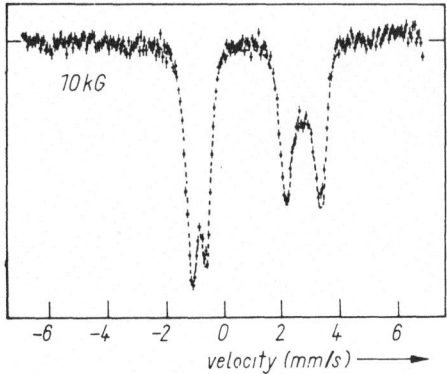

Figure 8. Mössbauer spectrum of ferrous fluosilicate hexahydrate (FeSiF$_6 \cdot$6H$_2$O), taken at 4.2°K in a 10kG parallel field.

take into account the finite linewidth Γ.   Champion [10]
has shown that the Δ's to be used are obtained by subtracting
approximately 0.7 Γ from the width at half height).   We now
turn to a more wicked compound, reduced cytochrome P450.

In Figures 9-12 we have displayed some representative
spectra taken on the high-spin ferrous form of P450.   Figure
9 shows a spectrum taken in zero-field.   An applied field
induces a polarisation of the electronic spin resulting in
spectra showing paramagnetic hyperfine interactions.   The
spectrum in Figure 10 was taken at 4.2°K in a parallel field
of 8.6 kG.   Notice that the quadrupole interaction is still
dominant although the magnetic hyperfine interaction is
sizable.   For a 25 kG field the internal field has become
dominant (see Figure 11).   From the field and temperature
dependence of the low temperature Mössbauer spectra and from
magnetic susceptibility data it emerged that D > +15°K and
E/D < 0.15.   Thus the hyperfine field in P450 is essentially
in the x-y plane of the zero-field splitting tensor.

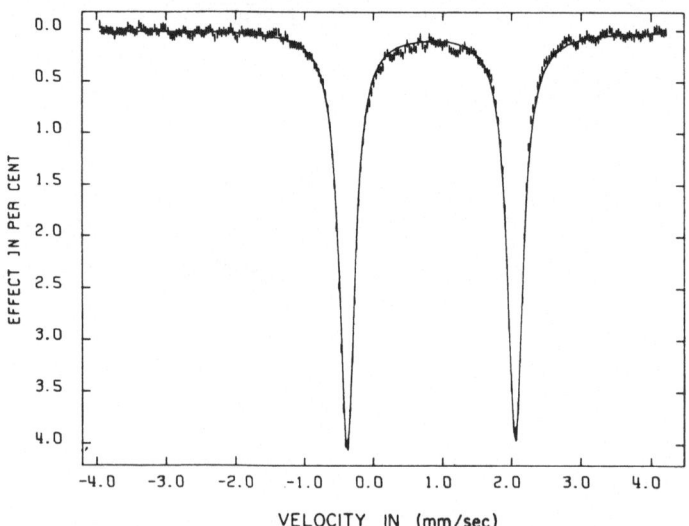

Figure 9.   Mössbauer spectrum of reduced cytochrome P450
taken in zero field at 4.2°K.   Note the clean quadrupole
doublet; Mössbauer spectra of biomolecules are not necessarily
messy!

With this knowledge we can inspect the magnetically perturbed spectrum shown in Figure 10 and utilize the effective g-value diagram in Figure 3. Figure 10 shows that the ratio of the widths of the magnetically perturbed lines is closer to 1 than to 2; this excludes $\eta = 0$. Figure 3 shows that we can broaden the $\pm 3/2$ line by increasing η. For large values of η the splitting of the $\pm 3/2$ states is essentially given by g_x and g_y (solid lines), the splitting for the $\pm 1/2$ states results from g_y (dashed line). However, even for $\eta = 1$ the $\pm 3/2$ line comes out too narrow. We can improve the situation by requiring that the hyperfine field acts in the y-z plane of the EFG tensor, i.e. we rotate the EFG tensor by 90° around the y axis. The magnetic splittings are now determined by g_z (solid line) and by g_y (dashed line). This arrangement allows that both absorption lines in Figure 10 have similar width. The recognition of a rotated field gradient was the crucial step in analysing the intricate high-field spectra of reduced cytochrome P450.

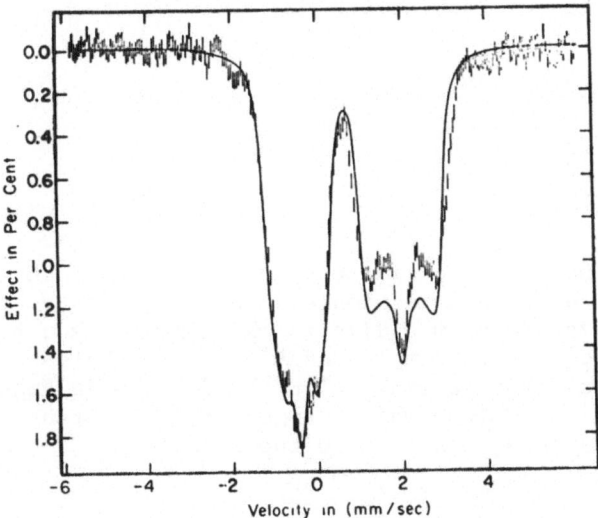

Figure 10. Spectrum of reduced P450 taken at 4.2°K in a field of 8.6kG applied parallel to the transmitted γ-rays. The solid lines in Figures 10-12 are the results of extensive computer simulations. The parameters used for the computation of the spectra are quoted at the end of this Chapter.

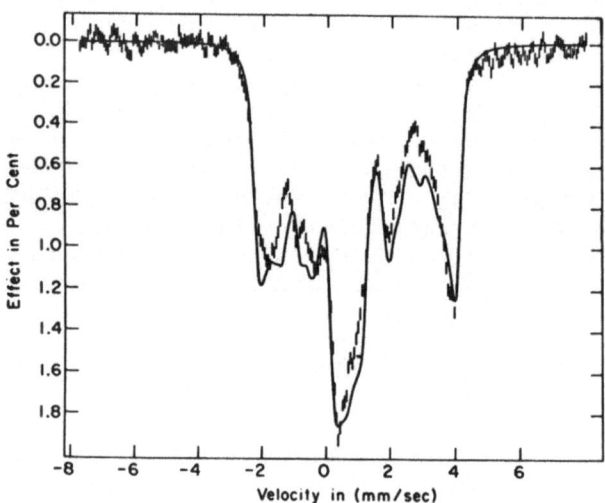

Figure 11. Mössbauer spectrum of reduced P450 taken in a parallel field of 25kG at 4.2°K.

We do not want to give the impression that the procedure just described is simple. On the contrary, a fool-
proof utilization of the diagram in Figure 3 requires some
insight into the problem (D and E!). Nevertheless, even at
a crude state of data evaluation the diagram in Figure 3
can provide some clues which, together with extensive computer simulations, can lead to the solution of a complex
problem.

The solid lines in Figures 9-12 show the results of
simulations by means of a computer program that generates
Mössbauer spectra of polycrystalline specimen from Eqs (7)
and (8). The solutions shown reflect a set of parameters

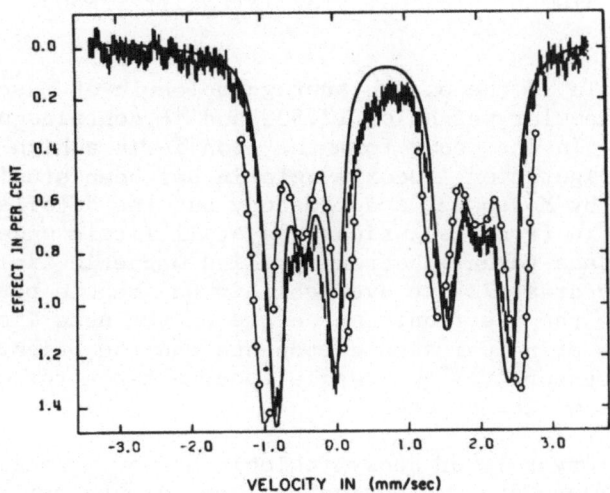

Figure 12. Spectrum of reduced P450 taken in a 45kG field
at 180°K. The solid line represents a computer simulation
utilizing the same parameters as used for the computation
of the low-temperature spectra shown in Figures 10 and 11.
The circled line uses the same parameters, but assumes dia-
magnetism, i.e. $\langle \vec{S} \rangle_T = 0$. The circled line has been·added
to show the sizable effects of the internal magnetic field
even at 180°K. These anisotropic fields have to be taken
into account if η is to be determined from such data.

consistent with the spectra obtained for a variety of applied
fields (5 - 45kG) and temperatures (1.5°K - 180°K). The
simulations shown correspond to the following set of parameters: D =+20°K, E/D = 0.15, g_x = 2.24, g_y = 2.32,
g_z = 2.00, A_x = -180kG, A_y = -125kG, A_z = -150kG,
ΔE_Q = +2.42 mm/s, and η = 0.8. The Euler angles which take
the zero-field splitting tensor into the frame of the EFG
tensor were α = 60°, β = 70°, and γ = 0. For details we
refer the reader to the original publication [9].

IV. SINGLE CRYSTALS. THE INTENSITY TENSOR

    Myoglobin is the oxygen storage molecule of muscle.
It has a molecular weight of 17,500 and it contains one
heme group. In its deoxy form the iron is in a high-spin
ferrous configuration. Deoxymyoglobin has been studied
extensively by Mössbauer spectroscopy but the details of
the spectra in frozen solutions are still little understood,
especially data taken in strong applied magnetic fields.
Since single crystals are available it has become possible
to elucidate the electronic structure of the heme iron by
deducing the principal axes components and the orientation
of the EFG tensor from quadrupole spectra taken for different
orientations of the crystal.

    Single crystals of deoxymyoglobin are of monoclinic
symmetry. There are two equivalent heme groups per unit
cell which can be transformed into each other by a 180° rotation and a translation. In the past, quadrupole spectra
of monoclinic single crystals have been evaluated by a
method given by Zory [16]. Little attention, however, has
been paid to the fact that the solutions are not unique for
monoclinic crystals. In other words, it is not possible to
determine the principal axes values and the orientation of
the EFG tensors of two equivalent lattice sites unless some
independent information can be secured by other methods.
In the following we will discuss this matter (which applies
to all substances of monoclinic crystallographic symmetry).

    In the presence of an EFG at the nucleus the ^{57}Fe
Mössbauer line is split in two quadrupolar lines, the intensities of which depend on the orientation of the single
crystal relative to the γ-rays. To mathematically describe
the angular dependence of the absorption intensity we utilize
that it traces an ellipsoid in three dimensional space [17],
isotropic Debye Waller factor, thin absorber, and non-polar-
ized γ-beam being implied.* An ellipsoid can be described
by a tensor. Thus if $I^{(h)}$ and $I^{(1)}$ are the intensities of

---

*This description can also be applied in case of combined
electric and magnetic hyperfine interactions [18]. It rests
on the fact that the ^{57}Fe transition is dipolar in
character.

the lines at high and low velocity, respectively, the reduced intensity of the high velocity line is given by

$$I^{(h)}/I^{(tot)} = \sum_{p,q = x,y,z} \overline{I^{(h)}_{pq} e_p e_q} \qquad (12)$$

where $I^{(tot)} = I^{(h)} + I^{(1)}$ is the total intensity, e_x, e_y, e_z
are the direction cosines of the γ-rays with
respect to some arbitrary frame, and $I^{(h)}_{pq}$ are the components
of an intensity tensor, which is real, symmetric, and of
second rank. Since intensities are additive Eq (12) holds
even for more than one equivalent lattice site per unit
cell. For a derivation of Eq (12) the reader is referred
to reference [17].

For a random powder the two quadrupolar lines have
equal intensity. Since the powder mean value, as calculated
from Eq (12), is just 1/3 of the trace of the intensity
tensor the latter has the trace 3/2, i.e.

$$I^{(h)}_{xx} + I^{(h)}_{yy} + I^{(h)}_{zz} = 3/2 \qquad (13)$$

Thus the intensity tensor has five independent components.
Once these components have been determined experimentally
we have collected the total information which can be extracted from the angular dependence of the line intensities.
The determination of all possible solutions for the local
EFG's at each site is now a matter of analyzing the intensity tensor.

The evaluation of the tensor $I^{(h)}_{pq}$ is simplified by the
introduction of the traceless macroscopic intensity
tensor.

$$I^{(m)}_{pq} = I^{(h)}_{pq} - 1/2 \; \delta_{pq} \qquad (14)$$

$p,q = x,y,z$. The adjective "macroscopic" has been given to
$I^{(m)}_{pq}$ to emphasize that it results, like $I^{(h)}_{pq}$, from the super-
position of the intensities of each lattice site.
In a similar way we can define a traceless local intensity
tensor which belongs to the local EFG associated with each

lattice site.* It is easily seen that the macroscopic intensity tensor is just the mean value of the local intensity
tensors. Using this and the proportionality of the EFG
tensor with the traceless intensity tensor (see Eq (9) of
reference [17]), we can write

$$eQV_{pq}^{(m)} = 8 \ |\Delta E_Q| \ I_{pq}^{(m)} \tag{15}$$

where $V_{pq}^{(m)}$ is the mean value of the local EFG tensors. If
N is the number of equivalent lattice sites and
$V_{pq}^{(loc,i)}$ the local EFG tensor associated with the i-th
lattice site, $V_{pq}^{(m)}$ is given by

$$V_{pq}^{(m)} = \frac{1}{N} \sum_{i=1}^{N} V_{pq}^{(loc,i)} \tag{16}$$

Though usually the macroscopic EFG tensor $V_{pq}^{(m)}$ does not have
much meaning in the discussion of Mössbauer spectra, its
importance for single crystal work is quite obvious. The
macroscopic EFG tensor is the quantity, which can be uniquely
obtained from the measurements (at least up to the factor
$eQ/|\Delta E_Q|$).   The local EFG tensors are determined only in
so far as they lead to the same macroscopic EFG tensor.
Before discussing monoclinic crystals we discuss some consequences of Eq (15) for crystals with only one lattice site
per unit cell.

     In the case of only one lattice site per unit cell the
local EFG tensor is identical with the macroscopic EFG tensor. Omitting the distinction (m) and (loc) in the super-
scripts of Eq (15) we may write

$$eQV_{pq} = 8 \ |\Delta E_Q| I_{pq} \tag{17}$$

---

*For equivalent lattice sites, the local EFG's transform
into each other by the symmetry transformations of the
crystal. The corresponding local intensity ellipsoids have
identical shape, but are differently oriented.

The traceless intensity tensor and the EFG tensor are connected by a positive proportionality constant. Thus the traceless intensity tensor has the important properties:

(i)   The principal axes system of the EFG, $O_{\hat{x}\hat{y}\hat{z}}$ with $|V_{\hat{z}\hat{z}}| \geq |V_{\hat{y}\hat{y}}| \geq |V_{\hat{x}\hat{x}}|$ is identical to the principal axes system of the intensity tensor, with $|I_{\hat{z}\hat{z}}| \geq |I_{\hat{y}\hat{y}}| \geq |I_{\hat{x}\hat{x}}|$

(ii)   The asymmetry parameter of the EFG, η, can be obtained from

$$\eta = \frac{I_{\hat{x}\hat{x}} - I_{\hat{y}\hat{y}}}{I_{\hat{z}\hat{z}}}$$

(iii)   The sign of the quadrupole splitting is obtained from the sign of the principal component of maximum absolute value, $I_{\hat{z}\hat{z}}$

$$\text{sign } (\Delta E_0) = \text{sign } (I_{\hat{z}\hat{z}})$$

(iv)   The quantity I_Δ, defined analogously to the square of the quadrupole splitting, has a constant value

$$I_\Delta = 16 \; I_{\hat{z}\hat{z}}^2 \; \sqrt{(1 + 1/3 \; \eta^2)} = 1$$

The first three properties facilitate the determination of the EFG from the reduced intensities. The fourth property is something like a check that one is really concerned with identical EFG's. Disregarding blackness effects or anisotropic Debye-Waller factors it can be shown that the quantity I_Δ is only equal to 1 if the EFG tensors are identical, otherwise it is smaller than one [17]. Thus single crystal Mössbauer spectra in zero field can differentiate between crystals of one or more lattice sites per unit cell.

We now return to deoxymyoglobin as an example of monoclinic symmetry. There are two equivalent hemes per unit cell which can be transformed into each other by a rotation of 180° around the b axis of the unit cell. The additional translation is not relevant in this context. The symmetry of the crystal requires that the macroscopic EFG tensor (and similarly the macroscopic intensity tensor) have one principal axis along b. This can be understood as follows:

Let z be along the b axis. With $V_{pq}^{(loc,1)}$ and $V_{pq}^{(loc,2)}$ being the local EFG tensors at site 1 and site 2, respectively, the two-fold symmetry $(x \to -x, \; y \to -y, \; z \to z)$ implies the following relations

$$V_{pp}^{(loc,2)} = V_{pp}^{(loc,1)}$$

$$V_{xy}^{(loc,2)} = V_{xy}^{(loc,1)}$$

$$V_{xz}^{(loc,2)} = -V_{xz}^{(loc,1)} \tag{18}$$

$$V_{yz}^{(loc,2)} = -V_{yz}^{(loc,1)}$$

The components $V_{xz}^{(loc)}$ and $V_{yz}^{(loc)}$ change their sign upon a two-fold symmetry transformation. Thus, when we calculate the macroscopic EFG tensor according to Eq (16) these two components cancel and we have

$$V_{pp}^{(m)} = V_{pp}^{(loc,1)}$$

$$V_{xy}^{(m)} = V_{xy}^{(loc,1)}$$

$$V_{xz}^{(m)} = 0 \tag{19}$$

$$V_{yz}^{(m)} = 0$$

The cancellation of the two components of $V_{pq}^{(loc)}$ on one hand allows the macroscopic EFG tensor to reflect the symmetry of the crystal (one principal axis along b = z). On the other hand, this cancellation shows why the local EFG tensors cannot be determined uniquely from zero-field spectra of monoclinic single crystals.

Many aspects of the preceeding discussion would apply for the susceptibility tensor also. There is, however, an important difference in the treatment of both tensors. The local intensity tensor has to obey condition (iv)*. Writing

---

*It is a tensor of actually only four independent parameters. The EFG has, of course, five independent parameters. The fifth parameter is hidden in the proportionality factor between the EFG and the traceless intensity tensor.

this equation for a nondiagonal local EFG tensor and solving for the nondiagonal elements $V_{xz}^{(loc)}$ and $V_{yz}^{(loc)}$ we obtain [17]

$$[eQV_{xz}^{(loc)}]^2 + [eQV_{yz}^{(loc)}]^2 = [eQV_L]^2 . \qquad (20)$$

$[eQV_L]^2$ is determined by the components of the macroscopic intensity tensor (i.e. experimentally),

$$[eQV_L]^2 = 3|\Delta E_Q|^2 (1-I_\Delta^{(m)})$$

with $I_\Delta^{(m)}$ being the I_Δ value of the macroscopic intensity tensor

$$I_\Delta^{(m)} = 16[(I_{zz}^{(m)})^2 + 1/3(I_{xx}^{(m)} - I_{yy}^{(m)})^2 + 4/3(I_{xy}^{(m)})^2] .$$

Using Eq (20) to reduce the number of parameters, only one unknown is left. To describe this unknown we define an angle

$$\psi = \arctan (V_{yz}^{(loc,1)}/V_{xz}^{(loc,1)}) \qquad (21)$$

The local EFG tensor at one lattice site is finally given by

$$V_{pq}^{(loc,1)} = \begin{pmatrix} V_{xx}^{(m)} & V_{xy}^{(m)} & V_L\cos\psi \\ V_{xy}^{(m)} & V_{yy}^{(m)} & V_L\sin\psi \\ V_L\cos\psi & V_L\sin\psi & V_{zz}^{(m)} \end{pmatrix} \qquad (22)$$

For this tensor all quantities with the exception of ψ can be determined from the macroscopic EFG tensor. To each value of ψ ($0 < \psi < 2\pi$) there corresponds one possible solution. The problem of finding all possible local EFG's is now solved in principle. (For a convenient representation one diagonalizes the local EFG tensor (22), determines its asymmetry parameter η, the sign of ΔE_Q and the Euler angles α, β, γ describing the orientation of its principal axes system. Fortunately, the tensor in Eq (22) can be diagonalized analytically. For formulae the interested reader is referred to reference [17].)

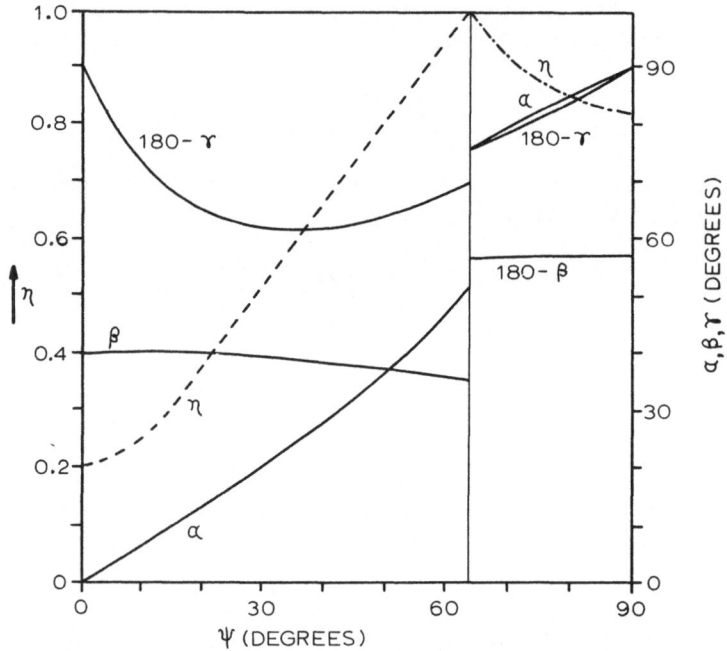

Figure 13.   Results of the single crystal evaluation of sperm
whale deoxymyoglobin at 77K.   Each value of ψ represents a
possible solution for the local EFG.   The latter is described
by the asymmetry parameter η (left hand scale), the sign of
ΔE_Q (positive, if the η-line is dashed, negative if it
is $-\cdot-\cdot$), and the Euler angles α,β,γ, which describe the
orientation of the principal axes system relative to that of
the macroscopic intensity tensor.   β and α correspond to the
polar and azimuthal angles of $V_{\hat{z}\hat{z}}$, respectively [17] (taken
from reference [20])

Comment on Figure 13.

The values of ψ in Figure 13 have been restricted to the first
quadrant, $0 \leq \psi \leq \pi/2$.   The solutions in the other 3 quadrants
are obtained by reflecting the local EFG at planes which contain two principal axes of the macroscopic intensity tensor.
The four solutions to each value of η and ΔE_Q result from
the invariance of the macroscopic intensity   tensor under
reflections at these three planes.

We now return to deoxymyoglobin. Single crystal measurements were first reported by Trautwein, Gonser, and collaborators [19]. Recently these researchers (Maeda et al [20])
have refined the evaluation of the deoxymyoglobin data by
using the above described method. Results of this evaluation are presented in Figure 13. The diagram shows all
solutions compatible with the experimental data for deoxymyoglobin and it illustrates some problems which may arise in
single crystal work. To each value of ψ there corresponds
a solution for the local EFG tensor. This tensor is charac-
terized by η, the sign of ΔE_Q and the Euler angles α, β and γ
which describe the orientation of the principal axes system.
It is obvious that we have to know η and the sign of ΔE_Q in
order to extract α, β and γ. Conversely, the knowledge of
the orientation of a local symmetry axis is sufficient to
determine the other two principal axes and η.

Most naturally one would attempt to determine η by
studying the protein in applied magnetic fields. For a diamagnetic, polycrystalline sample a determination of η is
fairly straightforward although this quantity can practically
not be determined to better than ± 0.2 for $\eta < 0.6$. In deoxymyoglobin the situation is more complex since the heme iron
is paramagnetic and the application of an external magnetic
field induces a sizable, and more importantly, an anistropic,
internal magnetic field. As an example we consider again
cytochrome P450. The data taken at $180°K$ in a 45kG field
show a residual hyperfine field of approximately 10kG (see
Fig. 12) making an accurate determination of η virtually
impossible unless the magnetic hyperfine tensor and the zero-
field splitting parameters are known from a low temperature
investigation.

If η cannot be extracted readily from measurements on
polycrystalline specimen we might study single crystals.
For large crystals, like $FeCl_2 \cdot 4H_2O$, such measurements usually give the necessary information [17]. For biological
compounds, however, some experimental problems arise since
teeny-weeny protein single crystals lead to very unfavorable
solid angles in standard spectrometers with superconducting
coils.

As we can see the single crystal investigations of deoxymyoglobin so far have not yielded the desired results. Because of the practical difficulties encountered in the determination of η Maeda et al [20] utilized some information
obtained from x-ray diffraction work on myoglobin. These

authors argue that the x-ray data, taken at 1.4 Å resolution, indicate the heme normal to be a local twofold axis, an information which they use to locate one principal axis of the EFG tensor.* With this assumption two solutions can be selected from Figure 13:

(i) $\psi \approx 20°$, $\eta \approx 0.4$, $\Delta E_Q > 0$, $V_{\hat{y}\hat{y}} \simeq$ parallel to the heme normal
(ii) $\psi \approx 340°$, $\eta \approx 0.4$, $\Delta E_Q > 0$, $V_{\hat{x}\hat{x}} \simeq$ parallel to the heme normal

$(|V_{\hat{z}\hat{z}}| \geq |V_{\hat{y}\hat{y}}| \geq |V_{\hat{x}\hat{x}}|)$. Both solutions have in common that $V_{\hat{z}\hat{z}}$ is positive and that it is located close to the heme plane. In discussing these solutions Maeda et al [20] have pointed out that three theoretical models [22-24] (which are also based on the assumption of C_{2v} symmetry) yield $0.3 \leq \eta \leq 0.5$ in reasonable agreement with both selected experimental solutions.

To conclude, the matters discussed in this Chapter point out some ambiguities associated with the evaluation of data taken on monoclinic single crystals. Therefore, in order to extract the desired information from often time consuming and difficult experiments one should make sure that some additional knowledge can be secured by other means. Crystallographic data, theoretical models or single crystal measurements in applied fields might furnish such information.

ACKNOWLEDGEMENTS

A successful Mössbauer investigation of a protein depends critically on the help of biochemists, and often enough on microbiologists. We enjoy a fruitful cooperation with various groups of the University of Wisconsin. In particular, we appreciate the collaboration with Dr. W. H. Orme-Johnson's group on the nitrogenase system. We wish to thank Dr. A. Trautwein for providing us with the manuscript of the myoglobin single crystal work (ref. 20) prior to publication. We wish to express our gratitude to Dr. J. Rossillon for many hours of stimulating discussions. This work is supported by a grant from the National Science Foundation and by a Research Career Development Award K04-GM 00057 (E.M.).

---

* In the literture, quite often the proximal histidine ligand is made responsible for lowering the symmetry to C_{2v}. Interestingly, EPR work [21] on ferrimyoglobin shows that the iron atom resides in a site of tetragonal symmetry (E/D < 0.01).

REFERENCES

1. E. Münck, P.G. Debrunner, J.C.M. Tsibris, and I.C. Gunsalus, Biochemistry 11, 855 (1972).

2. W.R. Dunham, A.J. Bearden, I.T. Salmeen, G. Palmer, W.H. Orme-Johnson, and H. Beinert, Biochim. Biophys. Acta 253, 134 (1971).

3. C.W. Carter, Jr., J. Kraut, S.T. Freer, R.A. Alden, L.C. Adman, and L.H. Jensen, Proc. Nat. Acad. Sci 69, 3526 (1972).

4. W.H. Orme-Johnson, W.D. Hamilton, T. Ljones, M.Y.Tso, R.H. Burris, V.K. Shaw, and W.J. Brill, Proc. Nat. Acad. Sci. 69, 3142 (1972).

5. E. Münck, H. Rhodes, W.H. Orme-Johnson, L.C. Davis, W.J. Brill, and V.K. Shah, Biochim. Biophys. Acta 400, 32 (1975).

6. B.E. Smith and G. Lang, Biochem. J. 137, 169 (1974).

7. E. Münck, J.L. Groves, T.A. Tumollillo, and P.G. Debrunner, Computer Phys. Commun. 5, 225 (1973).

8. M. Sharrock, E. Münck, P.G. Debrunner, V. Marshall, J. Lipscomb, and I.C. Gunsalus, Biochemistry 12, 258 (1973).

9. P.M. Champion, J.D. Lipscomb, E. Münck, P. Debrunner, and I.C. Gunsalus, Biochemistry 14, 4151 (1975).

10. P.M. Champion, Thesis, University of Illinois (1975).

11. R. Zimmermann, H. Spiering, and G. Ritter, Chem. Phys. 4, 133 (1974).

12. G. Lang and W. Marshall, Proc. Phys. Soc., London 87, 3 (1966).

13. A. Abragam and B. Bleaney, Electron Paramagnetic Resonance of Transition Ions, Chapter 19.1, Clarendon Press, Oxford (1970).

14. R. Zimmermann, G. Ritter, H. Spiering and D.L. Nagy, Suppl. au J. de Physique (France), Fasc. 12, Colloque N°6, 439 (1974).

15. H. Spiering, R. Zimmermann, and G. Ritter, phys. stat. sol. b 62, 123 (1974).

16. P. Zory, Phys. Rev. 140, A1401 (1965).

17. R. Zimmermann, Nucl. Inst. and Meth. 128, 537 (1975).

18. R. Zimmermann, Chem. Phys. Lett. 34, 416 (1975).

19. A. Trautwein, Y. Maeda, U. Gonser, F. Parak and H. Formanek, Proc. of the 5th Intern. Conf. on Mössbauer Spectroscopy, Bratislava (CSSR), Sept. 1973; U. Gonser, Y. Maeda, A. Trautwein, F. Parak and H. Formanek, Z. f. Naturforschung 29b, 241 (1974).

20. Y. Maeda, T. Harami, A. Trautwein, U. Gonser, Z. f. Naturforschung B (in press).

21. E.F. Slade and R.H. Farrow, Biochim. Biophys. Acta 278, 450 (1972).

22. B.H. Huynh, G.C. Papaefthymiou, C.S. Yen, J.L. Groves and C.S. Wu, J. Chem. Phys. 61, 375o (1974).

23. A. Trautwein, R. Zimmermann, F.E. Harris, Theor. Chim. Acta 37, 89 (1975).

24. H. Eicher, D. Bade and F. Parak (to be published).

MÖSSBAUER PARAMETERS OF RUBREDOXIN,

A ONE-IRON-SULFUR PROTEIN

P. Debrunner and C. Schulz

Department of Physics
University of Illinois at Urbana-Champaign
Urbana, IL 61801

Rubredoxins are red, iron-containing redox proteins. They are found in many bacteria and are among the simplest iron proteins known. [1] They typically consist of a polypeptide chain of 50-60 amino acids which binds a single iron atom by means of four characteristically located cysteine residues. The three-dimensional structure of rubredoxin from Clostridium pasteurianum, the particular protein under study here, has been determined by x-ray diffraction. [2] It is found that the thiolate sulfurs of the four cysteine residues bind to the iron in a roughly tetrahedral array. The iron-sulfur distances range from 2.05 Å to 2.34 Å according to the x-ray data [2], and the local symmetry at the iron site is C_1 only. Recent EXAFS studies suggest smaller differences in the Fe-S distances [3], but the symmetry is still expected to be very low.

The iron is clearly the active center of the protein; it changes charge state in the redox reaction from Fe^{3+} to Fe^{2+}. The protein can easily be enriched in ^{57}Fe by chemical reconstitution of the apo protein. [4]

A detailed Mössbauer study of rubredoxin is of interest for a number of reasons. (i) Rubredoxin is the prototype of the large class of iron-sulfur proteins [5] which all

Supported by a grant from the U.S. Public Health Service, USPH GM 16406.

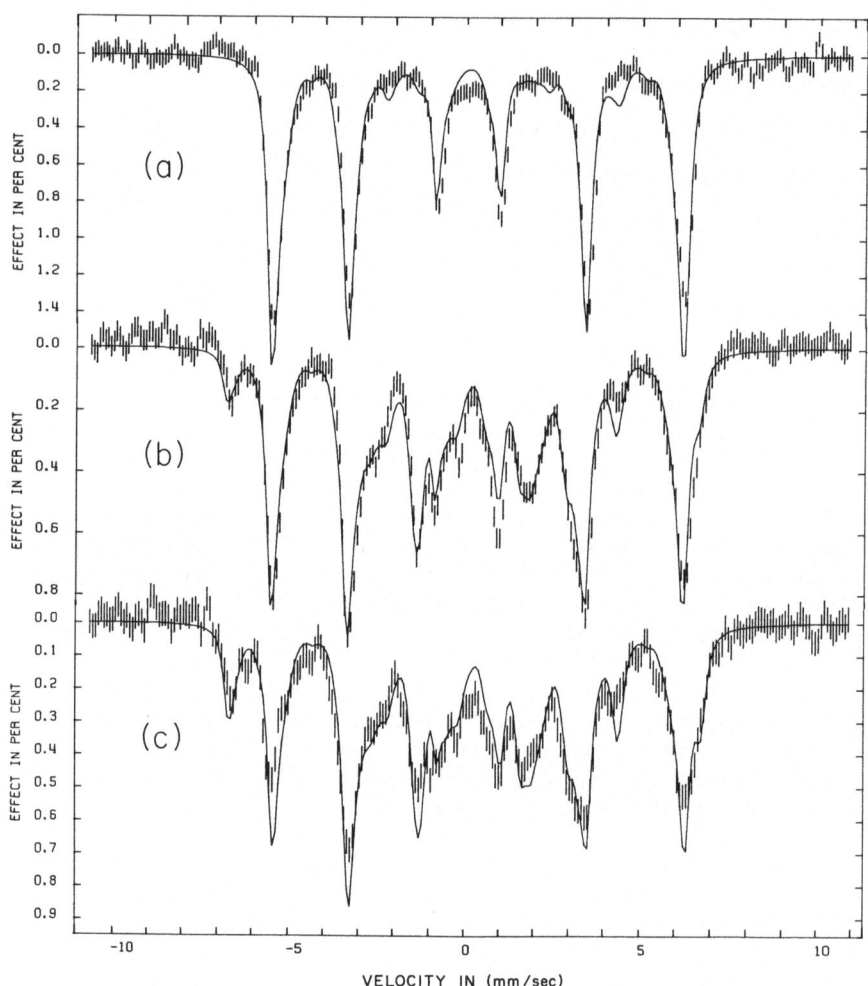

Fig. 1 Mössbauer spectra of a frozen solution of oxidized
        rubredoxin from Clostridium pasteurianum in a per-
        pendicular field of 1.3 kG at 1.5 K(a), 10 K(b) and
        20 K(c).   The solid lines represent simulations
        obtained with the parameters of Table I.

contain clusters of the basic FeS_4 tetrahedron found in the
active site of rubredoxin. Specifically, the 2Fe-2S* proteins contain two FeS_4 complexes sharing two sulfide ions
S*, while the 4Fe-4S* proteins contain a cube-like cluster
of four of the basic tetrahedra with four sulfides S* common
to three complexes each. (ii) Model complexes for all three
types of iron-sulfur proteins have been synthesized recently,
and it is important to check the extent to which the models
duplicate the spectroscopic properties of the actual proteins
and how they are affected by chemical modifications. [6]
(iii) Earlier Mössbauer measurement revealed well resolved
electric and magnetic hyperfine splittings for both oxidized and reduced rubredoxins[7,8], and previous experience
with similar spectra suggested the feasibility of a quantitative analysis in terms of a spin Hamiltonian. (iv) Experimentally determined Mössbauer parameters can be compared
with molecular orbital calculations[9,10] based on the
known structure of the active center in rubredoxins[2] and
its model complex. [11]

OXIDIZED RUBREDOXIN

Figure 1 shows the Mössbauer spectra of a frozen
solution of oxidized rubredoxin from Clostridium pasteurianum*) measured in a transverse magnetic field of 440 G at
1.5 K(a), 10 K(b) and 20 K(c). Spectra virtually identical
with Fig. 1a have been published by Rao et al.[8] Careful
inspection of the data shows that they represent a super-
position of three distinct component spectra with intensities determined by different Boltzmann factors. At 1.5 K,
Fig. 1a, a simple six-line pattern is observed that can
be characterized by an internal field of 360 kG and a small
positive quadrupole interaction. The second and third
component spectra can be obtained from a detailed analysis
only. The simulations shown as solid lines in Fig. 1 give
the results and Fig. 2 illustrates the three component
spectra.

The spectral simulations are based on a spin Hamiltonian that proved successful in the interpretation of the
EPR spectra of oxidized rubredoxin.[12] The spin Hamiltonian formalism has been applied to Mössbauer spectroscopy by Wickmann[13], Lang[14], Spartalian and Oosterhuis[15],
and others, and the reader is referred to these authors for
details.
_____
*The samples were prepared by Dr. W. Lovenberg, N.I.H.,
Bethesda, MD.

TABLE I

Parameters used in the simulation of the Mössbauer spectra, Fig. 1-3, of rubredoxin from Clostridium pasteurianum and comparable parameters of a typical 2Fe-2S* protein.   The numbers in parentheses are the uncertainties in units of the last significant digit.

	Rubredoxin		Putidaredoxin[a]	
	oxidized	reduced	Fe^{3+}	Fe^{2+}
$D(°K)$	2.5(5)	12		
λ	0.17(2)	0.12		
μ	0.3(1)	0.33		
δ_{Fe} @4.2K (mm/sec)	0.33(5)	0.70(2)	0.27(2)	0.64
ΔE_Q (mm/sec)	-0.5(1)	-3.25(1)	0.6(1)	-2.7(1)
η	0.2(1)	0.65	0.5	-3
A_x (MHz)	-22.7(15)	-17.4	-24	-11
A_y (MHz)	-21.5(4)	-12.1	-21.4	-16
A_z (MHz)	-23.5(4)	-38.5	-18.5	-26
Γ (mm/sec)	0.31	0.28	0.31	0.31

a) E. Münck et al., Biochemistry 11, 855 (1972).

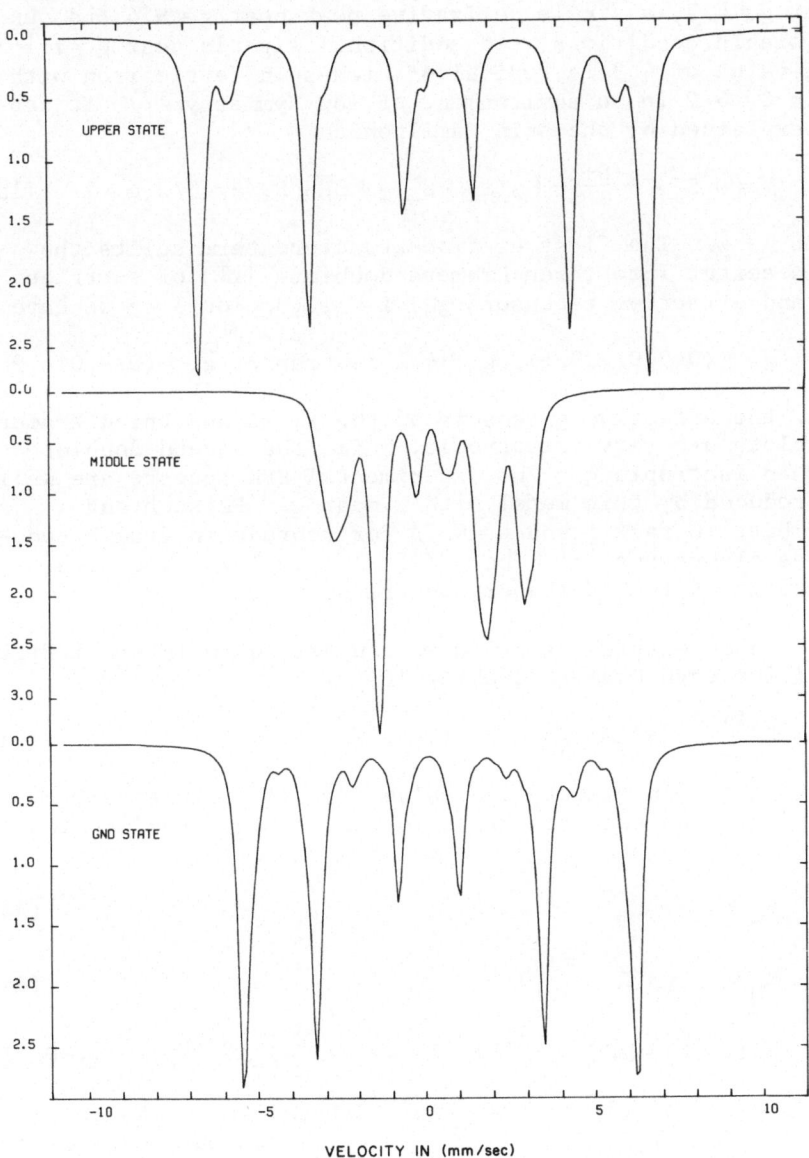

Fig. 2 Component spectra of the three Kramers doublets
calculated with the parameters of Table I.

Oxidized rubredoxins show a characteristic EPR signal near g = 4.3, a single derivative peak near g = 9.4 and, under favorable conditions, two additional signals near g = 1.[12,16] A g-value of 4.3 is typical of high-spin ferric iron with spin S = 5/2 in an environment of low symmetry.[17] It can be explained by the spin Hamiltonian

$$\mathcal{H} = D[S_z^2 - \frac{S(S+1)}{3} + \lambda(S_x^2 - S_y^2)] + \beta\vec{H}g\vec{S}, \quad S = 5/2, g = 2 \quad (1a)$$

with $\lambda \simeq \frac{1}{3}$. The first or fine-structure term splits the spin sextet into three Kramers doublets $|i^{\pm}\rangle$ of energies E_i and effective g-tensors $\tilde{g}_i$, $i = 1, 2, 3$. For $\lambda = \frac{1}{3}$ we have

$$\tilde{g}_1 \simeq (0.9, 9.7, 0.6), \quad \tilde{g}_2 \simeq 4.3 \text{ isotropic}, \quad \tilde{g}_3 = (0.9, 0.6, 9.7),$$

i. e. the effective g-tensors of the first and third Kramers doublets are very anisotropic, while the second doublet has an isotropic g. The experimental EPR spectra are well reproduced by this model with λ near $\frac{1}{3}$. Peisach and Blumberg in fact found $\lambda = 0.28$ for rubredoxin from Pseudomonas oleovorans.[12] The model also explains the main features of the Mössbauer spectra quite well.

In an external field $\vec{H}$ we can define an internal field $\vec{H}_{int}^{(i)}$ for each Kramers doublet $|i\rangle$,

$$\vec{H}_{int}^{(i)} = -\langle\vec{S}\rangle_i A / g_N \beta_N \quad (S = 5/2) \quad (2a)$$

where the spin expectation value $\langle\vec{S}\rangle_i$ is calculated from Eq. (1a). We can then write for the nuclear Hamiltonian $\mathcal{H}_N$

$$\mathcal{H}_N = \mathcal{H}_M + \mathcal{H}_Q \quad (3a)$$

$$\mathcal{H}_M = -\beta_N g_N (\vec{H} + \vec{H}_{int}^{(i)}) \cdot \vec{I} \quad (3b)$$

$$\mathcal{H}_Q(I = \tfrac{3}{2}) = (eQV_{zz}/12)[3I_z^2 - 15/4 + \eta(I_x^2 - I_y^2)], \quad \mathcal{H}_Q(I = \tfrac{1}{2}) = 0. \quad (3c)$$

The spectra of Fig. 1 are mainly determined by the internal field, since $\mathcal{H}_M \gg \mathcal{H}_Q$ and $|\vec{H}_{int}| \gg |\vec{H}|$, and our model of a rhombic fine structure, $\lambda = 1/3$ in Eq. (1a) makes the following specific predictions about $\vec{H}_{int}^{(i)}$ for the three Kramers doublets: (i) For the two anisotropic doublets $|1^{\pm}\rangle$ and $|3^{\pm}\rangle$ the internal field essentially follows the largest

component of the g-tensor, i.e. the y-direction for the
ground doublet and the z-direction for the highest doublet.
(ii) For the isotropic middle doublet $|2^{\pm}\rangle$ the internal
field follows the direction of the applied field. (iii) The
magnitudes of $H_{int}^{(i)}$ for the three doublets are essentially
in the ratio $g_{1,y} : g_2 \approx 4.3 : g_{3,z}$.

Inspection of Fig. 2 bears out most of these predic-
tions: the ground state and upper state show relatively
sharp spectra, since the internal field is practically fixed
within the molecule. The quadrupole interaction mainly de-
pends on the component $V_{\ell\ell}$ along $\vec{H}_{int}^{(i)}$ and is positive for
the ground doublet, $V_{yy} > 0$, and negative for the upper
doublet, $V_{zz} < 0$. The middle Kramers doublet has a smaller
overall splitting, depends strongly on the direction of the
applied field and shows broad lines since $\vec{H}_{int}^{(2)}$ assumes all
directions relative to the molecule-fixed quadrupole tensor.

A detailed calculation shows, however, that the model
described so far is inadequate in two respects:

(i) The internal field $H_{int,z}^{(3)}$ is 15% larger than $H_{int,y}^{(1)}$
and, from the condition $H_{int,y}^{(1)} : H_{int,z}^{(3)} = g_{1,y} : g_{3,z}$ and
the known value of $g_{1,y} = 9.4$,[8] one finds $g_{3,z} \approx 1.15 \cdot g_{1,y}$
=10.8. A g-value in excess of 10 is not possible, though,
if the high-spin ferric iron has no orbital angular momen-
tum, as we had tacitly assumed in setting g of Eq. (1a) (in
the S = 5/2 representation) equal to its spin-only value,
$g = g_s = 2.0023$, and writing A in Eq. (2a) as a scalar. The
problem can be solved by either making A in Eq. (2a) or g
in Eq. (1a) a tensor, or both. We have chosen the first
approach, keeping Eq. (1a) with g = 2 but substituting a
tensor $\tilde{A}$ in Eq. (2).

(ii) The experimental Boltzmann factors of the Kramers
doublets $|2^{\pm}\rangle$ and $|3^{\pm}\rangle$ imply an energy ratio, setting $E_1 = 0$,
of $E_3/E_2 = 3.3 \pm 0.5$. Eq. (1a) with $\lambda = \frac{1}{3}$, however, yields a
ratio $E_3/E_2 = 2$ which is incompatible with the data. We
follow Spartalian, Oosterhuis and Neilands[15] and introduce an
additional fine structure term of the form

$$(D\mu/6)[S_x^4 + S_y^4 + S_z^4 - S(S+1)(3S^2 + 3S - 1)/5]. \tag{1b}$$

With the two parameters λ and μ of Eqs. (1a) and (1b) it is possible to obtain the proper energy ratio while maintaining an isotropic g-value of $g = 30/7$[15] for the second Kramers doublet.

The parameters used in the simulations of Figs. 1 and 2 are listed in Table I. They reproduce the data quite well but it remains to be seen whether a simultaneous fit of Mössbauer and EPR spectra can be obtained. Table I also lists the Mössbauer parameters derived for the high-spin ferric iron in the paramagnetic Fe^{2+}-Fe^{3+} cluster of a reduced 2Fe-2S* protein, putidaredoxin.[18] The similarities are striking, particularly in the anisotropy of the magnetic hyperfine tensor $\widetilde{A}$. Model calculations[9] predict $|A_y| > |A_x| > |A_z|$ in agreement with experiment if a coordinate system is chosen such that $g_{1,y} \simeq g_{3,z}$ are the large components of the effective g-tensor.

REDUCED RUBREDOXIN

The quadrupole splitting of reduced rubredoxin is $|\Delta E_Q| = 3.25$ mm/sec at 4.2K and decreases slightly to $|\Delta E_Q| = 3.21$ mm/sec at 200K. The isomer shift, $\delta_{Fe} = 0.70$ mm/sec at 4.2K, is relatively small for high-spin ferrous iron but is typical for a complex with tetrahedral sulfur coordination.[19] Rao et al.[8] found that the low-temperature, high-field spectra of rubredoxin from Chloropseudomonas ethylica show a well resolved line pattern with an overall magnetic splitting that saturates at relatively modest applied fields. The spectra of rubredoxin from Clostridium pasteurianum, Fig. 3, are quite similar and, as the solid lines in Fig. 3 indicate, they can be will reproduced by a formalism based on the spin Hamiltonian, Eq. (1).

We assume that the orbital ground state of the high-spin ferrous ion, $S = 2$, is far removed in energy from all higher orbital states, and that the five-fold spin degeneracy of the orbital ground state is lifted by spin-orbit interaction leading to the fine structure term of Eq. (1a). In an applied field $\vec{H}$ an internal field

$$\vec{H}_{int} = -\langle \vec{S} \rangle \, \widetilde{A}/g_N \beta_N \qquad (2b)$$

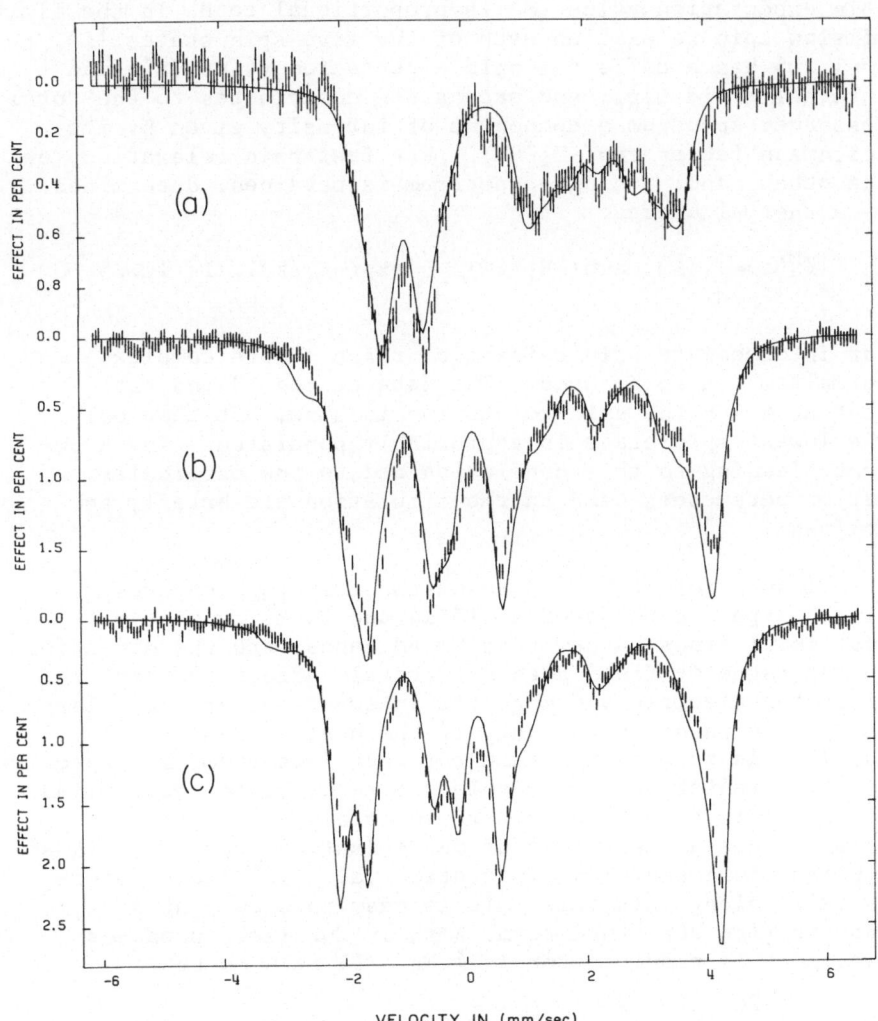

Fig. 3 Mössbauer spectra of reduced rubredoxin from C.
pasteurianum at 4.2 K in parallel fields of
6.6 kG(a), 15 kG(b) and 24 kG(c). The simulations
shown as solid lines were obtained with the parameters of Table I in the slow relaxation limit.

can be defined as in the case of Fe^{3+}, but for small H the spin expectation value $\langle \vec{S} \rangle$ is proportional to H. In the limit of slow spin relaxation each of the five spin states $|i\rangle$, $i = 1 \ldots 5$ has a different spin expectation value $\langle \vec{S} \rangle_i$ and internal field $H_{int}^{(i)}$, and each state contributes to the total Mössbauer spectrum a component of intensity given by its Boltzmann factor $\exp(-E_i/kT)$. For fast spin relaxation, on the other hand, a single spectrum is obtained, determined by a thermal average

$$\langle \vec{S} \rangle_T = \sum_i \langle \vec{S} \rangle_i \exp(-E_i/kT) / \sum_i \exp(-E_i/kT), \quad i = 1 \ldots 5.$$

For intermediate spin relaxation rates a more complex formalism has to be used. The data of Fig. 3 indicate that at 4.2 K the spin relaxation is slow, but that only the lowest spin state is appreciably populated. The arguments leading to this conclusion and to the particular set of parameters used in the simulation are briefly as follows:

Inspection of Fig. 3 shows two striking features. (i) The spectra measured at 15 kG and 24 kG consist of individual lines rather than broad bands, and the direction of the magnetic field does not greatly affect the pattern. Since the electric and magnetic interactions are both large, sharp spectra can arise only if the nuclear Hamiltonian, Eq. (3), is roughly the same for each iron atom, irrespective of the orientation of the molecule relative to the applied field. This situation in turn is obtained only if each molecule has an easy axis of magnetization, such that in an applied field the spin expectation value $\langle \vec{S} \rangle$ lies preferentially along this axis. It is easy to show that a rhombic fine structure term, $\lambda \simeq \frac{1}{3}$ in Eq. (1a), produces exactly this result. For $\lambda = \frac{1}{3}$ Eq. (1) reduces to

$$\mathcal{H} = \frac{2D}{3} (S_z^2 - S_y^2) + \beta \, \vec{H} \cdot \widetilde{g} \cdot \vec{S}$$

and the easy axis of magnetization lies in the y-direction. (ii) The internal field H_{int}, Eq. (2b), saturates for moderate applied fields. This implies that the spin expectation value $\langle S_y \rangle$ along the easy axis of magnetization, which is the dominant component of $\langle \vec{S} \rangle$ in the model just discussed, must have the proper saturation behavior. With $\lambda = \frac{1}{3}$ one finds for the ground state $\langle S_y \rangle = (\text{const}) y (1+y^2)^{-\frac{1}{2}}$,

where $y \propto H_y \beta g_{yy}/D$ is proportional to the Zeeman interaction
in the y-direction in units of D.   This model reproduces
the data quite well if the slow relaxation limit is assumed.
Fast relaxation can be ruled out since it leads to saturation at larger applied fields than was observed experimentally.

With these considerations in mind the parameters λ,
A_x, A_y, A_z, η and sign (ΔE_Q) were optimized, and the additional
fine structure parameter μ, Eq. (1b), was added for the
final simulations.   To minimize the number of variables all
the tensors were assumed to have the same principal axes
system as the fine structure term.   Furthermore, $\tilde{g}$ was
calculated from D and λ using second order perturbation
theory.   The final parameters used for the simulations of
Fig. 3 are listed in Table I.   They should be considered
as tentative in view of the approximations made, and more
work is required to establish the uniqueness of the
solution, limits in the parameter values, etc.   A number of
interesting conclusions, however, can be drawn at this
stage already.
(i) The spin Hamiltonian, Eqs. (1a,b)   permits a satisfactory simulation of the data in the slow relaxation limit.
(ii) The fine structure term is large, $D \simeq 12$ K, and has low
symmetry.   In fact, the "rhombic" case, $\lambda \simeq \frac{1}{3}$, $\mu = 0$, provides
a good approximation for both ferric and ferrous rubredoxin.
(iii) The quadrupole interaction has a large asymmetry
parameter, i.e. there is a large negative as well as a
large positive component ($eQV_{zz}/2 = -3$ mm/sec, $eQV_{yy}/2 = 2.5$
mm/sec).
(iv) The magnetic hyperfine interaction is negative in all
its components and is quite anisotropic.

The Mössbauer parameters of reduced rubredoxin can be
compared with those of the ferrous iron in the paramagnetic
$Fe^{3+}-Fe^{2+}$ cluster of reduced 2Fe-2S* proteins.   Table I
shows that the Fe^{2+} site indeed is quite similar in the two
types of proteins as far as the electric and magnetic hyper-
find interactions are concerned.   Another very interesting
comparison can be made with one of the recently synthesized
model complexes of rubredoxin. The similarity of the high-
field, low-temperature spectrum of this complex[20] with the
spectrum of Fig. 3b suggests that it not only has comparable
hyperfine interaction but also a fine structure term of the
same symmetry and magnitude as rubredoxin.

REFERENCES

1) W. A. Eaton and W. Lovenberg in "Iron-Sulfur Proteins" II, p. 131. W. Lovenberg, ed. Academic Press, New York (1973).

2) K. D. Watenpaugh, L. C. Sieker, J. R. Herriott and L. H. Jensen, Acta Cryst. B$\underline{29}$, 943 (1973).

3) R. G. Shulman, P. Eisenberger, W. E. Blumberg and N. A. Stombaugh, Proc. Nat. Acad. Sci. USA $\underline{72}$, 4003 (1975).

4) W. Lovenberg and W. M. Williams, Biochemistry $\underline{8}$, 141 (1969).

5) Iron-Sulfur Proteins, W. Lovenberg, ed. Academic Press New York (1973).

6) R. H. Holm in "Iron-Sulfur Proteins" III, W. Lovenberg, ed. Academic Press, New York (1976).

7) W. D. Phillip, M. Poe, J. F. Weiher, C. C. McDonald and W. Lovenberg, Nature $\underline{227}$, 574 (1970).

8) K. K. Rao, M. C. W. Evans, R. Cammack, D. O. Hall, C. L. Thompson, P. J. Jackson and C. E. Johnson, Biochem. J. $\underline{129}$, 1063 (1972).

9) G. H. Loew et al., Theor. Chim. Acta (Berl.) $\underline{32}$, 217 (1974); $\underline{33}$, 125, 137, 147 (1974).

10) J. G. Norman, Jr. and S. C. Jackels, J. Amer. Chem. Soc. $\underline{97}$, 3833 (1975).

11) R W. Lane, J. A. Ibers, R. B. Frankel and R. H. Holm, Proc. Nat. Acad. Sci. USA $\underline{72}$, 2868 (1975).

12) J. Peisach, W. E. Blumberg, E. T. Lode and M. J. Coon, J. Biol. Chem. $\underline{246}$, 5877 (1971).

13) H. H. Wickman, M. P. Klein and D. A. Shirley, J. Chem. Phys. $\underline{42}$, 2113 (1965).

14) G. Lang, R. Aasa, K. Garbett and R. J. P. Williams, J. Chem. Phys. $\underline{55}$, 4539 (1971).

15) K. Spartalian, W. T. Oosterhuis and J. B. Neilands, J. Chem. Phys. 62, 3538 (1975).

16) W. E. Blumberg and J. Peisach, Ann. New York Acad. Sci. 222, 539 (1973).

17) T. Castner, Jr., G. S. Newell, W. C. Holton and C. P. Slichter, J. Chem. Phys. 32, 668 (1960).

18) E. Münck, P. G. Debrunner, J. C. M. Tsibris and I. C. Gunsalus, Biochemistry 11, 853 (1972).

19) W. M. Reiff, "Mössbauer Effect Methodology" 8, I. J. Gruverman and C. W. Seidel, ed. p. 89, Plenum Press, New York (1973).

20) A. Kostikas, V. Petrouleas, A. Simopoulos, D. Coucouvanis and D. G. Holah, submitted to Chem. Phys. Letters.

ROLE OF OXYGEN MOTION IN THE TEMPERATURE DEPENDENCE OF ΔE

IN OXYHEMOGLOBIN AND MODEL COMPOUNDS

G. Lang and K. Spartalian

Department of Physics
The Pennsylvania State University
University Park, Pennsylvania   16802

I.   INTRODUCTION

Hemoglobin is the most intensively studied of the biological macromolecules; it has served and continues to serve as the proving ground for techniques and ideas which are leading toward a detailed understanding of life processes in general. In view of the fact that reversible combination with oxygen is the primary function of this molecule, it is a sobering thought to realize that the nature of the attachment of the oxygen is only poorly understood. There is no general agreement on the disposition of bonding electrons. General agreement as to the geometric arrangement of the oxygen and the iron atom of the binding site has been achieved only recently. The oxygenated form of hemoglobin, having spin-paired electrons, is EPR-silent. Fortunately, however, a site-specific probe is available in the form of ^{57}Fe Mössbauer spectroscopy.

The deoxy- and oxy- forms of hemoglobin were the first proteins examined by Mössbauer spectroscopy.[1-4] The former exhibited quadrupole-split spectra which were typical of high-spin ferrous materials in respect to both isomer shift and the presence of large and temperature-dependent quadrupole splitting. In oxyhemoglobin the quadrupole splitting was again found to be large and temperature dependent. If this zero-spin complex were to be considered as low-spin ferrous, the large quadrupole interaction would imply the presence of large asymmetric 3d electron delocalization through covalent bonding. The observed negative and approximately axial electric field gradient[5] would be compatible

169

with loss of almost one electron charge from a planar t_{2g} orbital.

A more curious feature of the oxyhemoglobin Mössbauer spectrum is the temperature dependence of the quadrupole splitting. Such behavior is usually associated with thermal excitation of electrons into empty low-lying orbital states, and is common in high-spin ferrous and low-spin ferric complexes. However, oxyhemoglobin is diamagnetic and remains so up to room temperature, indicating that its electrons remain paired and consequently that electronic excitation is not involved. An attractive alternative is conformational excitation, involving the motion or displacement of some part of the molecular structure in the neighborhood of the iron. This could redistribute charge without unpairing electrons. The relative flexibility of protein originally led to a suggestion that distortion of the peptide chain near the heme might be involved.[6] This explanation of the temperature dependence of ΔE was, however, almost certainly ruled out by the Mössbauer emission measurements of Münck, et al.,[7] who observed the transient oxygenated iron imidazole heme formed in the decay of the corresponding ^{57}Co complex, and found a quadrupole-split spectrum whose behavior closely followed that of oxyhemoglobin. This finding seems to implicate either the oxygen or the heme, probably the former since no other ligand shows such effects. The detailed nature of the implied oxygen motion is of interest because of the light it could throw on the electronic structure of the complex and because of its possible relevance to biological function.

The study of relatively small non-protein models has traditionally been of importance in efforts to understand the behavior of the large biological molecules. A recent successful simulation of the oxygen binding sites of hemoglobin is described in reference [8], where the synthesis and characterization of a series of compounds of iron, meso-tetra ($\alpha,\alpha,\alpha,\alpha,$-o-pivalamidophenyl) porphyrin, are presented. With a variety of bases attached to the iron on the side opposite the 'picket fence' structure, these are capable of binding O_2 reversibly as the sixth iron ligand. Our Mössbauer measurements of the oxygenated model indicate a close similarity to oxyhemoglobin with respect to isomer shift, quadrupole splitting, and lineshape over a

range of temperature.   X-ray studies of the N-Me-imid
oxygenated complex show a statistical disorder in oxygen
position at room temperature.[9] We believe this is a
dynamic thermal distribution and that it is possible to
account for the observed Mössbauer spectra in terms of it.
In the present paper we support this contention with quantitative treatment, and then discuss its relevance to the
oxygen binding site in hemoglobin.

II.   EXPERIMENTAL

The spectra at 4.2K were recorded by keeping the sample
immersed in liquid helium in a cryostat of a design described elsewhere.[6]   The temperatures between 4.2 and 100K
were achieved with a variable temperature insert, in which
the sample was mounted inside a vacuum can immersed in the
helium bath.   The temperature was maintained by a temperature controller of a design described in reference [10].
A thermocouple (0.03 at. % iron-gold vs. chromel) was used
to record the temperature.   The temperature of 178K was
reached by using a frozen acetone slush bath in a liquid-
nitrogen type experimental cryostat described in reference
[6], and 195K was obtained by using dry-ice in the same
dewar.   The powder samples were contained in 3/4 inch diameter lucite sample holders which were stored in liquid
nitrogen when not in use.

High external magnetic fields were achieved with a
split-coil superconducting magnet (American Magnetics)
which provided a maximum of 6.0T transverse to the γ ray
beam.

The Mössbauer spectra were taken in horizontal transmission geometry using a constant acceleration spectrometer
operated in connection with a 256 channel analyzer in the
time scale mode.   The source was kept at room temperature
and consisted of 50 millicuries of [57]Co diffused in rhodium
foil.   The spectrometer was calibrated against metallic
iron foil and zero velocity was taken as the centroid of
its room temperature Mössbauer spectrum.   In calibration
spectra linewidths of about 0.23 mm/sec were normally
observed.

III.   RESULTS

Mössbauer spectra of a powdered sample of [Fe(O_2)
(N-Me-imid) ($\alpha,\alpha,\alpha,\alpha$-TpivPP)], measured at a variety of
temperatures, are shown in figure 1.   In a first attempt to
characterize them numerically, least-squares fits were made,
assuming a symmetric pair of lorentzian lines in each case
and allowing the linewidth to vary.   The results are shown
in Table I.   In many cases the fits were less than satisfactory, but the table does indicate general trends.   The

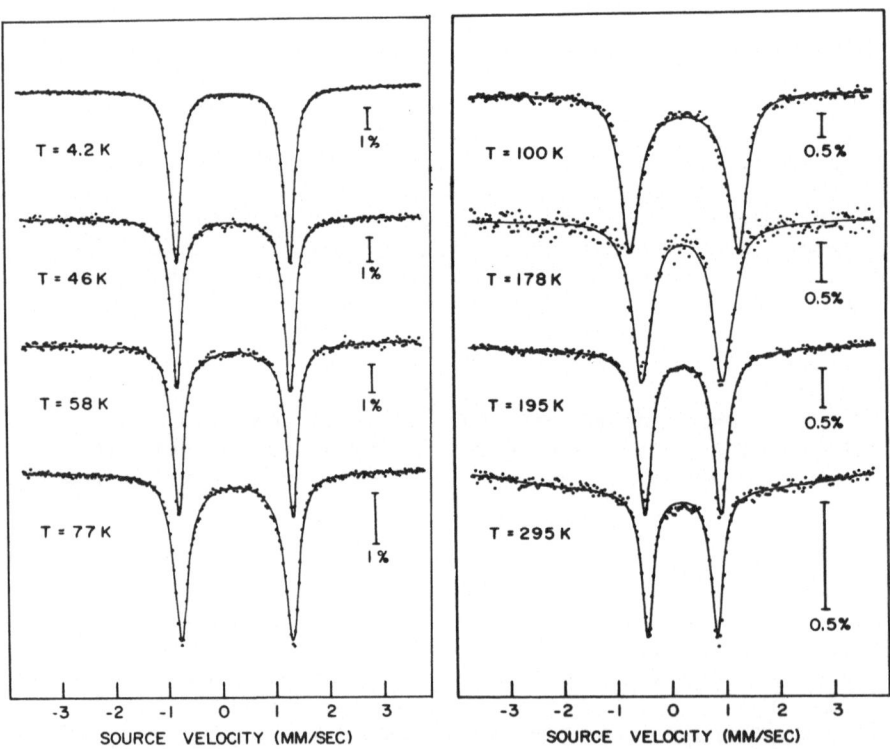

Figure 1.   Zero-field Mössbauer spectra of FeO$_2$(N-Me-imid)P
at various temperatures as shown.   The solid
curves are theoretical least-squares fits as
described in the text.

Table I. Zero-field Mössbauer parameters determined by fitting lorentzians to the data from sample FeO_2(N-Me-imid)P.[a]

T(K)	Γ(mm/sec)	δ^b(mm/sec)	ΔE(mm/sec)
4.2	0.236	0.280	2.106
46	0.244	0.276	2.109
58	0.253	0.274	2.099
77	0.304	0.271	2.042
100	0.421	0.267	1.947
178	0.430	0.248	1.488
195	0.330	0.243	1.389
295	0.248	0.195	1.288

[a]P = ($\alpha,\alpha,\alpha,\alpha$-TpivPP). Statistical uncertainties are ±0.005 mm/sec for all Mössbauer parameters.

[b]With respect to metallic iron.

essential features are a quadrupole splitting which decreases with increasing temperature, and an effective linewidth which is maximum at an intermediate temperature. At 4.2K and 295K this linewidth is approximately the same as observed in calibration runs using thin iron foils. Only the full curve at 4.2K shown in figure 1 is a lorentzian fit. The others are based upon a detailed relaxation model to be described below. In order to determine the sign of the electric quadrupole interaction a measurement was made with the sample at 4.2K in a magnetic field of 6.0T. A straightforward theoretical fit[11] indicated a nearly axial efg with negative principal component and an asymmetry parameter of 0.23.

IV. DISCUSSION

At 4.2K the [Fe(O_2)(N-Me-imid)($\alpha,\alpha,\alpha,\alpha$-TpivPP)] sample exhibits a typical quadrupole splitting, resulting from interaction of the iron nucleus with a local electric field gradient according to the following Hamiltonian:

$$H = \frac{QV_{z'z'}}{4} \left[I_{z'}^2 - \frac{5}{4} + \frac{\eta'}{3} (I_{x'}^2 - I_{y'}^2) \right] \quad ;$$

$$\eta' = \frac{V_{x'x'} - V_{y'y'}}{V_{z'z'}} \quad . \tag{1}$$

Here I is the nuclear spin operator, Q is the nuclear quadrupole moment, and the V_{ii} are the second spatial derivatives of the electrostatic potential in the x'y'z' principal axis system, evaluated at the iron nucleus--the so-called electric field gradient tensor or efg. The asymmetry parameter is denoted by η', and we orient the coordinate system to make $V_{z'z'}$ the principal second derivative of largest magnitude. At this stage we have not determined the orientation of the efg relative to the molecular axes.

When no magnetic interactions are present, the electric quadrupole interaction gives rise to a pair of absorption lines with energy separation $\Delta E = 1/2 \; QV_{z'z'} \; \sqrt{1 + \eta'^2/3}$. Measurements of a powder sample yield the value of ΔE, but do not determine $V_{z'z'}$ and the asymmetry parameter η' separately. Application of a strong external magnetic field to a sample will normally perturb the spectrum so that $V_{z'z'}$ and η' may be separately determined. The situation is particularly clearcut in diamagnetic materials such as the present ones, for the complications of internal contributions to the magnetic field at the nucleus are avoided. Using an appropriate computer program,[11] we have allowed $V_{z'z'}$ and η' to vary in order to achieve a least-squares fit to our high-field 4.2K spectrum. The optimum values found were $QV_{z'z'}/2 = -2.088$ mm/sec (^{57}Fe Mössbauer energy units) and asymmetry parameter $\eta' = 0.23$. The residual sum of squares achieved was 320, to be compared with the value 248 which would be expected from the statistical uncertainty in the gamma-ray counts. It is a characteristic of the method that low values of η' such as the present one are determined with rather low accuracy, and we would roughly describe our value as lying between 0.1 and 0.3 with high probability. Thus the electrostatic potential at the iron nucleus has nearly axial symmetry about its principal direction, and a negative second derivative along that direction. Such a situation would be produced, for example, by starting with a set of six full

t_{2g} electron orbitals on the iron (low-spin ferrous configuration) and removing approximately 2/3 of an electron
charge from a single one of the planar orbitals.

We propose to explain the decrease in quadrupole
splitting of the [Fe(O$_2$)(N-Me-imid)($\alpha,\alpha,\alpha,\alpha$-TpivPP)] with
increasing temperature in terms of a conformational exci-
tation of the oxygen molecule. A brief consideration of
the molecular structure of the oxygenated complex is
essential to the understanding of our model. The structure
of [Fe(O$_2$)(N-Me-imid)($\alpha,\alpha,\alpha,\alpha$-TpivPP)] as determined by X-
ray diffraction,[9] is shown in figure 2. The imidazole plane
lies normal to the heme plane, and makes an angle of about
25° with the line joining the iron and a pyrrole nitrogen.
The line passing from the nearer oxygen through the iron to
the imidazole nitrogen is a 2-fold axis of the crystal, with
the imidazole group statistically disordered between the
symmetry related orientations. The oxygen was found to be
randomly distributed over four orientations as shown, with
the Fe-O-O plane bisecting the angles subtended by the
pyrrole nitrogens in each case. Thus the iron-oxygen plane
can lie either about 20° or 70° from the imidazole plane.
If we neglect the asymmetry produced by the imidazole

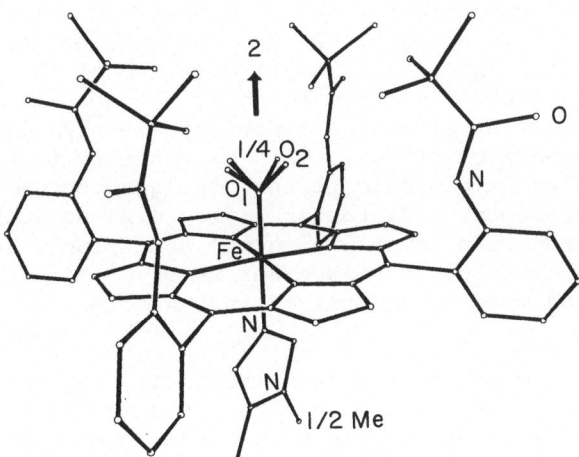

Figure 2.   The molecular structure of compound FeO$_2$(N-Me-
imid)P as determined from X-ray studies.

N-methyl group, this gives rise to just two types of molecular conformation. Within the experimental accuracy of the X-ray study, these are related by 90° rotation of the O_2 molecule about the heme normal, with no change in its altitude above the heme plane.

We seek a quantitative model consistent with the data and the known geometry of the complex. In order to keep it to manageable proportions it is necessary to make some assumptions about the efg orientation. We begin by assuming that for both oxygen conformations the heme plane is a principal plane of the efg tensor, and select the heme normal as the z axis of the principal axis system. We further assume that the field gradients produced by the two conformations have a common set of principal axes x,y lying in the heme plane. It is not necessary to specify their orientation within the plane, but it is reasonable to believe that they are largely determined by the iron-oxygen bond, and hence bisect the angles subtended by the pyrrole nitrogens. The oxygen motion does not involve atomic displacement in the z direction; we will make the additional assumption that it does not involve the displacement of any charge in this direction, and hence that V_{zz} is unaffected. We now must consider the relative orientation of the molecular xyz coordinate system and the system x'y'z' in which the low temperature high magnetic field results give $QV_{z'z'}/2 = -2.088$ and $\eta' = 0.23$. We can see immediately that z' cannot be along z, because the assumed constancy of V_{zz} would allow only a very small decrease of ΔE with temperature. In fact if z' were along z, the quadrupole splitting could vary by at most a factor of $\sqrt{1 + \eta^2/3}$ which for $\eta = 1$ has a maximum of 14%. This is contrary to our experimental results which indicate that the quadrupole splitting decreases by almost a factor of 2 from 4.2K to room temperature. Therefore z' must lie in the heme plane; let it be along y. The choice as to whether x' or y' lies along z is relevant, and must be determined in the detailed calculations.

We also assume that conformation (II) lies at an energy E_0, with $E_0/k \gg 4.2K$, and that the relative probabilities of the conformations are given by the Boltzmann factor, $p(II)/p(I) = \exp(-E_0/kT)$. If the conformational relaxation were very fast on the Mössbauer time scale (i.e., rate $\gg$ 2 MHz), the effective efg would be found by adding corresponding components of efg(I) and efg(II) with appropriate

Boltzmann weighting factors. The large quadrupole splitting
and narrow absorption lines observed at 4.2K result from
the interaction of the iron nucleus with a field gradient
produced by the oxygen in its low energy conformation. The
decrease of ΔE with increasing temperature occurs because
the contribution to the electric field gradient from the
high energy oxygen conformation tends to oppose that from
the low energy one. The narrow absorption lines observed
at room temperature indicate a conformational relaxation
which is fast compared with the nuclear lifetimes. At
intermediate temperatures, however, the absorption lines
are broad and asymmetric, indicating intermediate relaxation rates.

The problem of calculating the lineshape of Mössbauer
spectra when the nucleus finds itself in a fluctuating
electric field gradient has been solved by Tjon and Blume.[12]
The theoretical considerations as adapted to our case can

Table II. Mössbauer parameters found by fitting the spectra
from FeO_2(N-Me-imid)P to the relaxation model.

T(K)	δ^a (mm/sec)	W(MHz)
4.2	0.280	---[b]
46	0.276	1.4
58	0.274	1.5
77	0.271	3.7
100	0.270	5.9
178	0.248	26
195	0.243	58
295	0.197	173
$E_0/k = 147$ deg K	$\Gamma_{nat.} = 0.19$ mm/sec	
$\eta(II) = -1.22$	$\Gamma_{instr.} = 0.04$ mm/sec	

[a] With respect to metallic iron.

[b] W value not relevant because the high energy conformation
is not appreciably populated at 4.2K.

be found in reference [13] and will not be presented in
detail here. It suffices to say that the zero-field
Mössbauer spectra from the model compound were fit to com-
puter-generated calculations with the following adjustable
parameters:

    1. A choice of the orientation of the low temperature
        efg(I), viz the decision as to whether x' or y'
        should lie along z, the heme normal,
    2. The value of η which characterizes efg(II),
    3. The conformational excitation energy E_0,
    4. The relaxation rate at each temperature of obser-
        vation.

    The solid lines in figure 1 are calculations based on
the set of parameters presented in Table II. The electric
field gradients that correspond to these parameters for
each conformation are shown in figure 3. The value of our
model lies in its simplicity and in its agreement with
intuitive notions based on the molecular structure of the
sample studied. We have shown that, even with the con-
straints imposed by the model, there exists a satisfactory
choice of efg(I), efg(II) and E_0. This choice, combined
with a relaxation rate monotonically increasing with temperature, consistently accounts for the temperature dependence

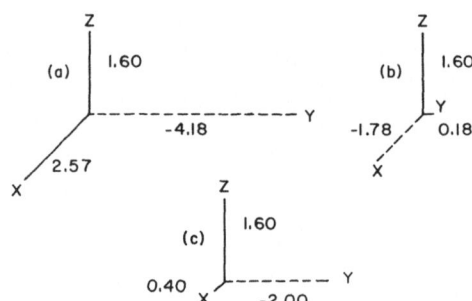

Figure 3. Schematic, showing the values of the efg tensors
          for the model compound. Numbers give values of
          principal components QV_{ii} in ^{57}Fe Mössbauer
          units of mm/sec. (a) Corresponds to the low
          energy conformation I; (b) corresponds to the
          high energy conformation II; (c) is their aver-
          age, i.e., the expected high temperature fast
          relaxation limit.

of the spectral lineshape and quadrupole splitting over the entire range observed. The generally good agreement between theory and data in figure 1, together with the small number of adjustable parameters, suggests that the model is highly plausible. The energy E_0 corresponds to a population ratio $p(II)/p(I) = 0.6$ at room temperature; the X-ray study found random orientation of oxygen, but it may be that the experimental uncertainty could accommodate our predicted population ratio.

We may now shift our discussion to oxyhemoglobin. Its zero field spectra exhibit the same qualitative behavior as the model compound. At low temperature two relatively narrow and symmetric lines are observed with a quadrupole separation slightly larger than the model compound (2.25 mm/sec). As the temperature is increased, the quadrupole separation decreases while the lines display the same asymmetry as the model compound at comparable temperatures. Moreover, the principal component of the efg ($V_{z'z'}$) is negative and the efg itself is nearly axial ($\eta = 0.2$). A noteworthy difference between oxyhemoglobin and the model compound is that in the former the quadrupole splitting seems to extrapolate to a larger high-temperature limit (~1.70 mm/sec) than the latter. Although the zero-field spectra from oxyhemoglobin have not been analyzed yet in terms of a relaxation model, it should be pointed out that the difference in the high-temperature limit of ΔE might reflect a difference in the oxygen motion between the model compound and oxyhemoglobin. Structural information and X-ray results[14] indicate that the O_2 molecule in oxyhemoglobin is sterically hindered from completing a full circle about the Fe-O axis; instead it may "wobble" within an angle of about 45°.

Finally, in an effort to establish possible differences in the relaxation characteristics of the different chains in our oxyhemoglobin, we have studied the Mössbauer spectra from samples having selectively enriched α and β chains with ^{57}Fe in the fully oxygenated tetramer. Preliminary analysis shows no dramatic difference between the two chains although the α chains at 4.2K have a quadrupole splitting ΔE which is slightly larger than the one observed from the β chains by 0.03 mm/sec.

We now speculate briefly on the sources of the electric field gradients, and propose the following as a working hypothesis. The heme plane has axial symmetry about the normal through the iron. Assume that the oxygen is primarily responsible for breaking this symmetry, and the effect of the imidazole is relatively small. Let the low-energy equilibrium conformation correspond to oxygen lying in the yz plane. An appropriate set of iron t_{2g} orbitals would then be (xz), (yz), and (x^2-y^2). (Note that x and y axes bisect N-Fe-N angles in our system.) The symmetry of one of the empty oxygen π^* orbitals would be appropriate to mix with (xz) causing a withdrawal of electronic charge. If this were the only source of efg, it would produce a potential function axial about y with V_{yy} negative. In the higher energy conformation this efg component would simply be rotated about z. Any rigorous defense of this picture would of course require an extensive molecular orbital calculation.

ACKNOWLEDGMENTS

    This work was supported by NIH Grant HL16860 from the National Heart and Lung Institute. The collaboration of J. P. Collman of Stanford University and T. Yonetani of the University of Pennsylvania is gratefully acknowledged.

REFERENCES

1. W. Karger, Z. Naturforsch. 17B, 137 (1962).
2. U. Gonser, R. W. Grant, and J. Kregzde, Appl. Phys. Lett. 3, 189 (1963).
3. J. Maling and M. Weissbluth, in "Electronic Aspects of Biochemistry," 93 (1963).
4. U. Gonser, R. W. Grant, and J. Kregzde, Science 143, 680 (1964).
5. G. Lang and W. Marshall, Proc. Phys. Soc. 87, 3 (1966).
6. G. Lang, Quart. Rev. Biophys. 3, 1 (1970).
7. L. Marchant, M. Sharrock, B. M. Hoffman, and E. Münck, Proc. Nat. Acad. Sci. USA 69, 2396 (1972).
8. J. P. Collman, R. R. Gagne, C. A. Reed, T. R. Halbert, G. Lang, W. T. Robinson, J. Amer. Chem. Soc. 97, 1427 (1975).
9. J. P. Collman, R. R. Gagne, C. A. Reed, W. T. Robinson, and G. A. Rodley, Proc. Nat. Acad. Sci. USA 71, 1326 (1974).
10. B. Window, J. Phys. E2, 894 (1969).
11. G. Lang and B. W. Dale, Nuc. Inst. and Meth. 116, 567 (1974).
12. J. A. Tjon and M. Blume, Phys. Rev. 165, 456 (1968).
13. K. Spartalian, G. Lang, J. P. Collman, R. R. Gagne, and C. A. Reed, J. Chem. Phys. 63, 5375 (1975).
14. M. Perutz (private communication).

MÖSSBAUER STUDIES OF INTERNALLY NITRIDED AND OXIDIZED

ALLOYS

G. P. Huffman and H. H. Podgurski

U. S. Steel Research Laboratory
Monroeville, Pennsylvania   15146

I.   INTRODUCTION

The reactions between mobile interstitial atoms and substitutional solute atoms in alloys is a topic of considerable metallurgical interest. Internally nitrided Fe base alloys frequently exhibit significantly improved mechanical properties, such as high hardness and yield strength values. Alloy oxidation (both internal and external), on the other hand, is generally undesirable, and the aim of most research in this area is to prevent or limit such oxidation. Both subjects have been extensively studied,[1,2] and many of the most interesting aspects of both problems involve internally formed nitride or oxide "phases" which are dispersed on an extremely fine scale, ranging from molecular clusters to fine platelets with thicknesses of the order of 10 to 20 Å. Mössbauer spectroscopy, with its high sensitivity to valence state and local atomic environment is an excellent method of studying such systems and in the current paper, some recent work in this area is summarized.

II.   EXPERIMENTAL PROCEDURE

All spectra were obtained using a constant acceleration Mössbauer spectrometer of standard design, utilizing a multichannel analyzer with dual input capability which allowed simultaneous accumulation of the unknown spectrum and a calibration spectrum, normally that of an Fe foil.

The source for the Fe^{57} spectra was Co^{57} in Pd (the source
strength ranged from approximately 30 to 80 mCi for the
various spectra reported here) and that for the Sn^{119}
spectra was approximately 20 mCi of Sn^{119m} in a $BaSnO_3$
matrix.* Low and high temperature spectra were obtained
using a continuous flow cryotip and a bifilarly wound
vacuum furnace.

Sample preparation for the internally nitrided and
oxidized alloys has been described in more detail
elsewhere,[3,4] and only a brief summary of the techniques
used will be given here. The nitrided alloys were
prepared at temperatures ranging from 400 to 600 C in
flowing NH_3-H_2 atmospheres in which the nitrogen activity,
defined by

$$a_N = P_{NH_3} / (P_{H_2})^{3/2} \ (atm^{-1/2}) \tag{1}$$

could be maintained constant for extended time intervals,
and was always held at levels too low to allow formation
of any Fe nitride phases. Nitrogen uptake was continuously
monitored gravimetrically.[3] Two Fe-Mo alloys containing
2.9 and 0.28% Mo and three Fe-Ti alloys containing 0.68,
1.29 and 2.21% Ti were studied.**

Most of the internal oxidation results reported are
for two Ag-Sn alloys which contained 0.33% Sn (enriched to
90% Sn^{119}) and 6.1% Sn (1% of which was enriched to 90%
Sn^{119}). Preliminary results are also reported for an Fe-
4.45% Mn alloy in which the Fe^{57} content was enriched to
approximately 35%. The Ag-Sn alloys were oxidized
at constant pressure (P_{O_2} = 100 to 450 torr, T = 300 and
400 C) and the oxygen uptake was measured volumetrically.[4]
The Fe-Mn alloy was oxidized in a flowing N_2- H_2 - H_2O gas
mixture at a constant oxygen potential too low to produce
any Fe oxides. For the Fe-Mn sample, the surface layers are
of most interest and electron re-emission Mössbauer (ERM)
spectroscopy was used.

---

*   All sources were obtained from New England Nuclear
     Corp.

**   All percentages referred to in this paper are atomic
     percentages.

III.   INTERNALLY NITRIDED ALLOYS

For many of the nitrided alloys discussed in this section, a more detailed report of the Mössbauer, nitriding kinetics, and electron microscopy data has been given elsewhere.[3] The primary objective of the current paper is to briefly summarize the different types of information that can be obtained from Mössbauer studies of these systems, and the correlations between this information and other types of data.

A typical nitrogen uptake curve is shown in Fig. 1 (Fe-0.68% Ti).

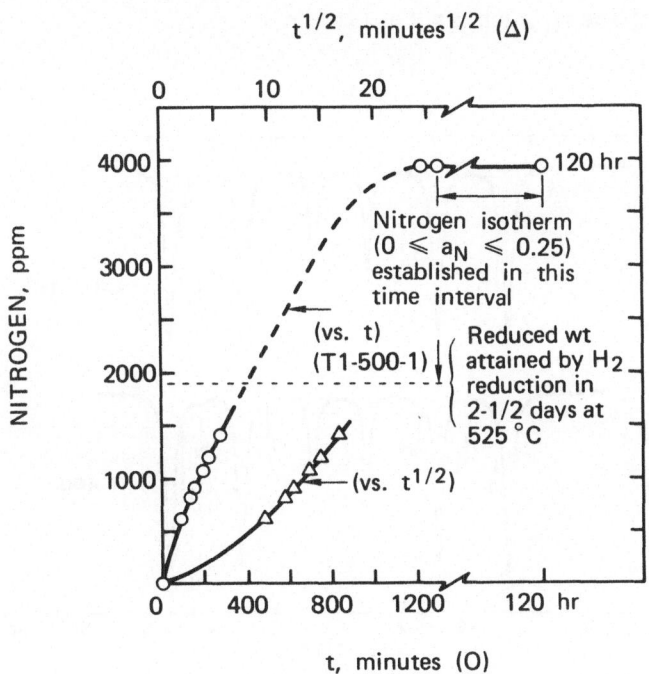

Fig. 1. Nitrogen uptake (ppm by weight) as a function of time for an Fe-0.68% Ti alloy (T= 500°C, a_N = 0.25 atm$^{-1/2}$).

Absorption isotherms for nitrided Fe-Ti alloys show that
the solubility of N in the ferrite matrix is over 2 and 1/2
times larger than that in pure Fe,[3] presumably as a
result of strain generated in the matrix by the highly
dispersed nitride phase. A large fraction of the absorbed
N in the as-nitrided Fe-Ti alloys is therefore present as
excess N in the ferrite matrix. A significant, but much
smaller, fraction of the N in as-nitrided Fe-Mo alloys is
also present in this form. Most of this excess lattice N
can be removed by H_2 reduction for short times at temperatures below about 500°C without modifying the solute-nitride
phase and this procedure was carried out for many of the
samples studied in this investigation. In completely
nitrided Fe-Ti alloys, removal of N by H_2 reduction at
temperatures from 400 to 800°C is observed to cease as the
Ti/N ratio approaches 1 (note sample T1-500-1 in Fig. 1).
This infers that the nitride phase has the stoichiometry
TiN, but does not rule out the possibility of more complex
nitride phases, such as Fe_nTiN (n > 0).

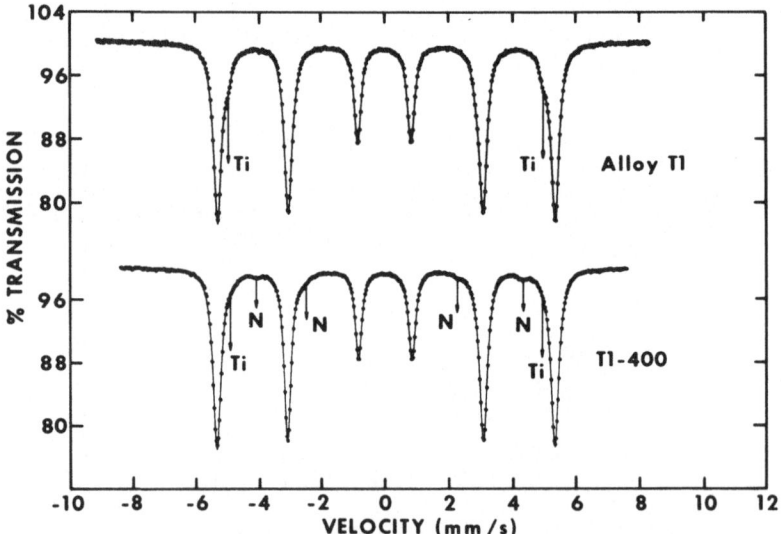

Fig. 2. Spectra of the Fe-0.68% Ti alloy before (top) and
after (bottom) nitriding at 400°C. Total N content of
T1-400 = 0.98%.

Typical spectra of the Fe-0.68% Ti and Fe-2.9% Mo
alloys before and after nitriding are shown in Figures 2 to
5. Spectra of the un-nitrided alloys appear at the top of
Figures 2 and 4. The satellite peaks labeled Ti in Fig. 2
arise from Fe atoms with one Ti nearest-neighbor (nn) or
next-nearest-neighbor (nnn) and those labeled (n,m) in
Fig. 4 from Fe atoms with n Mo nn and m Mo nnn; the
observed hyperfine fields and isomer shifts agree well with
the results of previous investigators.[5,6] As the nitriding
reaction proceeds, solute atoms are removed from solution
and the satellite peaks disappear. Simultaneously, a
rather broad ($\Gamma \approx 0.43$ mm/s) set of satellite peaks (labeled
N) arising from Fe atoms with one or more N nn in octahedral
interstitial positions appears; the Mössbauer parameters of
these peaks are given in Table I. There are two types of N
which can give rise to these N nn peaks; excess N in the
ferrite matrix and N in the surface planes of the nitride
platelets. In the nitrided Fe-0.68% Ti alloy, there is only
a small contribution from surface N, and the N nn peaks
arise mainly from excess N in the strained ferrite matrix
adjacent to the nitride particles. This is supported by
the absence of any detectable N nn satellites for sample
T1-500, 800 R (Fig. 3), which was nitrided to saturation
and then reduced sufficiently to remove all excess N.

Table I

Mössbauer Parameters of N nn Peaks in Nitrided
Fe - 0.68% Ti Alloys

T(K)	H(kG)	$\delta^{\dagger}$ (mm/s)	ε (mm/s)
77	276 ± 2	0.10 ± 0.04	0.08 ± 0.04
295	261 ± 2	-0.02 ± 0.04	0.10 ± 0.04
681	215 ± 4	-0.22 ± 0.08	0.09 ± 0.08

† All Fe^{57} isomer shifts are measured with respect to
metallic Fe at room temperature.

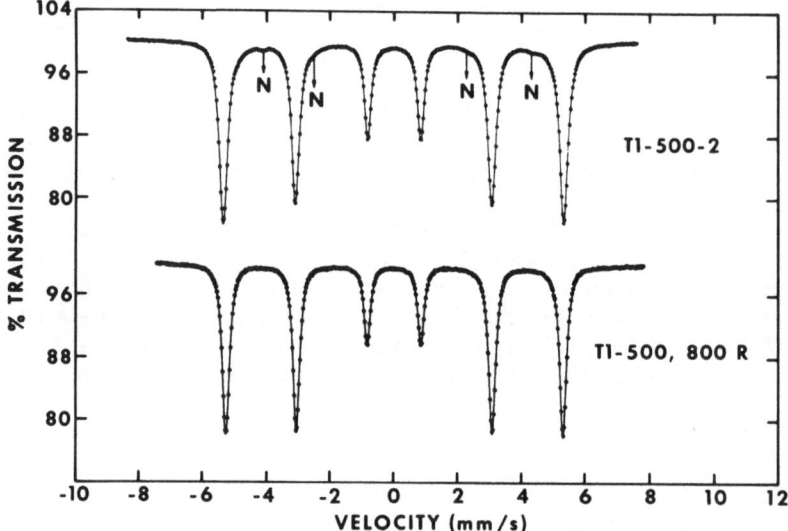

Fig. 3. Spectra of the Fe-0.68% Ti alloy after nitriding at 500°C (top, 0.98% N) and after nitriding at 500 and reducing at 800°C (bottom, 0.65% N).

In somewhat more concentrated alloys, satellite peaks caused by N in the nitride surface planes can be observed on removing the excess lattice N by reduction in H_2 at low temperatures (T $\lesssim$ 500°C) for short times, where little or no growth of the nitride precipitates occurs. For example, compare the spectra of the Fe-Mo-N samples 500-3 (bottom spectrum, Fig. 4) and 500-7 (top spectrum, Fig. 5). Sample

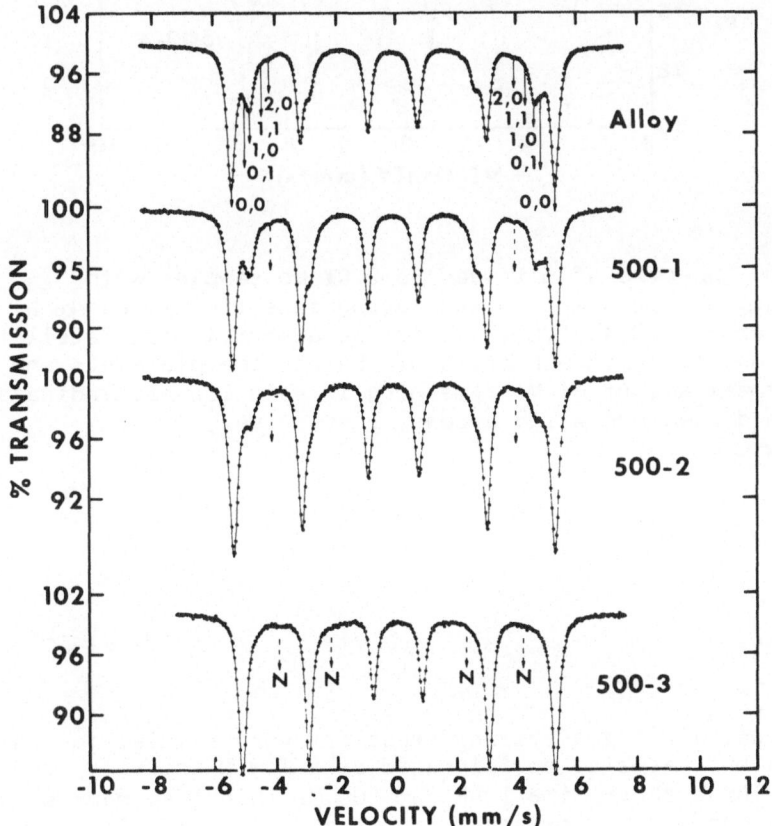

Fig. 4. Spectra of Fe-2.9% Mo alloy before (top) and after nitriding to various levels at 500°C. Total N content: 500-1--0.93%; 500-2--1.73%; 500-3--2.75%.

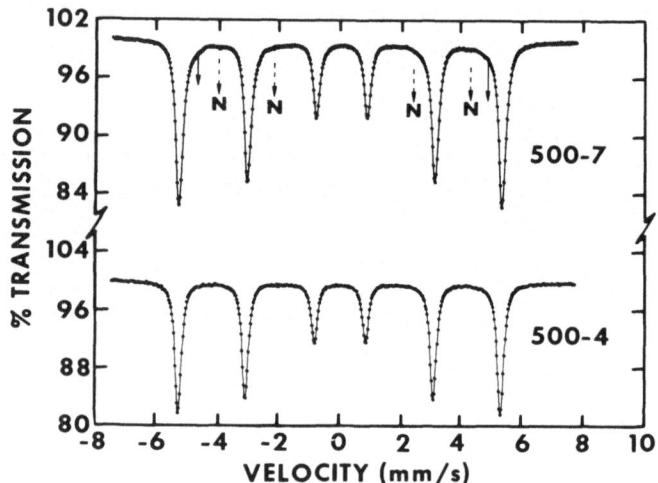

Fig. 5. Spectra of nitrided Fe-2.9% Mo samples with
negligible N in solution and having fine (500-7, 2.44% N)
and coarse (500-4, 2.81% N) nitride dispersions. Satellite
peaks due to surface N atoms of the nitride platelets and
to a small amount of Mo remaining in solution are indicated
by the dashed and solid arrows, respectively.

500-7 was given a nitriding treatment very similar to that of
500-3, but was also equilibrated at a low N activity and reduced for a short time in H_2 at 500 C; this treatment should
remove most of the excess N from the ferrite matrix and the
N nn peaks observed for 500-7 must therefore arise from
matrix Fe atoms at the nitride-ferrite interface. The most
prominent N nn satellite peaks for these samples are shown
in more detail in Fig. 6; the intensities of these peaks are
about 40% less for sample 500-7 than for 500-3 which tends
to confirm the above discussion.

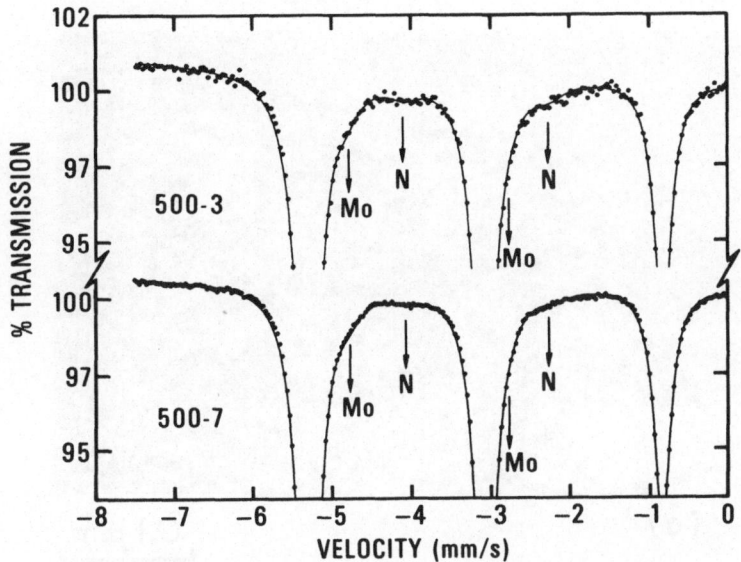

Fig. 6. Enlargement of the left-hand side of the spectra of samples 500-3 and 500-7.

Sample 500-4 is a case where all of the Mo has been nitrided and thermal treatment has caused growth of the nitride phase to a fairly coarse scale, so that the number of Fe atoms with N nn at interfaces or in solution is too small to be detectable. As seen (Fig.5), the spectrum is essentially identical to that of pure Fe. An electron micrograph of this sample is compared to that of an Fe-Mo-N sample having a finely dispersed nitride phase in Fig. 7. Fig. 7a is typical of nitrided alloys with a fine nitride dispersion and exhibits the characteristic tweed structure caused by the formation of thin (~ 10 to 20 Å thick) nitride platelets on (100) ferrite planes.[7-11]

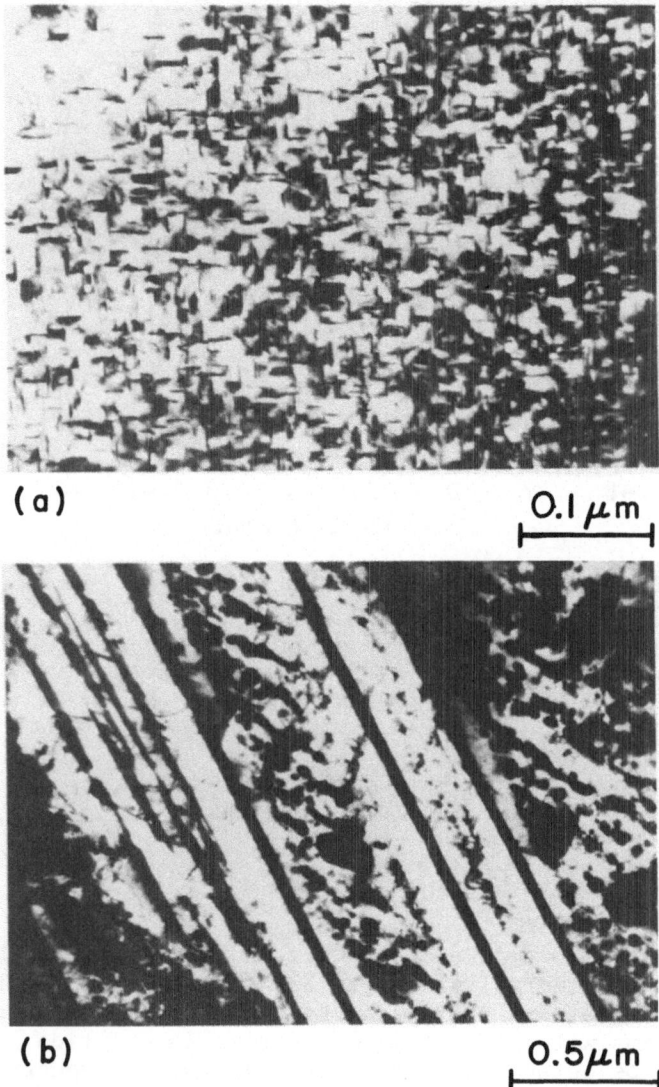

Fig. 7. Electron micrographs of nitrided Fe-2.9% Mo samples having fine (7a, sample 500-9, 1.11% N; see reference (3) for Mössbauer data) and coarse (7b, sample 500-4, 2.81% N, see Fig. 5) nitride dispersions.

It is worth noting that the nitride phase in the
Fe-Mo-N system is less stable than that in Fe-Ti-N alloys.
While short reductions at 500°C remove only the excess
lattice N, longer reductions at 600°C and above decompose
the nitride phase; an example is sample 600-R1, produced by
reduction of sample 500-4 at 600°C. As seen in Fig. 8, the
Mo satellite peaks begin to reappear and a non-magnetic
phase is observed; the isomer shift (-0.24 mm/s) and small
quadrupole splitting (0.16 mm/s) identify this phase
as Fe_2 Mo.[12] As nitrogen is removed from the nitride

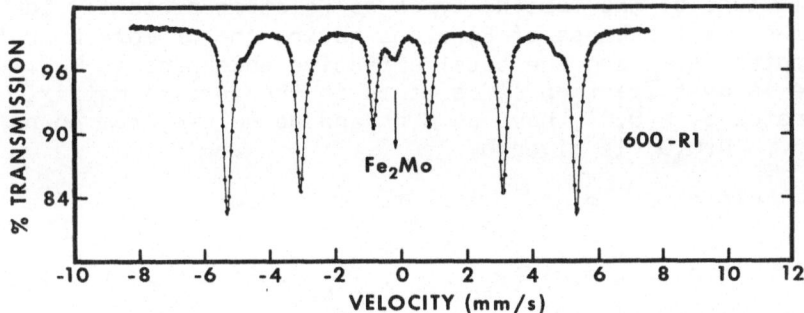

Fig. 8.   Spectrum of sample 600-R1 (total N ≈ 1.01%)
produced by rather extensive reduction of sample 500-4 at
600°C.

particles, it apparently leaves behind Mo-rich regions
which equilibrate with Fe to form Fe_2 Mo. A small amount
of a non-magnetic phase which was also tentatively
identified as Fe_2 Mo is observed to form during nitrogenation at 600°C. Nitrogenation of Fe-Mo at 600°C takes place
in a significantly different fashion than at 500°C, and we
refer the reader to reference (3) for a discussion of this
point.

Previous investigators of Fe-Mo-N alloys have reported
a nitride phase containing equal amounts of Fe and Mo
(Fe_3Mo_3N or $Fe_3Mo_3N_2$);[7-10] however, such a phase would
almost certainly be non-magnetic and should therefore give
a component near the center of the Mössbauer spectra. For

the Fe-Mo samples nitrided at 500°C, no detectable amount of any such phase was observed. To escape detection by Mössbauer spectroscopy, a nitride phase of this type would have to contain $\leq$ 0.3 percent of Fe and Mo, or less than about 10 percent of the total Mo content of the Fe-2.9% Mo alloy.

By combining the measured total N content with the Mössbauer data, the stoichiometry of the internal nitride phases may be determined. The Mössbauer data analysis program converts the absorption area associated with each component of the spectra to an effective thickness,[13,14] and the effective thicknesses are used to calculate the fractions of the various types of Fe atoms present. The effective thickness of Fe atoms having one or more N nn is denoted as x_N and the total effective thickness as x_{tot}. The largest fraction of Fe atoms in the ferrite matrix, denoted by $P(0,0)$, have no N nn and no solute atom nn or nnn. $P(0,0)$ is given by

$$P(0,0) = x(0,0)/(x_{tot}-x_N)---(\text{Fe-Ti-N});$$

$$P(0,0) = x(0,0)/(x_{tot}-x_N-x_{Fe_2Mo})---(\text{Fe-Mo-N}) \quad (2)$$

where $x(0,0)$ is the total effective thickness of the six main peaks in Figures 2-8. Assuming that solute atoms further away than the nnn distance contribute only to line broadening,

$$P(0,0) = (1 - C_s/100)^{14} \tag{3}$$

or

$$C_s = 100 \left\{ 1 - \exp \left[\frac{\ln P(0,0)}{14} \right] \right\} \tag{4}$$

where C_s is the solute atom percentage in solution in the ferrite matrix. Application of Eq.(4) to the un-nitrided alloys gives good agreement with the results of chemical analysis. For nitrided samples, the percentage of solute in the nitride phase, C_{Nit}, is

$$C_{Nit} = C_{tot} - C_s \text{ --- (Fe-Ti-N)}$$

$$C_{Nit} = C_{tot} - C_s - C_{Fe_2Mo} \text{ --- (Fe-Mo-N)}$$

$$(5)$$

where C_{tot} is the total solute atom percentage and C_{Fe_2Mo} is the Mo percentage in Fe_2Mo.

The relevant N percentages are denoted as follows: α_N = total N percentage; α_s = percentage of excess N in the ferrite lattice; and α_{Nit} = percentage of N in the nitride phase. The fraction of Fe atoms having one or more N nn is

$$P_N = x_N / x_{tot} . \qquad (6)$$

For dilute alloys (such as the nitrided 0.68% Ti sample), the N nn peaks arise mainly from excess lattice N and, to a first approximation,

$$P_N \approx 1 - (1 - \alpha_s / 100)^6 \approx 6 \alpha_s / 100 \qquad (7)$$

for small values of α_s. For more concentrated alloys (such as the nitrided 2.9% Mo and 2.2% Ti alloys), N atoms in the surface planes of the nitride platelets can make a significant contribution to P_N and it is preferable to work with samples in which the excess lattice N has been removed by reduction. The importance of the satellite peaks from nitride surface N with regard to mechanical properties will be discussed below.

The percentage of nitrogen in the nitride phase is

$$\alpha_{Nit} = \alpha_N - \alpha_s$$

and the solute atom to N ratio in the nitride phase, Me/N (Me = Ti or Mo), is

$$Me/N = C_{Nit} / \alpha_{Nit}. \qquad (9)$$

The results are summarized in Table II. Column 1 gives the percentage of Ti or Mo in the alloy, column 2 gives the temperature at which the principal nitriding reaction was

carried out,[*] column 3 gives the number of different
samples nitrided at each temperature, column 4 gives the
gravimetrically measured total N content, and column 5
gives the range and average values of the Me/N ratios
determined as described above. It is seen that the

Table II

Summary of the Nitride Stoichiometries Indicated by the
Mössbauer Data and Measured N Content. The Standard
Deviation of the Me/N Ratio was about ± 0.1 to 0.15 for
Most Samples

Alloy	Nitriding Temp. (°C) [*]	No. of Samples	Total N Content (at.%)	Me/N (Me = Ti or Mo)
0.68% Ti	400	1	0.98	1.04
0.68% Ti	500	3	0.65-0.98	0.98-1.15; average = 1.06
1.29% Ti	500	1	1.30	0.99
2.21% Ti	400	1	2.19	1.01
0.28% Mo	500	1	0.32	0.88 ± 0.2
2.9% Mo	500	12	0.70-2.88	0.97-1.06; average = 1.02
2.9% Mo	600	4	0.79-1.92	1.09-1.16; average = 1.12

dominant nitride stoichiometries in the two systems are
simply TiN and MoN. However, the average Mo/N ratio of
1.12 observed for Fe-2.9% Mo nitrided at 600°C may indicate
the formation of a small amount of Mo_2N at that temperature.

---

[*] For many of the samples in Table II, a more complete
summary of the Mössbauer results and nitriding treatments
(which were usually more complex than a simple nitrogenation
at a single temperature) is given in reference (3).

Additionally, as noted earlier, a non-magnetic phase
containing approximately 1% of the total Fe is observed to
form in Fe-Mo samples <u>during</u> nitrogenation at 600°C.
Although the Mössbauer parameters of this phase indicate
that it is Fe_2Mo,[12] the possibility that a non-magnetic
Fe-Mo nitride phase[7-10] might give a peak in nearly the
same location as Fe_2Mo cannot be ruled out at this point.
However, the assumption that the observed non-magnetic
phase is Fe_2Mo * gives results consistent with those
obtained for all other nitrided alloys we have studied and
a reasonable mechanism for the formation of this phase
during nitrogenation at 600°C has been discussed in
reference (3).

Before concluding this section, it is of interest to
consider in more detail the satellite peaks arising from
Fe atoms which are nn to N atoms in the surface planes of
the nitride platelets. As noted earlier, these surface N
satellite peaks can be observed in nitrided alloys whose
excess lattice N has been removed by H_2 reduction (see
Figure 6) and they are directly related to mechanical
properties. Our results are consistent with an atomistic
model proposed by D. H. Jack,[15] in which the (100) plane
of the nitride platelets consists of N atoms in octahedral
interstices surrounded by 4 planar solute (Ti or Mo) atoms,
as illustrated schematically in Figure 9. Each surface N
atom has an Fe nn in the dipolar position on this model and
two quantities of interest may be defined: (i) the nitride
surface area per unit volume, A_N, given approximately by

$$A_N \approx P_N N_{Fe} a_{Fe}^2 \tag{10}$$

where P_N is the Mössbauer determined fraction of Fe atoms
with surface N nn, N_{Fe} is the number of Fe atoms per unit
volume, and a_{Fe} is the Fe lattice constant; (ii) the

---

* It is worth noting that the Fe_2Mo observed in these
samples is not necessarily identical to the massive phase,
but the local atomic environment of the Fe atoms must be
very similar to that in massive Fe_2Mo.

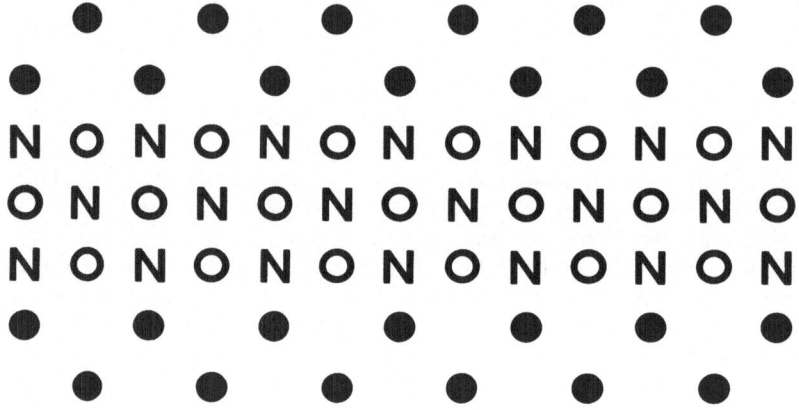

Fig. 9.   Schematic edge-on view of a 3 atomic layer nitride
platelet formed on (100) ferrite planes. The N's represent
N atoms, the white balls are solute atoms (Mo or Ti), and
the black balls are Fe.

average thickness of the nitride platelets in the alloy in
number of atomic layers, n_N, given by

$$n_N \approx 2 \, C_{Nit} / (P_N \times 10^2) \qquad (11)$$

On the basis of very simple considerations,[16] the yield
stress and hardness would be expected to decrease as P_N and
A_N decrease, and n_N increases. Such an effect is shown in
Figure 10, where the hardness values[17] of a partially
nitrided alloy (Fe-2.9% Mo-1.75% N) are plotted against A_N;
the values of n_N are also indicated. After nitriding at
500°C, this sample was first reduced in H_2 at 500°C for two
hours to remove most of the excess lattice N, and was then
annealed for progressively longer times in vacuum at 600°C
to promote thickening of the nitride platelets.[17] The
point for the sample annealed for 168 hours is somewhat
dubious, since the N nn peaks were barely detectable for
that sample, but otherwise the behavior shown in Figure 10
seems reasonable.

H.V. (kg/mm^2)

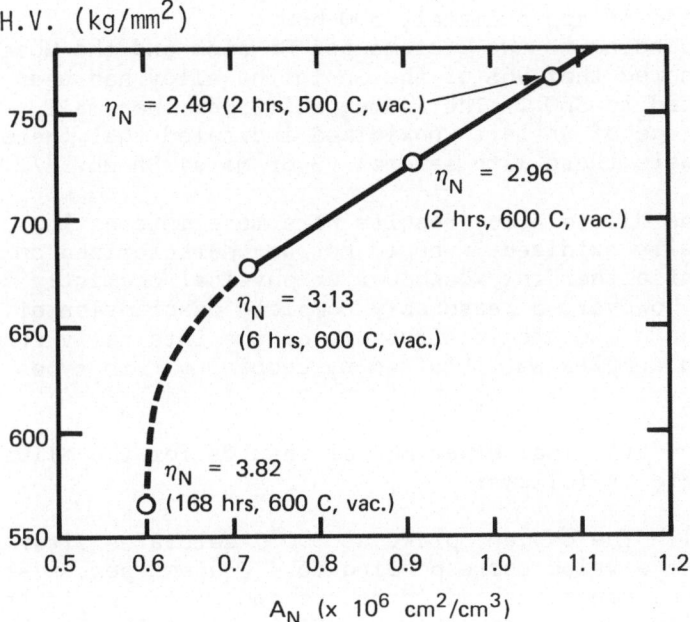

Fig. 10. Hardness values as a function of nitride platelet surface area per unit volume.

IV.   INTERNALLY OXIDIZED ALLOYS[*]

Two Ag-Sn alloys containing 0.33% and 6.1% Sn were subjected to internal oxidation treatments, as described in Section II (T = 300 and 400°C, P_{O_2} = 450 torr). The oxygen uptake was measured volumetrically and had a typical saturation behavior, similar to the nitrogen uptake curve in Figure (1).

For the more concentrated alloy, the data interpretation was fairly straightforward. The oxygen uptake

[*] A more detailed summary of the results for the internally oxidized Ag-Sn alloys appears in reference (4).

saturated in approximately 300 hours at a value corresponding to two O atoms per Sn atom and the Mössbauer data showed that 96% of the Sn in the alloy had been converted to SnO_2. The isomer shift for the small percentage of Sn left unoxidized indicated that these Sn atoms were those with several (3 or more) Sn nn.[4]

The dilute alloy results were more interesting. The internally oxidized Sn could not be characterized on the basis of either the Mössbauer or physical chemistry data alone; however, a reasonably complete description of the electronic and atomic structure of the internally formed Sn-O-Ag complex was obtained by combining both types of data.

The principal experimental results for the dilute alloy are as follows:

(1) The oxygen uptake at 300°C saturates after 120 hours at a value corresponding to 4 O atoms per Sn atom.[†] Mössbauer spectra obtained before and after the internal oxidation process are shown in Figure 11. The isomer shift[*] (δ = -1.77 mm/s) and small quadrupole splitting (ε = 0.53 mm/s) show that the internally oxidized Sn ions have an ionization state close to +4, but are slightly less positive than the Sn ions in SnO_2 (δ_{SnO_2} = - 1.95 mm/s). The isomer shift formula of Lees and Flinn[18] indicates approximately 0.20 to 0.40 electrons in the 5s and 5p shells of the internally oxidized Sn ions, as compared with 0.13 to 0.26 electrons in the 5s and 5p shells of Sn in SnO_2.

(2) Spectra of the dilute alloy after oxidizing approximately halfway to saturation (1.9 moles of O per mole of Sn) are shown in Figure 12. The spectra show no detectable amount of any intermediate oxidation state analogous, for example, to SnO. Only Sn atoms which have been totally oxidized (δ = -1.77 mm/s) and unoxidized Sn atoms (δ = 0.13 mm/s) are observed. This indicates that the oxidation process proceeds into the alloy as a front, forming a totally internally oxidized subscale.

---

† A negligible amount of this oxygen is in solid solution.[4]

* All isomer Sn^{119} shifts are measured with respect to Mg_2Sn at 77 K.

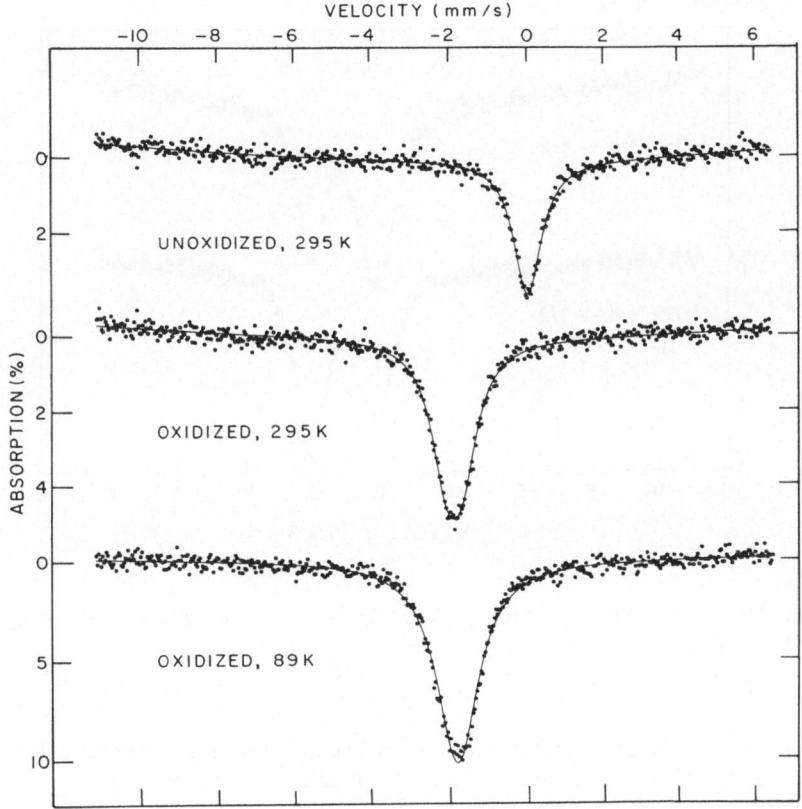

Fig. 11. Spectra of the Ag-0.33% Sn alloy before (top) and after oxidizing to saturation (middle and lower spectra); the absorber temperatures are indicated.

(3) On subjecting the totally internally oxidized alloy to hydrogenation (P_{H_2} = 100 torr, T = 400°C), approximately 2 H atoms per Sn atom are absorbed and the resulting isomer shift (δ = -1.93 mm/s) indicates that the Sn ions have been driven closer to a +4 charge state.

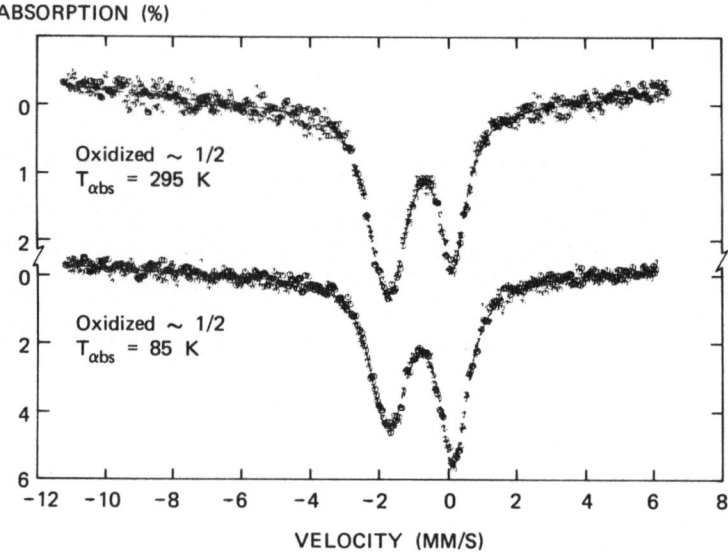

Fig. 12.   Spectra of the Ag-0.33% Sn alloy after internally oxidizing approximately halfway to saturation.

(4)   Isotope exchange experiments were performed by enclosing the internally oxidized sample in $^{36}O_2(P_{36_{O_2}} = 100$ torr) and following the $^{32}O_2$ and $^{16}O^{18}O$ contents of the gas phase to an apparent exchange equilibrium.   At 400°C, approximately 3 of the 4 O atoms held by each Sn ion could be exchanged. * Following hydrogenation of the same alloy at 400°C, it was found that for each mole of H_2 taken up, one additional mole of O was made unavailable for exchange.

All of the above results are consistent with a model shown in Figure 13.   An oxygen atom enters one of the tetrahedral or octahedral interstices adjacent to the

_____

*   An interesting di-atom diffusion process observed in isotope exchange experiments at 300°C has been described previously.[19]

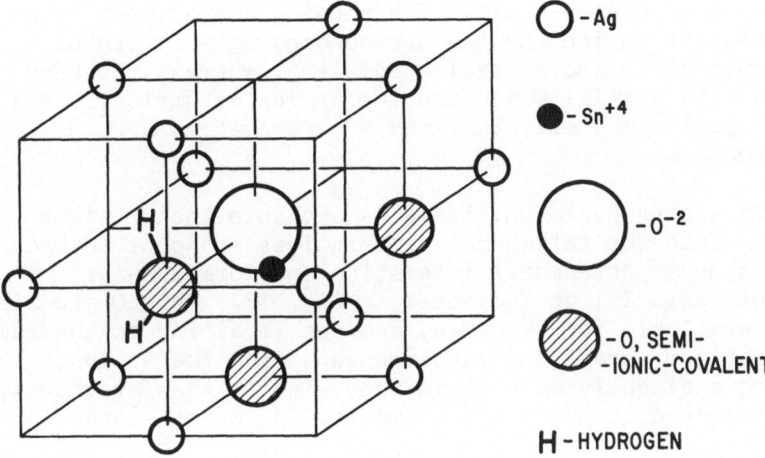

Fig. 13. Possible arrangement of the ions in the Sn-O-Ag complex. The sizes of the balls indicate schematically the relative sizes of the ions involved in the complex.

neutral Sn atom and immediately acquires two electrons from it and forms an ionic bond with it. While the Sn and O are essentially a single ionic molecule, the large size of the O ion and smaller size of the Sn ion would probably cause the O to occupy the original lattice site while the Sn might be displaced in the direction of an adjacent tetrahedral interstice. The presence of this large ionic molecule causes expansion of the Ag lattice around it, probably enlarging the three octahedral interstices nearest the Sn ion. The absence of intermediate oxidation states in the partially oxidized sample shows that the Sn ions strongly prefer to be in the +4 state; therefore, additional oxygens very quickly move into the three enlarged octahedral interstices. These three oxygens share the approximately 2 electrons the Sn ion must give up to put it into a +4 state; additionally they will probably accept some electrons from the surrounding Ag's since oxygen

prefers a -2 charge state. The bonding among these three
oxygens, the Sn ion and the surrounding Ag's is probably
partially ionic and partially covalent, whereas the bond
between the initial O ion and the Sn ion is strictly ionic.
This explains why only 3 of the 4 oxygen atoms are
exchangeable.

On adding hydrogen, it seems probable that H atoms
diffuse into the tetrahedral interstices adjacent to one of
the O's on an octahedral interstice and form a water
molecule with it, as indicated in Fig. 13. This O atom is
now unavailable for exchange, and the remaining octahedral
O's must now accept one electron each from the Sn ion,
forming a slightly more ionic bond with it than previously,
and thereby driving it closer to the +4 charge state.

Finally, an example of the use of Mössbauer
spectroscopy in studies of internal oxidation processes
involving Fe base alloys is given in Figure 14. These
electron re-emission Mössbauer (ERM) spectra were obtained
from an Fe-4.45% Mn alloy (enriched to 35% Fe^{57}) before
(bottom) and after (top) internal oxidation in a flowing
N_2-H_2-H_2O gas mixture (P_{H_2O} / $P_{H_2} \approx 0.09$, $T \simeq 700^oC$). As
in the case of the nitrided Fe base alloys discussed
earlier, the Mn in the top 3000 Å is totally removed from
solution during the oxidation process, and concentrated
into (Fe-Mn)O precipitates of the wustite type. More
extended studies of the nature of the reaction products
and subscale depth profile as a function of oxygen activity,
temperature, and time are in progress.

V.   SUMMARY AND CONCLUSIONS

In this paper, we have summarized some of the applications of Mössbauer spectroscopy in the study of gas-
alloy reactions. The technique is particularly valuable
when the internally formed solute reaction products
are dispersed on an extremely fine scale, ranging from
molecular clusters to thin platelets only 2 or 3 atomic
layers thick. With regard to the internally nitrided and
oxidized alloys discussed in this paper, the principal
information which can be determined from Mössbauer
spectroscopy is as follows: (1) the fraction of solute
which has been reacted and the depth of the reacted subscale;

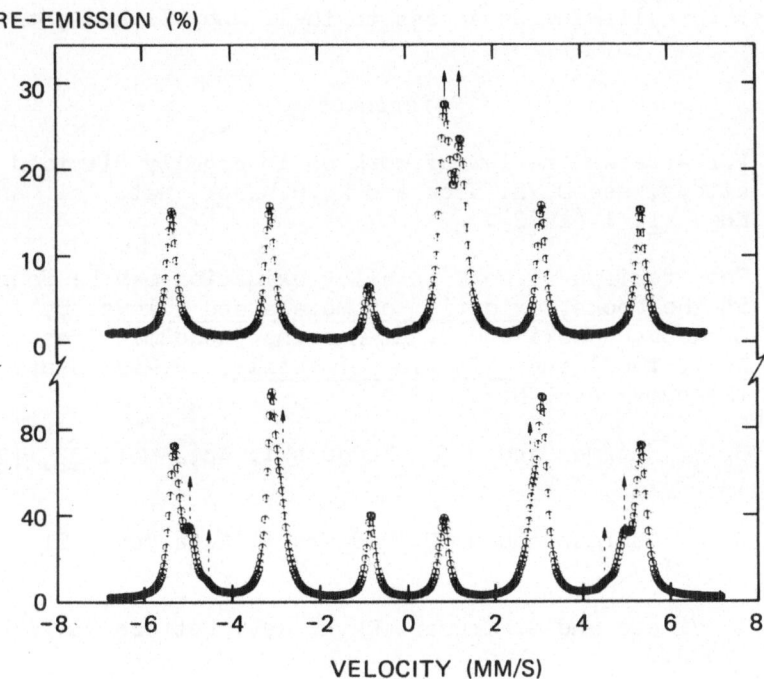

Fig. 14. Electron re-emission Mössbauer spectra of an Fe-4.45% Mn alloy before (bottom) and after (top) the oxidation treatment described in the text. The dashed arrows indicate the satellite peaks due to Mn in solution and the solid arrows, the (Fe-Mn)O peaks.

(2) the stoichiometry of the internally formed compounds and the electronic state of the reacted solute atoms; (3) the average dimensions of the fine internal precipitates; and (4) the detection and identification of any secondary phases. These techniques are clearly not limited to oxidizing and nitriding reactions and should also be useful for the study of reactions between other interstitials (C and H) and solute atoms in alloys.

Acknowledgements

We are grateful to Dr. L. J. Cuddy and Dr. Richard
Wagner for allowing us access to their unpublished data.

References

1.   For a recent review of work on internally nitrided
     alloys, see D. H. Jack and K. H. Jack, Mat. Sci. and
     Eng. 11, 1 (1973).

2.   Good reviews of work on alloy oxidation may be found
     in the books, Oxidation of Metals and Alloys, by
     O. Kubaschewski and B. E. Hopkins (Academic Press,
     N.Y., 1962) and Oxidation of Metals, by Karl Hauffe
     (Plenum Press, N.Y., 1965).

3.   G. P. Huffman and H. H. Podgurski, Acta Met. 23, 1367
     (1975).

4.   G. P. Huffman and H. H. Podgurski, Acta Met., 21, 449
     (1973).

5.   I. Vincze and G. Grumer, Phys. Rev. Letters 28, 178
     (1972).

6.   A. Asano and L. H. Schwartz, Proc. 19th Annual Conf.
     on Magnetism and Magnetic Materials-1973, part I,
     p.262, (A.I.P. Conf. Proc. No. 18, 1974).

7.   D. L. Speirs, W. Roberts, P. Grieveson and K. H. Jack,
     Proc. Second Int. Conf. on Strength of Metals,
     Monterey, Cal., 1970, A.S.M., P.601 (1970).

8.   J. H. Driver, D. C. Unthank and K. H. Jack, Phil. Mag.
     26, 1227 (1972).

9.   S. S. Brenner and S. R. Goodman, Scripta Met. 5, 865
     (1971).

10. J. H. Driver and J. M. Papazian, Acta Met. 21, 1139
     (1973).

11. L. J. Cuddy and H. H. Podgurski, unpublished research.

12. H. L. Marcus, M. E. Fine and L. H. Schwartz, J. Appl.
     Phys. <u>38</u>, 4750 (1967).

13. P. Z. Hien and V. L. Shpinel, Sov. Phys. <u>JETP</u> 17, 268
     (1963).

14. G. P. Huffman, F. C. Schwerer, R. M. Fisher and
     T. Nagata; Proc. Fifth Lunar Sci. Conf., Geochim.
     Cosmochim. Acta, Suppl. 5, Vol. 3, p. 2779 (Pergamon,
     1974).

15. D. H. Jack, personal communication.

16. J. Friedel, <u>Dislocations</u>, Ch.15 (Addison-Wesley, 1964).

17. Measured by Richard Wagner, personal communication.

18. J. K. Lees and P. A. Flinn, J. Chem. Phys. <u>48</u>, 882
     (1968).

19. H. H. Podgurski and G. P. Huffman, Nature, Phys. Sci.
     <u>237</u>, 77 (1972).

THEORETICAL EXPRESSIONS FOR THE ANALYSIS OF MULTI-LAYER

SURFACE FILMS BY ELECTRON RE-EMISSION MÖSSBAUER SPECTROSCOPY

G. P. Huffman

U. S. Steel Research Laboratory
Monroeville, Pennsylvania 15146

A theory of electron re-emission Mossbauer (ERM) spectroscopy has been derived in which analytic expressions are obtained for all quantities of interest in the interpretation of ERM spectra from multi-layer surface films. Only a very brief summary of the principal assumptions and results will be given here; a more detailed treatment will be published elsewhere.[1]

For convenience, all numerical values referred to will be those appropriate for Fe^{57}; the problem is illustrated in Fig. 1. A beam of 14.4 keV gamma rays emitted from a source moving with velocity v is incident on a multi-layer absorber consisting of layers of thickness, $Z_1, Z_2, \cdots Z_N$, $\cdots$. We wish to calculate the rate at which internal conversion and Auger electrons, which result from de-excitation of Fe^{57} nuclei following Mossbauer absorption in the Nth layer, emerge from the surface of the sample. The total escape distance of the electrons is much less than the sample radius so that θ effectively ranges from 0 to $\pi/2$.

It is assumed that the electrons are emitted isotropically. This should be equivalent to the condition of randomly diffusing electrons which Bothe[2] has shown leads to a total electron attenuation of the exponential form, $e^{-\rho z}$. Exponential attenuation has been experimentally verified for such isotropic electron distributions, both in thick-film electron transmission experiments[3] and in studies of internal conversion electron emission following radioactive

209

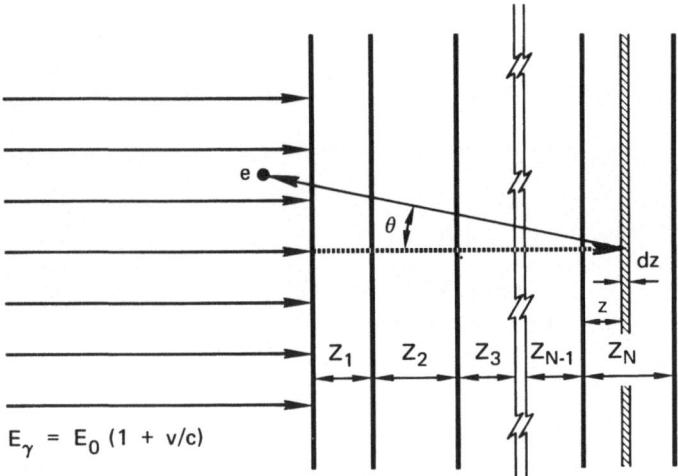

Fig. 1 - Schematic illustration of the multi-layer ERM
         problem.

ion implantation.[4] Therefore, the total fraction of the
internal conversion electrons produced at depth z in layer
N which are emitted from the sample surface is simply taken
to be

$$e^{-\rho_N z}\, e^{-D_{N-1}}/2; \quad D_{N-1} = \sum_{K=1}^{N-1} \rho_K Z_K \qquad (1)$$

where ρ_K is the attenuation coefficient[2,3] for 7.3 keV
electrons in layer K. It may then be shown[1] that the
normalized internal conversion electron counting rate*
arising from the Nth layer at velocity v is given by

$$N_C^{(N)}(v,T_N) = Ce^{-(B_{N-1}+D_{N-1})}\int_0^{T_N} dt\, e^{-(a_N+b_N)t}\left(\frac{-d\mathbf{I}^{(N)}(t,s)}{dt}\right) (2)$$

and the internal conversion electron contribution to the
Nth layer peak area is

$$A_C^{(N)}(T_N) = \frac{C\, e^{-(B_{N-1}+D_{N-1})}}{N_{tot}(\infty)}\int_0^{T_N} dt\, e^{-(a_N+b_N)t}\int_{-\infty}^{\infty} ds\left(\frac{-d\mathbf{I}^{(N)}(t,s)}{dt}\right) (3)$$

---

*The equations for Auger electrons are identical in form.
 All counting rates are normalized by dividing by the number of 14.4 keV gamma rays incident on the sample surface
per second.

where $I^{(N)}(t,s)$ is the fraction of recoilless gamma rays transmitted to depth z in layer N at source velocity v,

$$I^{(N)}(t,s) = \frac{1}{\pi} \int_{-\infty}^{\infty} \frac{dy \, \exp\left\{-\sum_{K=1}^{N-1} T_K/[(y-s_K)^2+1]\right\}\exp\left\{-t/[(y-s_N)^2+1]\right\}}{(y-s)^2 + 1} \tag{4}$$

For simplicity, the calculations are limited to single-level absorbers, and the various quantities appearing in Eq's. (2) to (4) are defined as follows:

$$C = f_s \alpha/2(1+\alpha); \qquad B_{N-1} = \sum_{K=1}^{N-1} \mu_K z_K;$$

$$T_K = t_o^{(K)} z_K; \qquad t = t_o^{(N)} z; \qquad a_N = \mu_N/t_o^{(N)};$$

$$b_N = \rho_N/t_o^{(N)}; \qquad s = 2v/\Gamma_o; \qquad s_K = 2v_K/\Gamma_o.$$

Here, f_s is the recoilless fraction of the source, $N_{tot}(\infty)$ is the background electron counting rate, α and Γ_o are the internal conversion coefficient and natural level width (8.18 and 0.097 mm/s for Fe^{57}), and μ_K, $t_o^{(K)}$, T_K, and v_K are, respectively, the gamma ray attenuation coefficient, effective thickness per unit length, total effective thickness, and resonance velocity for layer K (see reference (1) for more precise definitions).

The method of solution for the area integral of Eq. (3) is as follows: (i) $(-dI^{(N)}/dt)$ is obtained from Eq. (4) and inserted into Eq. (3); (ii) the integral over s is performed; (iii) the change of variables $u = y-s_N = [(1-x)/(1+x)]^{1/2}$ is made and the integral over dt is carried out; and (iv) the exponential function specifying the attenuation of gamma rays incident on layer N due to Mossbauer absorption in higher layers is replaced by an average form and removed from the integral. The result is,

$$A_C^{(N)}(Y_N) = C_C e^{-(B_{N-1}+D_{N-1})} < F_N(Y_K, \delta_{KN}) > H(\beta_N, Y_N); \tag{5}$$

$$<F_N(Y_K,\delta_{KN})> = \exp\{-\sum_{K=1}^{N-1} Y_K/(1+w(Y_K)\delta_{KN}^2)\}; \tag{6}$$

$$w(Y_K) = \{Y_K/[\ln(2)-\ln(1+e^{-Y_K})] - 1\}/4; \tag{7}$$

$$H(\beta_N,Y_N) = \frac{1}{\pi}\int_{-1}^{1} \frac{dx(1-e^{-(\beta_N+x)Y_N})}{(\beta_N+x)(1-x^2)^{1/2}}, \tag{8}$$

where $C_C = 2\pi C/N_{tot}(\infty)$, $Y_N=T_N/2$, $\delta_{KN} = 2(v_K-v_N)/\Gamma_o$ and

$$\beta_N = (2(\mu_N+\rho_N)/t_o^{(N)}) + 1. \tag{9}$$

β_N typically ranges from about 4 to 400[1], and a differential expansion of H in powers of $(1/\beta_N)$ can be made;

$$H(\beta_N,Y_N) = \frac{1}{\sqrt{\beta_N^2-1}} - \frac{e^{-\beta_N Y_N}}{\beta_N}\sum_{n=0}^{\infty}\frac{1}{(\beta_N)^n}\frac{d^n I_o(Y_N)}{dY^n} \tag{10}$$

where I_o is the modified Bessel function of order 0. It is shown in reference (1) that this reduces to

$$H(\beta_N,Y_N) = H_o(\beta_N,Y_N) - S(\beta_N,Y_N), \tag{11}$$

where $\qquad H_o(\beta_N,Y_N) = (1-e^{-\beta_N Y_N})/\sqrt{\beta_N^2-1}, \tag{12}$

and S is a very rapidly converging series whose total value is typically ≤ 0.1 to 0.5 percent of H_o. It is therefore a good approximation to simply take $H \approx H_o$; however, in the current work the S series has been summed to 6 terms, which gives results identical to those obtained by numerical integration of Eq.(8).[1] The resulting formula for peak area (Eq.(5) combined with Eq's.(6),(7),(11) and (12)) is exact for the top layer and, for lower layers, gives excellent agreement with the results found by numerical integration of the exact integral form. The Auger electron peak area is identical in form to Eq.(5), but is multiplied by 0.7 (for Fe[57], 70% of the internal conversion electrons are followed by Auger electrons) and the attenuation coefficients for 7.3 keV electrons are replaced by those for 5.4 keV electrons $(\rho_N \rightarrow \gamma_N, \beta_N \rightarrow \eta_N = 2(\mu_N + \gamma_N)/t_o^{(N)} + 1)$. The total ERM peak area is simply the sum of the internal conversion and Auger electron contributions.

In a somewhat similar fashion, it is shown in reference (1) that the internal conversion electron counting rate resulting from Mossbauer absorption in the Nth layer, Eq. (2), is given to an excellent approximation by the Lorentzian form,

$$N_C^{(N)}(v, Y_N) = C\, e^{-(B_{N-1} + D_{N-1})} \frac{<F_N(Y_K, \delta_{KN})>L(\beta_N, Y_N)}{1 + (s/r(\beta_N, Y_N))^2} \qquad (13)$$

where $s = 2(v - v_N)/\Gamma_o$;

$$L(\beta_N, Y_N) = (1 - e^{-\beta_N Y_N} I_o(Y_N)) - (\beta_N - 1) H(\beta_N, Y_N); \qquad (14)$$

$$r(\beta_N, Y_N) = \Gamma_C(\beta_N, Y_N)/2\Gamma_o = H(\beta_N, Y_N)/L(\beta_N, Y_N). \qquad (15)$$

The Auger electron counting rate is again identical in form (multiply Eq. (13) by 0.7, replace ρ_N by γ_N, β_N by η_N) and is added to Eq. (13) to obtain the total resonant electron counting rate from the Nth layer.

A detailed discussion of numerical results is given in reference (1), and here we give only a single example taken from that report. The problem considered is the growth of two separate oxide layers over an infinite substrate. The parameter values chosen for layers 1,2 and 3 (the substrate) are those of hematite, magnetite and metallic Fe (Fe^{57} abundance = 70%), but the condition that each layer gives a single peak* is maintained and the peak separation is arbitrarily taken to be $10\Gamma_o$. Parabolic growth is assumed ($Z_N = k_N \sqrt{time}$, $k_1 = k_2 = 8.83$ Å/(min)$^{1/2}$). The theoretical ERM spectrum (determined from Eq. (13)) after an oxidation time of 1500 minutes is shown in Fig.2, and the percentages of the total spectrum area (determined from Eq. (5)) contributed by layer 1 (P_1), layer 2 (P_2), and the substrate (P_3) are shown as a function of time in Fig.3. Although this example is hypothetical, the theory has been successfully applied in the determination of layer thicknesses in experimental studies of the oxidation of metallic Fe and Fe-base alloys and in Sn^{119} ERM studies of tinplate.[5]

*Modifications required for multi-level absorbers are discussed in reference (1).

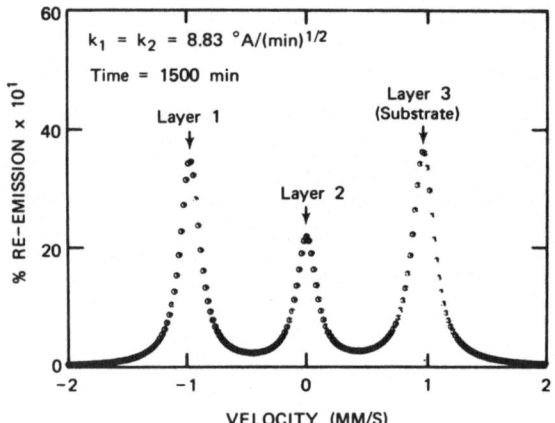

Fig. 2

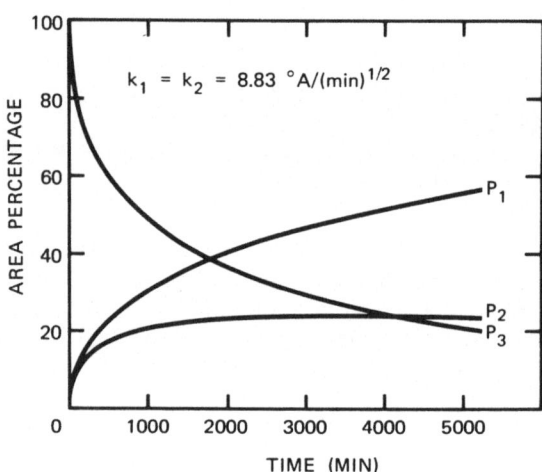

Fig. 3

References
1. G.P. Huffman, submitted to Nucl. Instrum. and Meth.
2. W. Bothe, Z.fur Physik 54, 161 (1929).
3. V.E. Cosslett and R.N. Thomas, Brit.J.Appl.Phys.15,883 (1964).
4. R.L. Graham, F. Brown, J.A. Davies and J.P.S. Pringle,
   Can.J. Phys. 41, 1686 (1963).
5. G.P. Huffman, H.H. Podgurski and G.R. Dunmyre, unpublished
   research.

APPLICATION OF MÖSSBAUER SPECTROSCOPY TO IN SITU STUDIES

OF THE ANODIC AND CATHODIC BEHAVIOR OF COBALT SURFACES

G. W. Simmons and H. Leidheiser, Jr.

Center for Surface and Coatings Research
Lehigh University
Bethlehem, Pennsylvania 18015

INTRODUCTION

The anodic behavior of metals is a complex phenomenon, particularly for experimental conditions that produce passivation of the metal surfaces. Because of the limited experimental techniques available for in situ studies of surface films, there remains considerable uncertainty about many aspects of passivation and anodic behavior of metals. Electrochemical methods have been useful in characterizing the anodic behavior of metals, but these techniques have not provided unambiguous information about the chemical composition and structure of anodically-produced surface films. The non-electrochemical approaches used thus far have also been of limited value. Electron diffraction investigations require that the specimen be removed from its initial environment, and reasonable objections to the interpretation of results can be raised because of possible changes in composition of the specimen surface, such as dehydration, that may occur during the diffraction analysis. Although ellipsometry is a highly surface sensitive technique and is applicable to in situ studies, the measured parameters of thin films, 10-50 A, lead only to information about the thickness. Mössbauer spectroscopy, on the other hand, offers the possibility of obtaining direct compositional and structural information about thin anodic films under in situ experimental conditions. The major objective of this paper is to present an evaluation of the emission Mössbauer spectroscopic technique for in situ determinations of the chemical composition and structure of cobalt surfaces associated with its electrochemical behavior. Cobalt was chosen because emission Mössbauer spectroscopy using the isotope Co^{57} is most suited to studies of this metal. Since the details of this research have been published elsewhere (1), only the salient features of the experimental techniques and of the results will be described.

BACKGROUND

Many electrochemical reactions are slow in comparison to the average lifetime of the nuclear excited state of Fe^{57} (~10^{-7}s), and thus the application of emission Mössbauer spectroscopy is well suited to studies of cobalt. In other words, the time between the formation of the excited Fe^{57} nucleus by the electron capture decay of Co^{57} and the emission of the gamma ray from Fe^{57} is so short that it is unlikely that the emission spectrum will contain any information about the chemical reactions of iron. The information derived from the emission spectra, therefore, depends principally upon chemical changes in the parent Co^{57} atoms, although the gamma ray emission is actually from the Fe^{57} daughter atoms.

The surface sensitivity required for studying passivation and anodic oxidation phenomena is obtained by controlling the amount of Co^{57} deposited on the specimen surface. The surface sensitivity of the emission method has been demonstrated in studies of surface lattice dynamics (2) and in investigations of the magnetic properties of thin films (3,4) in which surface sensitivities approaching submonolayer dimensions were obtained. The composition and structure of thin anodic films can, therefore, be elucidated from the emission Mössbauer spectra of Fe^{57} "probe" atoms that have been incorporated into the anodic films as Co^{57}.

In the interpretation of emission Mössbauer spectra, two important chemical effects must be considered. One is the chemical influences of the host lattice environment on the Fe^{57} impurity atom, and the other, termed "after effects" is the variety of chemical consequences produced by the electron capture decay of Co^{57} to Fe^{57}. Details of chemical effects and "after effects" in emission spectroscopy can be found in published reviews on this subject (5,6). The "after effects" are a result of the deexcitation by Auger electron emission of the vacant iron K level created after capture of a K electron by the Co^{57} nucleus. If the resulting highly ionized state of Fe^{57} has a sufficiently long lifetime, the immediate chemical environment of the "probe" atoms may be altered. The observable effects in the spectra depend strongly upon the nature of the host lattice and it has not been possible to predict a priori details of these effects. The charge state on the daughter iron atoms may be different from the parent cobalt for reasons that are completely independent of the Auger electron emission. In these cases the specific charge states may be stabilized by chemical equilibration of the iron ions with the host matrix. Detailed emission Mössbauer spectroscopic studies of the oxides, hydroxides and oxyhydroxides of cobalt are, therefore, necessary to provide a basis for the unambiguous interpretation of the results obtained from cathodically and anodically polarized cobalt surfaces.

EXPERIMENTAL

The cathodic and anodic behavior of cobalt polarized in deaerated borate solution, pH 8.5 was chosen for study. The polarization curve for cobalt was determined potentiostatically and the current at each potential (versus saturated calomel electrode, SCE) was recorded after the system reached steady state. Emission Mössbauer spectra were obtained at specific applied cathodic and anodic potentials in order to characterize the surface chemical species present during polarization.

Deposition of Co^{57} onto cobalt specimen surfaces was accomplished in a manner similar to that described by Dészi and Molnár (7). The thickness of the Co^{57} active layer was controlled by adding specific amounts of natural cobalt ($CoCl_2$) along with carrier free Co^{57} (in HCl) to the plating solution. Relatively thin deposits (20-30 A) were required to obtain sufficient sensitivity to study changes in the cobalt surface at low passivating potentials. For anodic potentials in the range from the middle of the passive region to beyond the transpassive region, thicker deposits (50-200 A) were used.

All emission spectra were obtained while the specimens were polarized under in situ conditions. The electrochemical cell was constructed from Plexiglass and was designed to minimize the amount of solution in the path of the emitted gamma rays. The specimen (emitter) was kept stationary, and the

Mössbauer spectra were generated in the usual manner by measuring count rate as a function of velocity of a Fe^{57}-enriched stainless steel absorber. Positive velocities correspond to the absorber moving toward the specimen.

At all polarization potentials used in this study, with the exception of low passivating potentials, the corrosion rate was sufficiently slow that the surface active region was not lost during the time required to obtain the Mössbauer spectra. For low passivating potentials, the radioactive surface was stabilized by quenching the specimen and solution to liquid nitrogen temperature, and interrupting the applied potential after the passive film had formed.

RESULTS AND DISCUSSION

Polarization of cobalt in the buffered borate solution produced a classical potential versus current dependence. An active-to-passive transition occurs between -500 mV and -300 mV and cobalt remains passive at higher anodic potentials up to +500 mV. Above +500 mV, thick anodic film formation and oxygen evolution occurred commensurate with an increase in anodic current. Five specific polarization potentials were chosen for study which represented different characteristic regions of the polarization curve. These potentials were as follows: (a) cathodic potential of -1100 mV, (b) low passivating potential at -100 mV which is about 200 mV above the onset of passivity, (c) potential of +200 mV near the center of the passive region, (d) potential of +500 mV near the onset of transpassive behavior, and (e) potential of +800 mV which is in the transpassive region. In the following sections typical results are presented for each of these polarization potentials.

Cathodic Polarization, -1100 mV

All of the specimens were polarized cathodically prior to the application of the anodic potentials. The emission Mössbauer spectra obtained during cathodic treatment was used to monitor the reduction of oxide or corrosion films that had formed subsequent to deposition of the Co^{57} layer or that had formed during previous anodic polarization of the specimen. The relatively intense line(s) at the center of the spectrum shown in Fig. 1a for a 100-200 A specimen is indicative of a

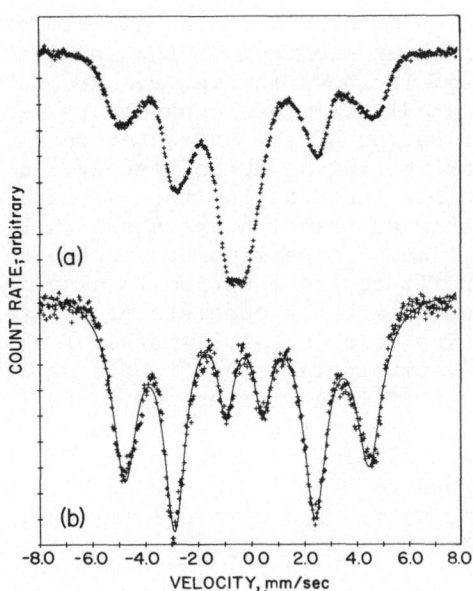

Figure 1 - (a) Emission spectrum of Fe^{57} in cobalt taken after deposition of 100-200 A Co^{57} layer. (b) Emission spectrum of same specimen taken during cathodic polarization at -1100 mV.

corrosion film that was present after preparation of the active Co^{57} layer. Cathodic polarization of the specimen at -1100 mV, current density 20 ma/cm^2, gave rise to the spectrum presented in Fig. 1b, which shows only the six lines expected from the magnetic hyperfine splitting of Fe^{57} in metallic cobalt. This result suggests that the corrosion film, which formed subsequent to the deposition of the active layer, was readily reduced by the cathodic treatment.

Anodic Polarization, -100 mV

The potential of -100 mV is just above the onset of passivity. The relatively high corrosion rate of cobalt at this potential did not permit _in situ_ emission Mössbauer spectra to be obtained at room temperature. The specimen and solution were quenched to low temperature after the formation of

the passive film as described in the experimental section.
The emission Mössbauer spectrum of the specimen after polarization at -100 mV is shown in Fig. 2. Resolution of the
spectrum of the anodic film was enhanced by removing the
background contribution of the unreacted metal. This was accomplished by subtracting an appropriately scaled spectrum
of the same specimen taken during the cathodic polarization.
The results of this subtraction are shown in Fig. 2a. The
simplest combination of Lorenzian curves that gave a satisfactory computer fit to the spectrum is also shown in Fig.
2b. The spectrum apparently consists of two sets of quadrupole doublets. The value of -1.31 mm/sec for the isomer
shift and the relatively large quadrupole splitting of 2.83
mm/sec for one set of lines correspond to Fe^{57} (+2) in the
anodic film. An isomer shift of -0.59 mm/sec and a relatively
small quadrupole splitting of 0.78 mm/sec for the other set
of lines correspond to Fe^{57} (+3). Because of the possibility
of Auger "after effects" and chemical effects, the Fe^{57} (+3)
daughter atoms may not have originally existed as Co^{57} (+3)

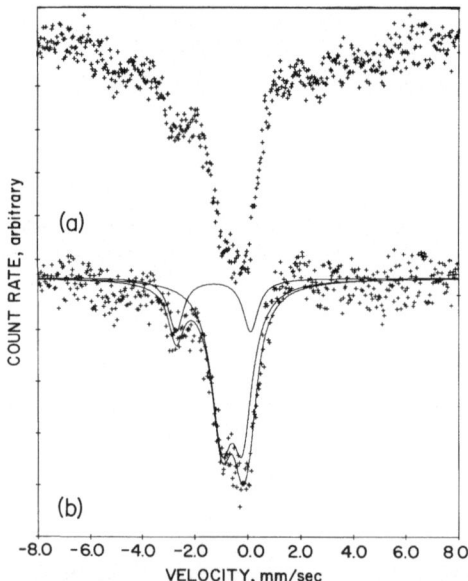

Figure 2 - Emission spectrum of Fe^{57} in cobalt anodically polarized at -100 mV, (a) initial spectrum and (b) spectrum
after subtraction of contribution from unreacted metal.

parent atoms. It is possible, therefore, that the passive
film formed at low passive potentials consists essentially
of a divalent oxide or hydroxide.

Anodic Polarization, +200 mV and +500 mV

The experimental results obtained at +200 mV and +500
mV were found to be similar, and these results are, therefore,
discussed together. The thickness of the Co^{57} layer on these
specimens was on the order of 100-200 A. The rate of anodic
dissolution was sufficiently slow at +200 and +500 mV that
the emission Mössbauer spectra could be readily obtained in
situ at room temperature without recourse to freezing. The
emission Mössbauer spectra taken at +200 and +500 mV are
shown in Figs. 3 and 4 respectively. The spectra taken at
these two potentials are shown before and after subtraction
of the unreacted metal background. Two Lorenzian lines of
equal intensity gave a satisfactory computer fit to the

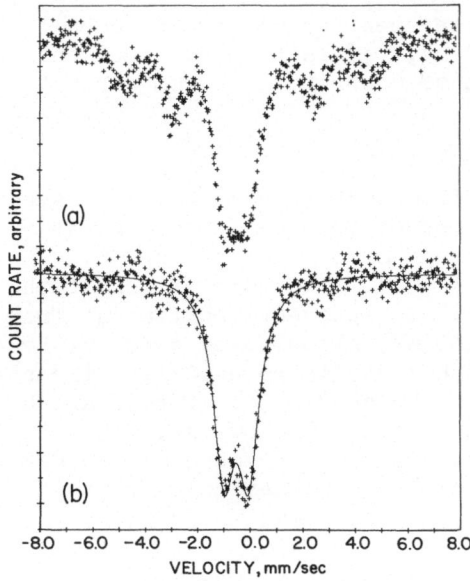

Figure 3 - Emission spectrum of Fe^{57} in cobalt anodically
polarized at +200 mV, (a) initial spectrum and (b) spectrum
after subtraction of contribution from unreacted metal.

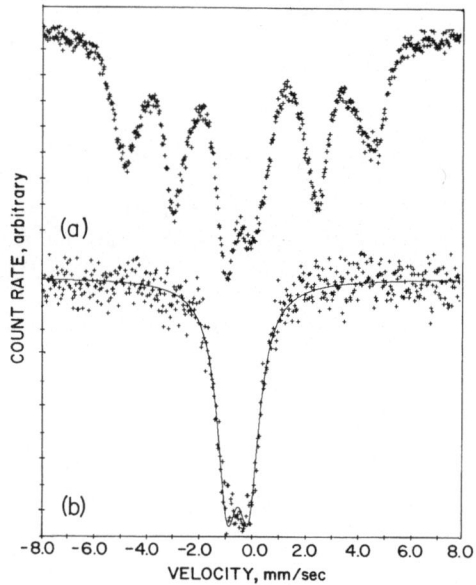

Figure 4 - Emission spectrum of Fe^{57} in cobalt anodically polarized at +500 mV, (a) initial spectrum and (b) spectrum after subtraction of contribution from unreacted metal.

spectra obtained from both specimens. The values of 0.054 mm/sec and -0.55 mm/sec for the isomer shifts and the small quadrupole splittings of 0.81 mm/sec and 0.94 mm/sec are indicative of Fe^{57} (+3). The composition of the passive films at +200 and +500 mV is apparently different from that of the film formed at -100 mV, since no evidence of Fe^{57} (+2) was found at these higher anodic potentials. The possible presence of Co (+2) in the anodic film formed at the higher potentials cannot be ruled out, however, since the daughter Fe^{57} (+2) ions could have been converted to Fe^{57} (+3) by Auger "after effects" of by chemical effects.

Anodic Polarization, +800 mV

The region of the polarization curve at which thick anodic films are formed and oxygen evolution occurs is referred to as transpassive. The potential of +800 mV is

approximately 300 mV above the onset of this transpassive
region. An emission Mössbauer spectrum obtained at this potential is shown in Fig. 5 before and after the subtraction
of the unreacted metal contribution. The spectrum of the
anodic film formed at +800 mV is similar to the spectra for
the surfaces anodically treated at +200 and +500 mV. The
asymmetry of the spectrum suggests, however, that anodic
polarization at transpassive potentials produces a characteristic change in the anodic film. Apparently the major
composition of the film, in this case, is the same as that
formed at the passive potentials, but part of the surface has
been converted to a higher oxidation state. A satisfactory
computer fit to the spectrum was obtained with a set of
quadrupole split lines, similar to those found for the passive film, and an additional line with an isomer shift of
0.14 mm/sec which indicates the presence of Fe^{57} (+4) in the
surface of the anodic film. In the computer fitting, the
hyperfine parameters of the major component of the spectrum
of the transpassive film were assumed to be identical with

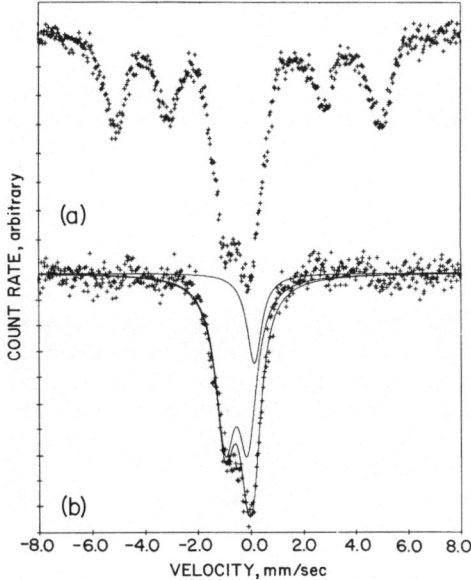

Figure 5 - Emission spectrum of Fe^{57} in cobalt anodically
polarized at +800 mV, (a) initial spectrum and (b) spectrum
after subtraction of contribution of unreacted metal.

TABLE I - Hyperfine parameters for Fe^{57} in anodically polarized cobalt surfaces doped with Co^{57}

Anodic Potential mV (SCE)	Isomer Shift[a] (mm/sec)	Quadrupole (mm/sec)
-100	-1.31[b] -0.59[b]	2.83[b] 0.78[b]
+200	-0.55	0.94
+500	-0.54	0.81
+800	-0.55[c] 0.14	0.85[c]

(a) Stainless Steel reference
(b) Measured at liquid nitrogen temperature
(c) Values constrained during computer fitting

parameters obtained from the passive films formed at +200 and +500 mV.

SUMMARY AND CONCLUSIONS

Emission Mössbauer spectroscopy technique was successfully demonstrated to be an effective method for in situ studies of changes in cobalt surfaces as a function of polarization. Despite the possible ambiguities introduced by effects associated with the emission technique, characteristic spectra were found for cobalt surfaces at specific applied potentials. The Mössbauer parameters obtained from the emission spectra of specimens for different polarization conditions are summarized in Table I. The following information has been derived from the results of this study: (a) cobalt was shown to be essentially free of a corrosion film during cathodic polarization (-1100 mV), (b) Resonance lines from both +2 and +3 oxidation states were found in the emission Mössbauer spectra of anodic films formed at low passivating

potentials (-100 mV), (c) At potentials in the passive region
of the polarization curve (+200 mV and +500 mV) the spectra
indicated that the passive film contained primarily +3 oxidation state, (d) The anodic film formed at transpassive potentials (+800 mV) was found to consist of +3 and +4 oxidation
states, and the +3 component of this film was shown to be
likely the same as that formed at the high passive potentials.
Further interpretation of the spectra obtained in this study
is possible with reference emission spectra for the oxides,
hydroxides and oxyhydroxides of cobalt. Progress on this
phase of the research has already been made, and a description
of the composition and structure of the anodic films formed
during anodic polarization of cobalt in buffered borate solution, pH 8.5, will be presented in the near future.

ACKNOWLEDGMENT

We acknowledge with appreciation support of this research
by the Army Research Office-Durham. Our special thanks go to
Dr. Henry Davis of this agency.

REFERENCES

1. G. W. Simmons, E. Kellerman and H. Leidheiser, Jr. (sub-
    mitted to J. Electrochem. Soc.).

2. J. W. Burton and R. P. Godwin, Phys. Rev., 158, 218 (1967).

3. M. N. Varma and R. W. Hoffman, J. Appl. Phys. 42, 1727
    (1971).

4. A. C. Zuppero and R. W. Hoffman, Vacuum Sci. and Tech. 7,
    118 (1969).

5. N. N. Greenwood and T. C. Gibb, Mössbauer Spectroscopy,
    Chapman and Hall Ltd., London (1971).

6. G. K. Wertheim, Accounts of Chemical Research, Nov. 1971,
    pp. 373-379.

7. I. Dézsi and B. Molnár, Nucl. Instr. and Meth. 54, 105
    (1967).

A ^{125}TE MÖSSBAUER STUDY OF ORGANOTELLURIUM COMPOUNDS

C.H.W. Jones

Simon Fraser University

British Columbia, Canada V5A 1S6

W.R. McWhinnie

The University of Aston in Birmingham

Birmingham B4 7ET, England

F.J. Berry*

Birkbeck College

University of London

London WC1E 7HX, England

* Present address:

        The University Chemical Laboratory
        Lensfield Road, Cambridge, England

INTRODUCTION

The investigation of tellurium compounds using the Mössbauer effect for ^{125}Te is of some interest since such experiments allow a comparison with similar studies of compounds of tin and antimony. In the past much of the work on ^{125}Te has been primarily concerned with the halides and oxides of tellurium. In recent years attention has turned to organo-tellurium compounds and a detailed study of Te(II) and Te(IV) complexes with thiourea and other sulphur containing ligands has been reported (1). Berry et al. (2) have reported ^{125}Te Mössbauer data for a number of aryl tellurides and ditellurides and aryltellurium dihalides and trihalides while Smith et al. (3) have investigated the dialkyltellurium dihalides. Overlapping with these latter experiments, Jones et al. (4) have also studied the aryl tellurides and ditellurides and the organotellurium halides together with derivatives of the latter with pyridine and tetramethylthiourea. The diaryltellurium diacetates and dibenzoates have also been investigated (5).

It is clearly timely, and of interest, to bring these many different results together and to review the present status of ^{125}Te Mössbauer studies of organo-tellurium compounds.

EXPERIMENTAL

The bulk of the results reported here were obtained from spectra recorded at Simon Fraser University. In preliminary studies spectra were recorded using a 5 mCi ^{125}I/Cu source (New England Nuclear) or a 1 mCi Pb^{125m}Te source prepared by neutron irradiation of a 100 mg Pb^{124}Te sample at Chalk River Nuclear Laboratories at a thermal neutron flux of ca. 10^{14} n.cm^{-2} s^{-1} for 1 month. However the considerable disadvantages of the relatively short half-lives of ^{125}Te and ^{125}I led us to use ^{125}Sb/Cu (New England Nuclear) as a source (10 mCi) in the present work. This source has been described in detail by Boolchand and is found to yield narrow line-widths and relatively large recoil-free fractions. It is interesting to note that Erickson and Maddock (7) used ^{125m}Te radiochemically milked from the parent ^{125}Sb and deposited on platinum, and

observed very large recoil-free fractions and narrow linewidths. This would appear to be the optimum arrangement
since the high-level background from the ^{125}Sb is then
avoided, while taking advantage of the long half-life of
the parent. However the radiochemical manipulation required is a significant disadvantage.

The spectra were recorded with the ^{125}Sb/Cu source
cooled to ca. 80K, using a cold-finger immersed in liquid
nitrogen, and the absorbers immersed in liquid helium in a
Harwell Instruments dewar. The absorbers were of natural
isotopic abundance and contained 5-10 mg cm^{-2} of ^{125}Te.

The 35.5 keV resonant γ-ray can be detected indirectly
through the escape peak, using a thin NaI(Tl) scintillation
detector or a Xe/CO$_2$ proportional counter, or directly using a high resolution Ge/Li or Si/Li γ-ray detector (8).
From time to time we have used all three and have found
that a 2 atm Xe/CO$_2$ proportional detector (Reuter Stokes),
coupled with a copper critical absorber to reduce background under the 6 keV escape peak, gives quite adequate
results with most tellurium absorbers.

The spectra (2 x 256 channels) were recorded in a
Nuclear Data 2200 series 1024 channel analyser equipped
with a dual input router. This allowed the simultaneous
accumulation of an Fe foil standard at the rear end of the
transducer. An NSEC AM1 constant acceleration drive system
was used. The drive was routinely calibrated assuming
standard parameters for the Fe foil spectrum (9). The
2 x 256 channel spectra were separately computer fitted to
Lorentzians and the parameters obtained averaged (10). The
computed errors in any one spectrum were found to be much
smaller than the reproducibility of independent measurements, and the overall errors were generally estimated to
be ± 0.08 mm s^{-1} in δ and ± 0.1 mm s^{-1} in Δ. The linewidths were found to lie in the range of ca. 5.2 to 6.5 mm
s^{-1}. The χ^2/N values for the computed fits lay in the
range 1.13 to 0.90. The per cent resonance absorptions
ranged from 2% (CH$_2$Te$_2$) to 9% (pyH$^+$[(p-EtOC$_6$H$_4$)TeCl$_4^-$]).
The isomer shifts are reported relative to I/Cu as a reference standard and this entailed adding + 0.15 mm s^{-1} to
δ observed relative to the ^{125}Sb/Cu source.

RESULTS AND DISCUSSION

The Organotellurides and Ditellurides

The results for the dimethyl and diaryl tellurides, the diaryl ditellurides, and the compounds $PhMe_2TeI$ and Ph_2MeTeI are shown in Table I. The agreement between the quadrupole splittings of the same or similar compounds as measured here and in reference 2 is reasonable. The agreement in the isomer shifts is also acceptable, except for $(p-MeC_6H_4)_2Te$ and $(p-MeC_6H_4)_2Te_2$. It must be noted that two sources, $^{125}I/Cu$ and $Pb^{125m}Te$, were used in reference 2 and the discrepancies in the δ values for similar compounds in that reference presumably arise from difficulties in the isomer shift calibration.

The structures of the tellurides, ditellurides and the triaryl iodides are shown in Figure 1. For the compound $(p-MeC_6H_4)_2Te$ the C-Te-C bond angle is 101.0 ± 2.7^o (11) while the C-Te-Te bond angle in $(p-ClC_6H_4)_2Te_2$ is 94.4^o (12) and in Ph_2Te_2 it is ca. 99° (13). The tri-aryl iodides are proposed to contain the Ar_3Te^+ cation on the basis of conductivity measurements in solution (14-16) and in analogy with $Ph_3Se^+Cl^-$, for which crystallographic data is available (17), and $Me_3Se^+Cl^-$ where the i.r. spectrum has been interpreted as evidence for the Me_3Se^+ cation (18).

The isomer shifts of the compounds reported in Table I, with the exceptions noted, are relatively small. The nuclear radius term $\Delta R/R$ is positive for the ^{125}Te transition (19) and thus δ becomes more positive as the s-electron density at the nucleus $|\psi s(o)|^2$ increases. Removal of 5p electron density from the tellurium will result in a deshielding of the s-electrons from the nucleus and an increase in $|\psi s(o)|^2$ while removal of s-electrons will have a more marked effect in directly decreasing $|\psi s(o)|^2$. The small isomer shifts of the tellurides, ditellurides and tri-aryl tellurium cation correspond to relatively small s-electron densities at the tellurium nucleus.

The large quadrupole splittings of the tellurides and ditellurides correspond to a considerable 5p orbital imbalance on the tellurium in each case. The indicates, not surprisingly, that the Te-C and Te-Te bonds have considerable covalent character and that the tellurium must have a significant number of holes in the 5p shell to give rise to such a large splitting. This alone would suggest that δ should be relatively large and positive. Thus the small isomer shifts necessarily imply some s-character in the Te-C and Te-Te bonds. This is consistent with the C-Te-C and C-Te-Te bond angles noted above which suggest

Table I

	δ^* (±0.08) mm s^{-1}	Δ (±0.1) mm s^{-1}	Ref†
Me$_2$Te	0.06	10.5	a
Ph$_2$Te	0.18	10.5	a
(p-MeOC$_6$H$_4$)$_2$Te	0.3	11.3	b
(p-MeC$_6$H$_4$)$_2$Te	0.7	10.1	b
Ph$_2$Te$_2$	0.37	10.7	a
(p-MeOC$_6$H$_4$)$_2$Te$_2$	0.42	10.1	a
	0.3	10.3	b
(p-EtOC$_6$H$_4$)$_2$Te$_2$	0.28	10.6	a
(p-PhOC$_6$H$_4$)$_2$Te$_2$	0.32	10.0	a
	0.3	10.3	b
(p-MeC$_6$H$_4$)$_2$Te$_2$	0.6	9.9	b
Ph$_2$MeTeI	0.38	5.6 (±0.2)	a
PhMe$_2$TeI	0.32	5.4	a
C$_2$H$_4$Te$_4$	0.42	9.3	a

* with respect to I/Cu. Source at 80K and absorbers at 4.2K.
† a) This work   b) Reference 2

some stereochemical activity of the 5s electrons and that the bonding in these compounds is intermediate between p^3 and sp^3.

The small quadrupole splittings for Ph$_2$MeTeI and PhMe$_2$TeI are also consistent with some s-character in the Te-C bond, since this implies some p-character of the lone pair. In the Ph$_2$MeTe$^+$ and PhMe$_2$Te$^+$ cations V_{zz} the principal component of the efg tensor would presumably lie along the approximate 3-fold axis, in the direction of the lone-pair. The isomer shifts for these compounds are more

positive than those of Me_2Te and Ph_2Te. This suggests
that the s-character in the Te-C bond must be relatively
small since the removal of s and p electron density from
the tellurium in the Te-C bond is such as to produce a net
deshielding of the remaining s-electrons and an increase
in $|\psi s(o)|^2$ in going from Ph_2Te to Ph_2MeTe^+ for example.

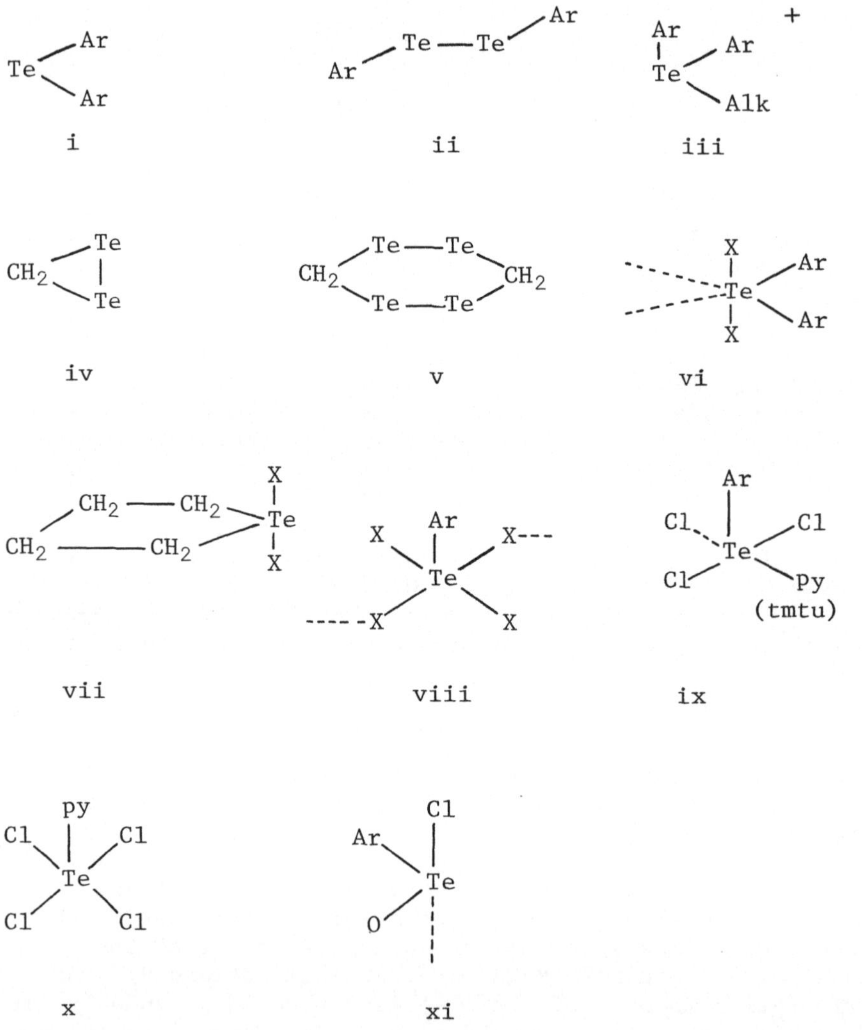

Figure 1

The quadrupole splitting for the proposed pyramidal Ph_2MeTe^+ and $PhMe_2Te^+$ cations are similar in magnitude to the splittings observed (20) for $TeCl_4$ (Δ=ca 6 mm s^{-1}) which contains pyramidal $TeCl_3^+$ ions with three distant bridging Cl$^-$ ions completing a distorted octahedron about the tellurium, and α-VTeO$_4$ (Δ=6.0 mm s^{-1}) and $CuTe_2O_5$ (Δ=6.6 mm s^{-1}) which contain pyramidal Te^{IV}O$_3$ units (21). In these different cases the efg presumably arises because of stereochemical activity of the lone pair. The compound Ph_3Sb may be considered to be iso-electronic and isostructural with Ph_3Te^+. The quadrupole splittings of ca. 5.5 mm s^{-1} for Ph_2MeTe^+ and $PhMe_2Te^+$ correspond to a coupling constant e^2qQ, assuming η=0, of ca. 320 MHz while the coupling constant for Ph_3Pb is 530 MHz (22). Taking the ratio of the principal components of field gradient tensor per 5p electron for Sb and Te to be 0.87 (23), the ratio of the coupling constants corresponds to Qqnd(^{121}Sb)/Qex(^{125}Te)~2 and this is not inconsistent with previous estimates of the magnitude of the quadrupole moments of these two states (22,24).

Morgan and Drew (25) prepared a compound with the stoichiometry CH_2Te_2 and proposed that this was the three-membered, cyclic ditelluromethane, Figure 1(iv). More recent work (26) has suggested, on the basis of mass spectral data, that the compound may be cyclo 1-4-dimethyl-enetetratellurium (Figure 1(v)). The Mössbauer parameters for this compound are similar to those of the other ditellurides although Δ is somewhat smaller. The data clearly indicates the presence of only one tellurium site in the compound and the absence of significant tellurium impurity; (the chemical analysis is hardly definitive in this latter regard for this compound). However the Mössbauer data does not distinguish between the two possible structures. In the structure of Figure 1(iv) if the tellurium utilised pd^2 hybrids in bonding, a more positive isomer shift would have been anticipated and a much smaller splitting, since the promotion of 5p electrons to 5d orbitals would place them further from the nucleus. However the bonding in Figure 1(iv) could be described as p^3 with the formation of bent bonds not directed along the interatomic axes. In this case parameters comparable with those for the structure of Figure 1(v) or for the other ditellurides would be expected. Nevertheless Δ is consistent with the presence of tellurium in the formal oxidation state of one.

The Organotellurium Halides

The data for the organotellurium halides are summar-
ised in Table II. With the exception of some of the
isomer shifts from reference 2, the agreement in the data
obtained in the different laboratories appears reasonable.
The errors quoted in our work are significantly greater
than those of reference 3.

For the dichlorides it is clear that the Me-, Ph-,
p-MeOC$_6$H$_4$-, and p-EtOC$_6$H$_4$- derivatives all have essentially
the same parameters. This appears to be generally true
within a given series of halides, although Me$_2$TeBr$_2$ and
Me$_2$TeI$_2$ appear to have somewhat larger Δ values than the
corresponding diaryltellurium dihalides.

Crystal structures of a number of dihalides (27-29)
show the coordination about the tellurium to be distorted
trigonal bipyramidal in which the halogens occupy trans
axial positions and there are distant bridging contacts in
the equatorial plane (Figure 1(vi)). The compound
ClCH$_2$CH$_2$.TeCl$_3$ has the coordination about tellurium shown
in Figure 1(vii) in which two of the halogens bonded to the
tellurium are terminal and two are bridging in a polymer
chain (30). A similar structure has been proposed on the
basis of i.r. and Raman data for other tellurium trihalides
(31) and the compound (pyH)$^+$((p-EtOC$_6$H$_4$)TeCl$_4^-$) would
appear to contain the square-based pyramidal anion (32).
The compounds (CH$_2$)$_4$TeX$_2$ (33) contain tellurium in a five-
membered ring, Figure 1(vii) while the compounds
ArTeCl$_3$.py (or tmtu) (34), TeCl$_4$.py (35) and ArTeOCl (32)
are proposed to have the structures shown in Figure 1(ix-
xi).

The Dihalides. The isomer shifts for the dihalides
are clearly more positive than those of the organotellur-
ides and this is consistent with the halogen ligands removing predominantly p electron density from the tellurium.
However within a given series of dihalides the isomer shift
does not change significantly with increasing electronegativity of the halogen. For example, the derivatives
(p-MeOC$_6$H$_4$)$_2$TeX$_2$ (X=F,· Cl, Br, I) all have the same isomer
shift within the errors. This may arise as a result of the

Table II

	δ^* mm s^{-1}	Δ mm s^{-1}	Ref[†]
Me_2TeCl_2	0.58	9.4	a
	0.64	9.97	b
Me_2TeBr_2	0.65	8.5	a
	0.61	8.76	b
Me_2TeI_2	0.52	7.6	a
	0.62	7.05	b
$(CH_2)_4TeBr_2$	0.54	7.1	a
$(CH_2)_4TeI_2$	0.64	7.1	a
Ph_2TeF_2	0.52	10.4	a
$(p-MeOC_6H_4)_2TeF_2$	0.54	10.0	a
$(p-EtOC_6H_4)_2TeF_2$	0.36	10.1	a
Ph_2TeCl_2	0.5	9.2	c
$(p-MeOC_6H_4)_2TeCl_2$	0.68	9.1	a
$(p-EtOC_6H_4)_2TeCl_2$	0.7	9.1	c
$(C_6H_4ClC=CH)_2TeCl_2$	0.57	8.0	a
Ph_2TeBr_2	0.60	7.9	a
	0.5	8.1	c
$(p-MeOC_6H_4)_2TeBr_2$	0.72	7.8	a
$(p-EtOC_6H_4)_2TeBr_2$	0.52	7.6	a
$(C_6F_5)_2TeBr_2$	0.73	5.3	a
$(p-MeOC_6H_4)_2TeI_2$	0.51	6.1	a
$(p-MeC_6H_4)_2TeI_2$	0.6	6.3	c
$PhMeTeI_2$	0.56	6.4	a
$(p-MeOC_6H_4)TeF_3$	0.56	8.6	a
$(p-EtOC_6H_4)TeF_3$	0.6	8.6	a
$(p-MeOC_6H_4)TeCl_3$	0.9	9.2	c
$(p-EtOC_6H_4)TeCl_3$	0.91	9.4	a
	1.1	9.2	c
$PhTeBr_3$	0.91	7.8	a
$(p-EtOC_6H_4)TeBr_3$	1.0	8.0	c
$PhTeI_3$	0.9	3.9	c
$(p-EtOC_6H_4)TeI_3$	1.0	5.2	c
$pyH^+(p-EtOC_6H_4)TeCl_4^-$	0.88	8.8	a
$\ddagger(p-MeOC_6H_4)TeCl_3py$	0.73	7.6	a
$\pm(p-MeOC_6H_4)TeCl_3tmtu$	0.85	7.9	a
$TeCl_4.py$	0.86	4.5	a
$(pEtOC_6H_4)TeOCl$	0.59	9.1	a

* δ with respect to I/Cu. Source at 80K and absorbers at 4.2K
† a This work   ‡ py = pyridine
   b Reference 3
   c Reference 2   ± tmtu = tetramethylthiourea

relatively small $\Delta R/R$ for ^{125}Te, and hence the small range
of isomer shifts. Another contributing factor may be small
changes in the s-p hybrid character of the bonds as the
electronegativity of the halogen changes.

The quadrupole splittings for the dihalides clearly
lie in the order F>Cl>Br>I for the (p-MeOC$_6$H$_4$) derivatives
for example. This is consistent with an increasing p-
orbital imbalance about the tellurium as the electronegativity of the halogen increases. If the bonding in these
compounds is predominantly p in character then V_{zz} the
principal component of the efg tensor would lie along
the axial X-Te-X bonding axis and would be expected to be
positive in sign. The positive isomer shift and the small
C-Te-C bond angle in Ph$_2$TeBr$_2$ are consistent with this proposal. The fact that (C$_6$F$_5$)$_2$TeBr$_2$ has a Δ which is significantly smaller than that of Ph$_2$TeBr$_2$ also supports such
an assignment. Thus C$_6$F$_5$ would be expected to be a poorer
σ-donor than Ph and in (C$_6$F$_5$)$_2$TeBr$_2$ the p-orbital imbalance
would be smaller than that in Ph$_2$TeBr$_2$, as is observed. If
the lone-pair exhibited considerable stereochemical activity in these compounds and the bonding could be described
formally as sp^{3}d, V_{zz} would lie through the lone-pair in
the equatorial plane and would be negative in sign. It is
not difficult to show that, in terms of Townes and Dailey
theory for example, Δ would then be expected to increase as
the σ-donor character of the organic ligand decreased.

While for the (p-MeOC$_6$H$_4$)$_2$TeX$_2$ derivatives Δ exhibits
a simple trend with the changing electronegativity of X,
within the dibromides for example Δ exhibits a range of
values such that Me->Ph-~MeOC$_6$H$_4$-~EtOC$_6$H$_4$->-（CH$_2$)$_4$-~C$_6$F$_5$-.
This may be explained by changes in the importance of
intermolecular bonding in the equatorial plane or by small
changes in the stereochemical activity of the lone-pair.

The compound PhMeTeI$_2$ has essentially the same Mössbauer parameters as (p-MeOC$_6$H$_4$)$_2$TeI$_2$ again suggesting that
the Ph-Te and Me-Te bonds must be very similar.

The Trihalides. For the trihalides, when X=Cl, Br, I
the isomer shifts are more positive than those of the dihalides, consistent with the removal of predominantly p-
character by the halogen. However the δ's for the difluorides and trifluorides are essentially the same and thus the
trifluorides are significantly different from the other
trihalides in this regard.

The quadrupole splittings for the trihalides when X=
Cl, Br, are the same, within the errors, as the corresponding dihalides. For the iodides the agreement is not quite
as close, although the two measurements on the tri-iodides
are themselves not in good agreement. For the fluorides
the data are consistent within themselves and show that Δ
for the trifluorides is significantly smaller than for the
difluorides. The fact that,for the chlorides and bromides,the
dihalides and trihalides have the same quadrupole splitting
can be readily understood in terms of the structures shown
in Figure 1   and assuming the bonding in these complexes
to be predominantly p in character. Thus if the X-Te-X and
R-Te bonds have constant covalent character in the two
cases, a simple additivity model based on the Townes and
Dailey theory for example would predict that e^2qQ would
have the same magnitude but be opposite in sign in the two
cases. It will be of interest to attempt to confirm the
signs of e^2qQ for these compounds and such experiments are
planned. It should be pointed out that if the lone-pair
exhibited considerable stereochemical activity in these molecules, Δ would not be expected to be the same in the two
cases.

The splittings for the iodides do not appear to fit
the simple picture, although the disagreement between the
two measurements for the tri-iodides itself raises some
questions about the data.

The relatively small splitting for the trifluorides
in comparison with the difluorides, together with the fact
that they have the same isomer shift, suggests that the

coordination about the tellurium in the trifluorides must
be significantly different from that in the other trihalides. The small isomer shift of the trifluorides suggests a greater stereochemical activity of the lone-pair
than in the other trihalides. However the relatively small
splitting suggests a more symmetrical environment about
the tellurium in the trifluorides than in the other trihalides. It may be noted that SbF_3 is significantly different in structure from the other antimony trihalides while
TeF_4 has a different structure than $TeCl_4$, the central
atom in both cases possessing one lone pair. It would
appear that $RTeF_3$ follows this general trend.

The compound $pyH^+[(p\text{-}EtOC_6H_4)TeCl_4{}^-]$ has the same
parameters, within the errors, as the aryltellurium tri-
chlorides and this suggests that the major contribution to
the electric field gradient in these compounds comes from
the valence shell and that lattice terms are negligible.

The compounds $TeCl_4.py$, $(p\text{-}MeOC_6H_4)TeCl_3.py$ and
$(p\text{-}MeOC_6H_4)TeCl_3.tmtu$ have small splittings which are consistent with the proposed structures for those compounds.
As will be shown towards the end of this paper a simple
additive rule for the quadrupole splittings appears to hold
and values of Δ for the pyridine and tetramethylthiourea
derivatives are in good agreement with those anticipated on
such a model.

Diaryltellurium Diacetates and Dibenzoates

The data for the diacetates and dibenzoates studied
are shown in Table III. Pant et al. (36) have interpreted
the i.r. spectra of a number of dicarboxylates as evidence
for a ψ trigonal bipyramidal geometry for those molecules
in which the carboxylate ligands occupy trans-axial positions. The Mössbauer data for these compounds are generally consistent with that proposal, exhibiting isomer
shifts and quadrupole splittings generally comparable with
those of the dichlorides. However the diphenyl compounds
exhibit much smaller splittings than the other derivatives.
The i.r. spectrum of $(p\text{-}EtOC_6H_4)_2Te(OCOCH_3)_2$ was
found to be more complex than that of $Ph_2Te(OCOCH_3)_2$ suggesting the possible presence of both uni- and bi-dentate
carboxylate ligands in the former case and only uni-dentate

in the latter. The different quadrupole splittings of (p-MeOC$_6$H$_4$)$_2$Te(OCOCH$_3$)$_2$ and Ph$_2$Te(OCOCH$_3$)$_2$ are consistent with a different environment about the tellurium in those two compounds. The smaller splitting for the latter compound

Table III

	δ* mm s^{-1}	Δ mm s^{-1}
Ph$_2$Te(OCOCH$_3$)$_2$	0.59	8.0
(p-MeOC$_6$H$_4$)$_2$Te(OCOCH$_3$)$_2$	0.50	9.3
(p-C$_7$H$_7$)$_2$Te(OCOCH$_3$)$_2$	0.60	9.5
Ph$_2$Te(OCOC$_6$H$_5$)$_2$	0.76	8.2
(p-MeOC$_6$H$_4$)$_2$Te(OCOC$_6$H$_5$)$_2$	0.52	9.5
(p-EtOC$_6$H$_4$)$_2$Te(OCOC$_6$H$_5$)$_2$	0.56	9.4

* δ with respect to I/Cu. Source at 80K and absorbers at 4.2K.

is indicative of a more symmetrical arrangement of the ligands about the tellurium.

The compound Ph$_2$Te(OCOC$_6$H$_5$)$_2$ was found to have a complex i.r. spectrum which was more similar to that of (p-EtOC$_6$H$_4$)$_2$Te(OCOCH$_2$)$_2$ rather than that of the diphenyl diacetate. However the Mössbauer data suggest that the tellurium environments in the diphenyl diacetate and diphenyl dibenzoate are very similar. Thus the i.r. and Mössbauer data are not wholly consistent but suggest that the structures of the diacetates and dibenzoates may show subtle differences with changes in the aryl ligands. A more full interpretation of the data must await structural studies on the carboxylates.

Other Compounds Studied

Table IV contains data for the compounds Ph$_3$Sn.TeR and the adducts of the aryltellurides with cuprous halides. The former compounds have isomer shifts comparable with those of the tellurides, while the quadrupole splittings are significantly smaller. Thus the Ph$_3$Sn moiety appears to be a better σ-donor than Ph or Ph-Te and the resulting p-orbital imbalance on the tellurium is smallest for Ph_3SnTeR.   As in the tellurides and ditellurides there would appear to be significant s-character in the Sn-Te bond.

The adducts of the aryltellurides with the cuprous halides (X=Cl, Br) appear to have surprisingly large isomer shifts comparable to those observed in the aryltellurium dihalides while the Δ values are comparable with the dichlorides. This suggests a significant coordination of the tellurium to the copper and the removal of considerable p-electron density from the tellurium. It would be of considerable interest to determine the sign of Δ in these compounds since it may well be negative, as proposed for the diaryl tellurium dihalides, rather than positive as in the parent telluride.

Table IV

	$\delta *$ mm s^{-1}	Δ mm s^{-1}
$p\text{-EtOC}_6H_4\text{TeSnPh}_3$	0.24	9.3
$p\text{-MeOC}_6H_4\text{TeSnPh}_3$	0.10	9.0
$Ph.TeSnPh_3$	0.15	9.2
$(p\text{-EtOC}_6H_4)_2Te.CuCl$	0.61	9.0
$(p\text{-}C_7H_7)_2Te.CuCl$	0.73	9.0
$(p\text{-EtOC}_6H_4)_2Te.CuBr$	0.68	8.8
$(p\text{-EtOC}_6H_4)_2Te.CuI$	0.15	9.2
$(p\text{-}C_7H_7)_2Te.CuI$	0.34	9.2

* δ with respect to I/Cu. Source at 80K and absorbers at
  4.2K.

An Additive Model for Isomer Shifts and
Quadrupole Splittings.

Additive models have been used from time to time to rationalise or parameterise isomer shifts and quadrupole splittings. The compounds studied here provide a very demanding test of such additive models, since the tellurium possesses one or two lone pairs. Where the 5s electrons are stereochemically active the additive rules would be expected to break down. A selection of isomer shift data is shown in Table V. Not surprisingly the isomer shifts are not additive over the whole range and the data suggests increasing lone-pair character of the 5s electrons in passing from R_2Te to $TeCl_6{}^{2-}$ i.e. with increasing coordination number.

The isomer shifts are sensitive to small changes in the relative s-p character of the lone-pair, whereas the

Table V

	$\delta*$ mm s^{-1}
Te^{2-}	−0.14
Ar_2Te	0.12
Ar_3Te^+	0.35
Ar_2TeCl_2	0.62
$ArTeCl_3$	0.91
$\dagger TeCl_4$	1.10
$TeCl_6{}^{2-}$	1.70

* Average values taken from the present work and reference 20.
† $TeCl_4$ contains $TeCl_3{}^+$ units with three distant bridging Cl^- contacts.

quadrupole splittings are not. Thus the Δ values appear to accord with a simple additive model expressed in terms of the Townes and Dailey theory. Assuming the bonding to be predominantly p in character and that a given bond has constant covalent character from one molecule to another, the Townes and Dailey theory allows the estimate of bond orbital populations as shown in Table VI. The signs of Δ have been assumed as shown and η is taken as zero in each case. Using the orbital populations so derived it is then possible to calculate Δ for $(p\text{-MeOC}_6H_4)TeCl_3py$ and $(p\text{-MeOC}_6H_4)TeCl_3.tmtu$ assuming structures of Figure 1 and the calculated values are both $(+)7.3$ mm s^{-1}, whereas the observed values are 7.6 and 7.9 mm s^{-1} respectively. Thus within the errors the splittings appear to be additive for this limited selection of compounds.

Table VI

Compound	Δ*	Up†	Bond-Orbital Population
Ar_2Te	(+) 10.5	−0.89	$U_c = 1.11$
Ar_2TeCl_2	(−) 9.1	+0.76	$U_{Cl-Te-Cl} = 0.35$
$ArTeCl_3$	(+) 9.2	−0.76	$U_{Cl-Te-Cl} = 0.35$
$TeCl_4 \cdot py$	(+) 4.5	−0.38	$U_{py} = 0.73$
‡$Te(thiourea)_2^{2+}$	(+) 15.6	−1.30	$U_s = 0.70$

* The sign of Δ is assumed, taking $Q(+3/2)$ as −0.2 barns.

† $Up = -e^2qQ$ and $e^2qQ_0 = +24$ mm s^{-1} for ^{125}Te

$$Up = \frac{-e^2qQ}{e^2qQ_0}$$

and $Up = -Uz + \dfrac{Ux+Uy}{2}$ where the U terms are p-orbital populations.

‡ From reference 24.

ACKNOWLEDGEMENTS

One of the authors (CHWJ) wishes to thank the National Research Council of Canada for a grant supporting this research. The assistance of P. Dobud, R.M. Cheyne, R. Shultz and N. Dance during various phases of this work are gratefully acknowledged. FJB wishes to acknowledge the support and encouragement of Dr. B.C. Smith, Birkbeck College.

REFERENCES

1. R.M. Cheyne, C.H.W. Jones and S. Husebye; Can. J. Chem.
    53, 1855 (1975).
2. F.J. Berry, E.H. Kustan and B.C. Smith; J. Chem. Soc.
    (Dalton), 1323 (1975).
3. K.V. Smith, J.S. Thayer and B.J. Zabransky, Inorg.
    Nucl. Chem. Lett. 11, 441 (1975).
4. C.H.W. Jones, R. Schultz, W.R. McWhinnie and N. Dance;
    submitted to Can. J. Chem.
5. F.J. Berry and C.H.W. Jones, unpublished results.
6. P. Boolchand; Nucl. Instr. and Methods 114, 159 (1974).
7. N.E. Erickson and A.G. Maddock; J. Chem. Soc. (A), 1665
    (1970).
8. C.E. Violet in The Mössbauer Effect and its Applica-
    tions in Chemistry. Advances in Chemistry Series of
    the American Chemical Society, Washington 1967, p 147.
9. Mössbauer Effect Data Index 1958-1965, Edited by A.M.
    Muir, K.J. Ando and H.M. Coogan, Interscience, New
    York, 1966, p 26.
10. A.J. Stone, appendix to G.M. Bancroft, W.K. Ong, A.G.
    Maddock, R.H.Prince and A.J. Stone; J. Chem. Soc. (A)
    1967 (1966).
11. W.R. Blackmore and S.C. Abrahams; Acta Cryst. 8, 317
    (1955).
12. F.H. Kruse, R.E. Marsh and J.D. McCullough; Acta Cryst.
    10, 201 (1957).
13. G. Llabres, O. Dideburg and L. Dupont; Acta Cryst. 288,
    2438 (1972).
14. D.A. Couch, P.S. Elmes, J.E. Ferguson, M.L. Greenfield
    and C.J. Wilkens; J. Chem. Soc. (A) 1813 (1967).
15. M.T. Chen and J.W. George; J. Amer. Chem. Soc. 90, 4580
    (1968).
16. K.J. Wynne and P.S. Pearson, Inorg. Chem. 9, 10 (1970).
17. J.D. McCullough and P.E. Marsh; J. Amer. Chem. Soc.,
    72, 4556 (1960).
18. K.J. Wynne and J.W. George; J. Amer. Chem. Soc. 91,
    1649 (1969).
19. B. Martin and R. Schule; Phys. Lett. 46B, 367 (1973).
20. J.J. Johnstone, C.H.W. Jones and P. Vasudev; Can. J.
    Chem. 50, 3037 (1972).
21. P. Dobud and C.H.W. Jones; J. Solid State Chem. (in
    press).

22. L.H. Bowen in Mössbauer Effect Data Index (1972),
    Edited by J.G. Stevens and V.E. Stephens, Plenum, New
    York, 1973 p. 71.
23. R.G. Barnes and W.V. Smith; Phys. Rev. 93, 95 (1954).
24. B.M. Cheyne, C.H.W. Jones and P. Vasudev; Can. J. Chem.
    50, 3677 (1972).
25. G.T. Morgan and H.D.K. Drew; J. Chem. Soc. 531 (1925).
26. F.J. Berry, B.C. Smith and C.H.W. Jones; submitted to
    Chem. Comm.
27. S.D. Christofferson and J.D. McCullough; Acta Cryst;
    11, 249 (1958).
28. G.D. Christofferson, R.A. Sparks and J.D. McCullough;
    Acta Cryst. 11, 782 (1958).
29. L.Y.Y. Chan and F.W.B. Einstein; J.C.S. (Dalton), 316
    (1972).
30. D. Kobelt and E.F. Paulus; Angewandte Chemie Inter-
    national Ed. 10, 73 (1971).
31. W.R. McWhinnie and P. Thavornyutikarn; J.C.S. Dalton,
    551 (1972).
32. P. Thavornyutikarn and W.R. McWhinnie; J. Organometal.
    Chem. 50, 135 (1973).
33. See for example J.D. McCullough, Inorganic Chemistry, 14,
    1142 (1975).
34. K.J. Wynne and P.S. Pearson; Inorg. Chem. 10, 2735
    (1971).
35. K.J. Wynne, A.J. Clark and M. Berg; J.C.S. Dalton 370,
    (1972).
36. B.C. Pant, W.R. McWhinnie and N.S. Dance; J. Organo-
    metal. Chem. 63, 305 (1973).

MÖSSBAUER SPECTROSCOPY OF RARE-GAS MATRIX ISOLATED ^{125}Te COMPOUNDS*

P. A. Montano

Department of Physics
West Virginia University
Morgantown, West Virginia 26506

and

P. H. Barrett and H. Micklitz⁺

Department of Physics
University of California
Santa Barbara, California 93106

INTRODUCTION

Rare-gas-matrix isolated (RGMI) atoms and molecules have been studied in the last twenty years by different experimental methods: These experiments indicate that the atoms and molecules trapped in the rare-gas-matrix have properties very similar to those of the free species, demonstrating that the weak binding in these solids does not change appreciably the atomic and molecular configuration. Matrix isolation techniques have been applied recently to Mössbauer spectroscopy. The first successful rare-gas-matrix isolation Mössbauer experiment was carried out with iodine molecules (I_2) imbedded in solid argon at 22K[1]. Since then this technique has been extended to ^{57}Fe,[2] ^{119}Sn [3], and ^{125}Te [4].

*Supported by the National Science Foundation.
⁺Permanent address: Department of Physics, Technische Universität München, 8046 Garching, Germany.

The use of the matrix isolation technique in Mössbauer spectroscopy offers a unique opportunity for studying "almost free" atoms and molecules. This is extremely useful in the interpretation and understanding of isomer shift [IS]. There is difficulty in correlating IS data and electron densities at the nucleus. This arises from the uncertainty in the application of free atom or free ion wave functions to solids and in the choice of electron configurations for particular reference compounds. The RGMI technique with atoms and ions has provided the possibility of obtaining IS values for Mössbauer nuclei where the free-atom or ion calculations can be applied. In the case of RGMI molecules self-consistent cluster calculation would be more useful in order to correlate IS data with electron densities at the nucleus.

We describe in the following paragraphs experiments in which the Mössbauer effect of ^{125}Te compounds is observed in rare-gas matrices of argon and krypton. ^{125}TeF$_6$ and TeCl$_4$ molecules were isolated in argon matrices and ^{125}Te dimers in argon and krypton matrices. The Te^0 monomer was obtained by photodissociation of H$_2$Te in argon. The IS measurements are discussed using Dirac-Fock electron density calculations for free atoms and ions.

EXPERIMENTAL

The samples are made in a liquid-helium cryostat, evacuated to a pressure below 10^{-7} torr. In figure 1 a general diagram of the matrix isolation system is given. The Mössbauer spectra were obtained with a conventional constant acceleration spectrometer. On one end of the velocity transducer was a (Cu-^{125}Sb) source cooled down to 4.2 K and at a distance of $\sim$2.5 cm from the absorber. On the other end, a (Pd-^{57}Co) source at room temperature was used for velocity calibration. The 5.8 keV escape peak of a Xe-filled proportional counter was used for detection purposes.

The TeF$_6$ was isolated in argon using a mixture of argon gas and with an Ar-to-TeF$_6$ ratio of 90 ± 5. The gas enters the cryostat through an adjustable needle valve and escapes from a tube at a distance of $\sim$ 6 cm from the depositing surface. The rare-gas-TeF$_6$ mixture condenses on a cooled beryllium disk which is thermally bonded to the bottom of

the liquid helium container. TeCl₄ was evaporated from an
8 cm long quartz crucible which was completely covered by Cu
foil. The bottom of the crucible was heated to 200°C and
the outlet of the crucible was around 90°C. These were
found to be the best conditions for deposition of relatively
good samples and not having too much condensation at the cool
end of the crucible. The TeCl₄ powder was handled in a dry
inert gas atmosphere before and during the filling of the
quartz crucible. The open end of the crucible was 3 cm
from the depositing surface (Be). The TeCl₄ deposition
rate was calibrated by a separate experiment by measuring
the attenuation of the 6 keV iron X-ray. Ar was used as a
matrix and it was deposited in the configuration shown in
Fig. 1a. The tellurium dimer matrix isolation experiments
were carried out by evaporating Te metal (95% enriched in
¹²⁵Te) from an alumina crucible inserted in a resistance
heated tantalum furnace. At 400°C the vapors of tellurium
are composed mainly of Te_2, no Te monomer is produced at
this temperature. The furnace temperature was measured
using a Chromel-Alumel thermocouple. The tellurium
deposition rate was calculated using the deposition
efficiency of the system (determined in a separate experiment) and by weighing the crucible before and after the
experiment. The rare-gas deposition rate was continuously
monitored by measuring the attenuation by the rare-gas of
the 35.5 keV γ-ray from the (Cu-¹²⁵Sb) source. We tried
to produce Te monomers from the metal vapor by using a
double Knudsen cell arrangement; in one cell vapors of Te_2
were produced at about 500°C, in the other cell the vapors
were heated up to 1500°C. We were unable to observe any
detectable amount of tellurium monomers in the matrix. Due
to the failure to obtain Te monomers from the metal vapor,
we decided to investigate the possibility of producing the
monomers by the dissociation of Te molecules. This type
of experimental technique has been widely applied in matrix
isolation work when the production of the desired species
from the vapor was not feasible[5]. We found H_2Te the most
appropriate molecule for our purpose. H_2Te is a highly
unstable molecule, even solid and liquid hydrogen telluride
are decomposed by light[6,7]. H_2Te is decomposed instantly
by dry oxygen at pressures below 10 cm of Hg, with the
formation of water and the deposition of tellurium metal;
if ignited in air it burns to form water and TeO_2. It is
very difficult to produce H_2Te by the action of hydrogen
on tellurium metal since high temperatures are necessary
and the compound is too unstable. Hydrogen telluride can

be prepared by the action of water or hydrochloric acid on
aluminum telluride; using 4N HCl one can obtain a yield of
80% H_2Te [6]. After several tests we decided to use the reaction of Al_2Te_3 and HCl as our source of H_2Te. In Fig. 1b
the two tight stainless steel containers used for the production of H_2Te are shown, they were connected through a
valve to the high vacuum system. The following procedure
was used for the production of H_2Te. Solid HCl was introduced in container A, the container was previously filled
with Al_2Te_3, 20% more than required by stoichiometry (the
handling of the substances was carried out in an inert
atmosphere). The container was sealed and pumped to a
vacuum better than 10^{-6} torr. Then the container was warmed
up to a temperature where the HCl liquified and H_2Te is a
gas (m.p. = 221.8 K, b.p. = 271 K) [7]. A_2Te_3 and HCl started
to react and H_2Te was produced and collected in container
B, which was previously evacuated. Then containers A and
B vacuum valves were closed, and container B was maintained
at a temperature of 223 K. Some contamination with water
in the container is still possible in our experiment; however,
there is not enough to dissociate all the H_2Te. The deposition for a pure H_2Te sample was carried out at a container
temperature of 223 K. The deposition rate was calculated
from the attenuation of the 6 keV X-ray of a ^{57}Co source by
the sample (in a separate experiment). The sample holder
was a Be disk at 78 K. The system was light tight so that
no photodissociation could occur. The above experiment was
repeated several times, it was observed that 90% of the
H_2Te was deposited in the first half hour.

        Tellurium monomers can be produced by photodissociation
of H_2Te isolated in a rare-gas solid. H_2Te molecules were
isolated in an argon matrix, the H_2Te gas temperature was
increased to 260 K. At 260 K a partial decomposition of
H_2Te occurs during deposition of H_2Te. A saphire disk was
used as a substrate and the sample was irradiated through
a side window of the cryostat with a tungsten lamp. The
sample was irradiated during deposition and for six hours
after deposition. The sample temperature was about 5 K.
The atomic ratio of rare-gas to tellurium in the sample
studied was 62 ± 5. Because of the fast deposition rate
(made necessary by the yield and instability of H_2Te) a
highly distorted matrix is obtained. After photodissociation
one can have recombination of hydrogen to H_2, with consequent
deformation of the environment of the Te atom.

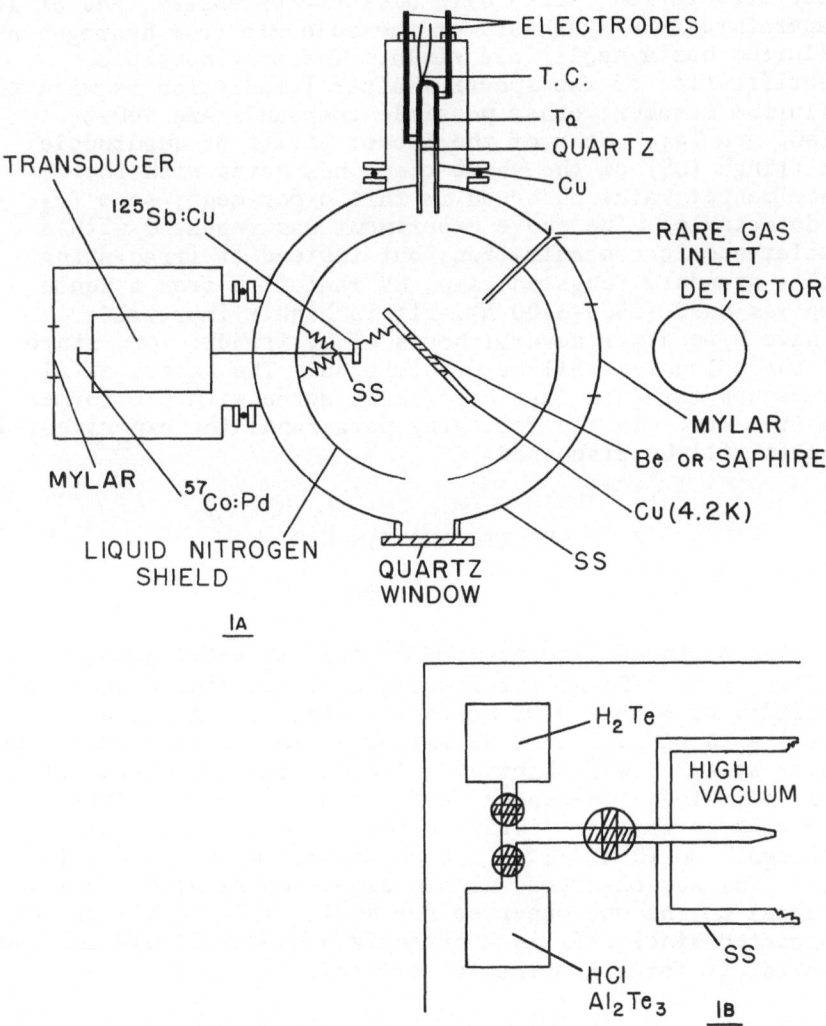

Fig. 1a: Experimental arrangement for the deposition of the Te compound-rare gas samples.

Fig. 1b: Experimental arrangement used for the production of H_2Te.

It is highly improbable to retain the H_2Te molecule after irradiation, since H_2Te dissociates easily, and at low temperatures the synthesis of the molecule from hydrogen and tellurium has a negligible yield. The only possible identification of the spectrum after irradiation is with the tellurium monomer; other possible compounds are TeO_2, H_2TeO_3 and Te_2. None of the isomer shifts or quadrupole splittings (QS) of the above compounds agree with the experimental value observed in this experiment; some Te_2 is detectable. The above experiment was repeated with a similar atomic concentration, but instead of irradiating with a standard tungsten lamp, UV radiation from a xenon lamp was used (2500-3100 Å). It is highly improbable to have H_2Te after several hours of UV irradiation, since all the molecules will be dissociated. The experimental parameters for this last experiment agree with the former measurements. In the following paragraphs the experimental results will be discussed.

RESULTS AND DISCUSSION

Ar - TeF$_6$

The Mössbauer spectrum of $^{125}TeF_6$ in solid argon ($Ar/TeF_6 = 90 \pm 5$) at 4.2 K shows a single line with a linewidth of 5.98 + 0.05 mm/sec. (Fig. 2). A large negative IS of -1.54 ± 0.05 mm/sec. referred to a ($Cu-^{125}Sb$) source at 4.2 K was observed. The Mössbauer spectrum of pure solid TeF_6 (thickness: 3.4 mg/cm^2 ^{125}Te) deposited on a cold surface at liquid-helium temperature was also obtained. An IS of -1.64 ± 0.05 mm/sec. with respect to ($Cu-^{125}Sb$) was observed; within experimental error this IS is equal to the one observed for RGMI - TeF_6. This is not unexpected since TeF_6 is a strongly bonded molecule and, as a solid, it forms a molecular crystal.

Ar - TeCl$_4$

The Mössbauer spectrum of $^{125}TeCl_4$ molecules isolated in solid argon ($Ar/TeCl_4 = 50 \pm 10$) at 4.2 K shows an un-resolved quadrupole doublet (Fig. 3). The solid line in Fig. 3 is the computer fit of the spectrum assuming Lorentzian lineshapes . From the fit, the values of the QS and IS were, respectively 3.5 ± 0.1 mm/sec. and

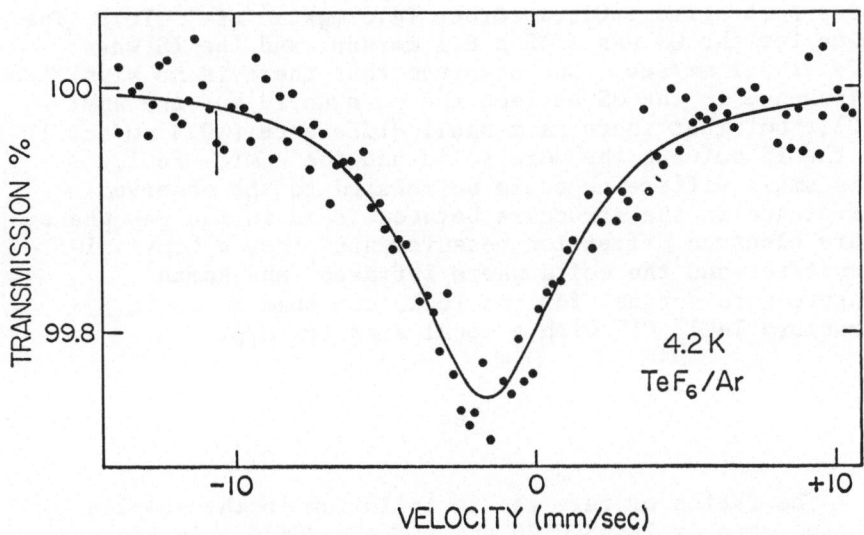

Fig. 2: Mössbauer spectrum of ^{125}TeF$_6$ in solid argon
(Ar/TeF$_6$ = 90 ± 5)

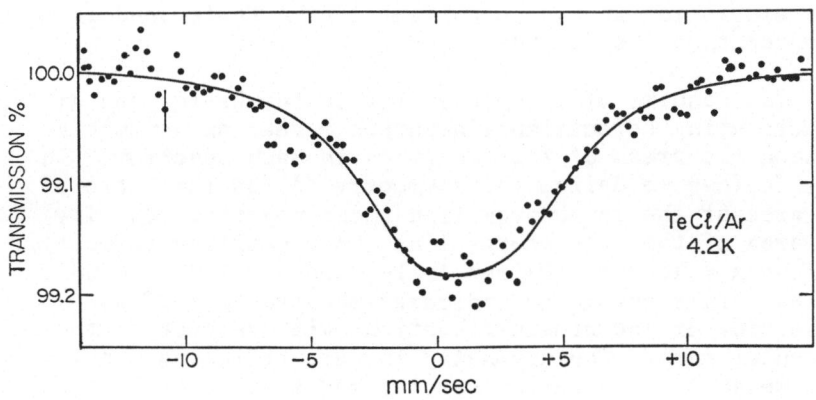

Fig. 3 : Mössbauer spectrum of ^{125}TeCl$_4$ molecules isolated
in solid argon (Ar/TeCl$_4$ = 50 ± 10)

1.0 ± 0.1 mm/sec. The Mössbauer spectrum was also obtained for a pure solid $TeCl_4$ absorber (4.6 mg/cm^2 of ^{125}Te). The value for the QS was 3.45 ± 0.1 mm/sec. and the IS was 0.74 ± 0.1 mm/sec. One observes that there is no significant difference in the QS between the pure solid and the RGMI - $TeCl_4$, but that there is a small difference ($\sim$0.3 mm/sec.) in the IS between the pure solid and the RGMI - $TeCl_4$. This small difference could be related to the observed difference in the structure between $TeCl_4$ in the gas phase, where electron difraction measurements[8] show a bipyramid structure, and the solid where infrared[9] and Raman[10] measurements suggest for the $TeCl_4$ compound an ionic structure $TeCl_3^+$ Cl^- with a local symmetry C_{3v}.

$$Te_2$$

The ratios of rare gas to tellurium in the samples studied were $Ar/Te_2 \simeq 500 \pm 20$ and $Kr/Te_2 \simeq 180 \pm 10$. In Fig. 4 the Mössbauer spectrum of $Kr-Te_2$ is shown. The main feature of this spectrum is the appearance of a well-resolved QS of 9.55±0.05 mm/sec. with an IS of +0.31±0.05 mm/sec.(referred to $Cu-^{125}Sb$). For $Ar-Te_2$ a similar spectrum gives a QS of 9.64±0.05 mm/sec. and IS of ±0.38±0.05 mm/sec. These results are consistent with the assumption that the distance between the Te atoms in the dimer is the same in krypton and argon. This assumption should be correct if the Te-Te bond is much stronger than the Te_2 host bond.

The results of a computer fit to the $Kr-Te_2$ and $Ar-Te_2$ spectra using Lorentzian lineshapes reveal an asymmetry between the areas of the two lines in each spectrum. In what follows we define the asymmetry (A) as the ratio of the area of the low-energy line (most negative velocity) to the area of the high-energy line (most positive velocity). We find A = 1.03 ± 0.02 for $Ar-Te_2$ and A = 1.07 ± 0.04 for $Kr-Te_2$. This asymmetry indicates the presence of an anisotropy of the atomic vibrations with respect to the molecular axis. The asymmetry in pure tellurium metal has been measured in powdered samples and a value of 1.05 was obtained[11]. This value is in the range of the anisotropy we observe in the tellurium dimers. For a crystal with axial symmetry Kagan[12] showed that the Mössbauer fraction can be written in the form

$$f = \exp\ [-k_\gamma^2(<x_\parallel^2>-<x_\perp^2>)\cos\phi-k_\gamma^2<x_\perp^2>] \qquad (1)$$

where $<x_\parallel^2>$ and $<x_\perp^2>$ are the mean squares of the projection of the amplitude of vibration of the absorbing nucleus along the symmetry axis and perpendicular to it, k_γ is the wave number of the Mössbauer radiation. The asymmetry for thin absorbers has been evaluated by Karyagin[13]. The values of A vs $\epsilon = k_\gamma^2(<x_\parallel^2>-<x_\perp^2>)$ were calculated by Flinn et al[14]. Using Flinn's calculation we obtain $\epsilon = k_\gamma^2(<x_\parallel^2>-<x_\perp^2>)\approx\pm\ 0.2$ for Kr and $\epsilon = \pm\ 0.5$ for Ar. The sign can be obtained by comparing with the results for tellurium metal and also by the assumption $<x_\parallel^2>><x_\perp^2>$. Thus $\epsilon>0$ for both Kr and Ar. This means that the transition $\pm\frac{1}{2}\rightarrow\pm\frac{1}{2}$ is lower in energy than the transition $\pm\frac{3}{2}\rightarrow\pm\frac{1}{2}$, i.e., $e^2qQ>0$. Since the quadrupole moment Q of the ^{125}Te nucleus for the $I = \frac{3}{2}$ level is negative[15], the electric field gradient (EFG) eq must be negative.

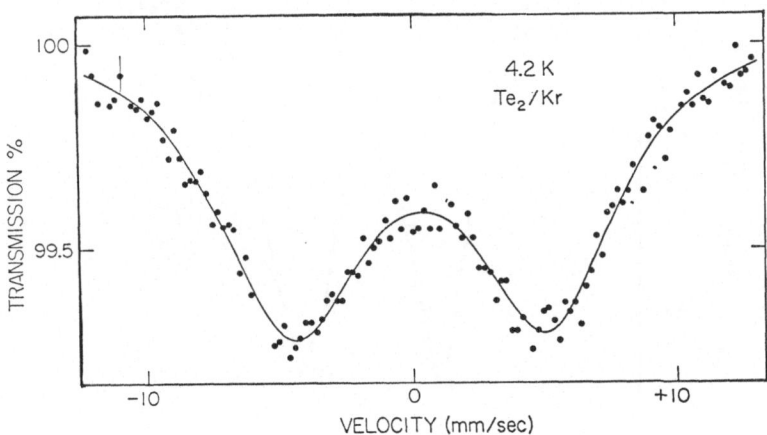

Fig. 4: Mössbauer spectrum of Kr-Te$_2$ (Kr/Te$_2$ = 180 ± 10)

The major contribution to the EFG in the Te dimers is expected to come from the aspherical electronic charge density around the ^{125}Te nucleus arising from the unbalanced p valence electrons. The number of electrons in the p_x and p_y orbitals (N_x and N_y) is the same for axial symmetry, i.e., $N_x = N_y$. In this case there are two possibilities for the occupation of the different p orbitals with the four 5p electrons: (a) $N_x = N_y = 1$, $N_z = 2$ (i.e., σ and π bonding) with eq = $-\frac{4}{5}e<1/r^3>$ or (b) $N_x = N_y = 2$, $N_z = 0$ (i.e., pure π bonding) with eq = $+\frac{8}{5}e<1/r^3>$. The sign of the EFG determined from the asymmetry in the quadrupole doublet is negative, therefore, (a) gives the occupation of the different 5p orbitals. Since Te$_2$ is a bound molecule, a mixed σ and π bonding seems more reasonable than the much weaker pure π bonding. For atomic tellurium $<1/r^3>$ = 15.0 a.u. according to Barnes and Smith[16]. Taking Q = -0.20 ± 0.03 [17] we calculate $\frac{1}{2}e^2qQ$ = -9.9 ± 1.5 mm/sec. This value agrees, within the experimental error in Q, with our measurement.

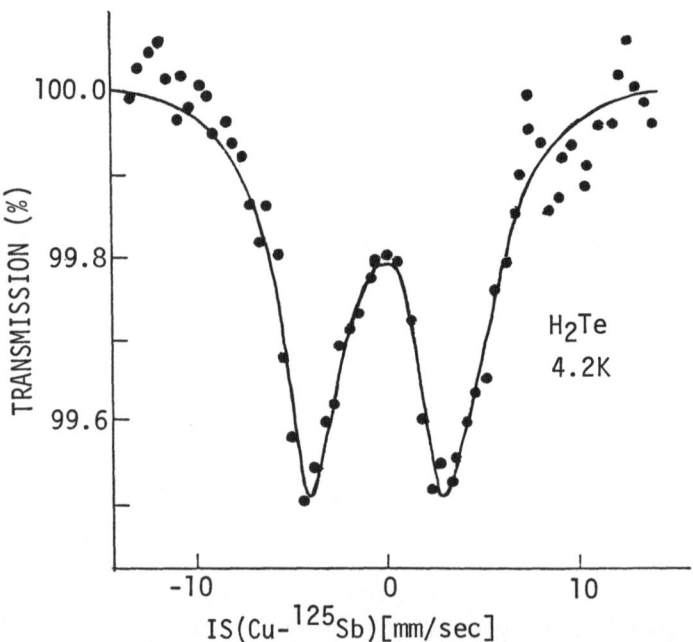

Fig. 5: Mössbauer absorption spectrum of H$_2$Te at 4.2 K.

H_2Te and Te^0

The Mössbauer spectrum of H_2Te at 4.2 K (Fig. 5) shows a QS of 7.2 ± 0.2 mm/sec. and an IS of -0.1 ± 0.2 mm/sec. (with respect to a Cu-[125]Sb source). The line width is 5.0 ± 0.2 mm/sec. and to the authors' knowledge the narrowest ever observed in a Te compound. The presence of a QS indicates that the symmetry of H_2Te at 4.2 K is lower than cubic. The isomorphous molecular compounds H_2Se and H_2S are cubic but H_2S has a stable tetragonal form below 103 K. No crystallographic information is available on H_2Te, but from the Mössbauer spectrum a tetragonal lattice at 4.2 K is the most probable. The tellurium-hydrogen bond length has been estimated at 1.7 Å, and the angle H-Te-H at 89°30' [18]. The QS observed in solid H_2Te is smaller than predicted for such a bond angle in the case of bonds formed with pure p-orbitals[19]. This QS is not unexpected for H_2Te since measurements of the quadrupole coupling constants of the H_2S molecule (bond angle 92°6') suggest that the bonding orbitals are spd hybrids[20]. The IS value observed for H_2Te is not far away from the expected value for Te^0 [4]. Due to similar electronegativities[21] of tellurium and hydrogen and the weak bonding of the molecule the tellurium atomic configuration will not be far from the neutral state.

The Te monomers were produced by photodissociation of H_2Te isolated in solid argon. One observes, after finishing the deposition and before irradiation with a tungsten light source, a Mössbauer spectrum with a QS of 8.9 ± 0.2 mm/sec, and IS of -0.2 ± 0.2 mm/sec, and a line width of 5.7 ± 0.2 mm/sec. After six hours of irradiation we observed a Mössbauer spectrum with a QS of 9.2 ± 0.2 mm/sec, an IS of 0.0 ± 0.2 mm/sec, and a line width of 5.0 ± 0.2 mm/sec. (Fig. 6). There is a slight asymmetry in the spectrum, this could be explained by the presence of some tellurium dimer. Since the atomic ratio is Ar/Te ≃ 62, there is about 10% probability of having Te-Te nearest neighbors. We interpret the pre-irradiation spectrum as produced by a mixture of H_2Te and Te monomers. The monomer being produced by dissociation in the H_2Te gas and photodissociation of H_2Te by room light entering the cryostat during deposition. The post-irradiation spectrum with its narrower line width is interpreted as being produced essentially by Te monomer. Due to the presence of light during all the experiment and and the weak bond and instability of H_2Te [6], we disregard

recombination of Te atoms with hydrogen in the matrix[6]. It
is more probable that hydrogen atoms recombine to form H_2
or combine with oxygen impurities and/or diffuse out of the
argon matrix. The asymmetry in the spectrum of Fig. 6, can
be explained by the presence of impurities, most probably
Te_2. Possible impurities in the matrix will have different
IS and QS: TeO_3 [22] (IS = -0.8 mm/sec.), TeO_2 [22] (IS = +0.9
mm/sec.) and H_2TeO_3 [22] (IS = +1.2 mm/sec.); all the IS are
given with respect to a $(Cu$-$^{125}Sb)$ source.

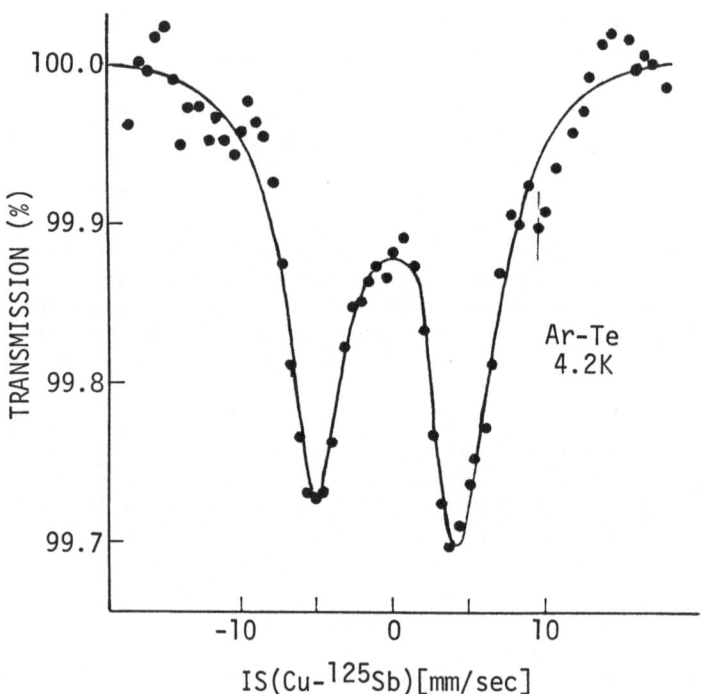

Fig. 6: Mössbauer absorption spectrum of ^{125}Te monomers
         in an argon matrix.

We repeated the above experiment with similar atomic
concentrations, but instead of irradiating with a standard
tungsten lamp, UV radiation from a xenon lamp was used
(2500-3100 Å). Under those circumstances it is impossible

to have any H_2Te left after irradiation. We observed a
Mössbauer spectrum with a QS of 9.1 ± 0.2 mm/sec. and IS of
-0.1 ± 0.2 mm/sec. in good agreement with the former measurements. The QS of the Te monomer is less than the one observed
for Te dimer[4]. The EFG at the ^{125}Te nucleus is produced by
the splitting of the p-levels in the weak non-cubic crystal
field of the matrix produced by lattice defects, due to a
fast deposition rate. A partial population of the excited
p-states is possible at 4.2 K, consequently reducing the
observed QS.

ISOMER SHIFT

The Mössbauer isomer shift is given by[24]

$$\delta V_{IS} = 6.105 \ (Z/E_\gamma)\Delta<r^2>\Delta\rho(o) \qquad (2)$$

where δV_{IS} is the isomer shift in mm/sec, $\Delta<r^2>$ is the
change in mean-square nuclear radius $\Delta<r^2>=<r^2>_{excited}$
$-<r^2>_{ground}$ and is given in units of fm^2, $\Delta\rho(o)$ is the
difference between electron densities at the nucleus (in
units of a_o^{-3}), E_γ is the transition energy in keV, and Z
is the atomic number. Sometimes the nuclear parameter used
is $\delta R/R$, where R is the nuclear charge equivalent radius
(R = 1.2 $A^{1/3}$ fm) and δR is the change in the charge radius
between the excited and ground states of the nucleus, the
intercomparison is $R^2(\delta R/R)=\frac{5}{6}\Delta<r^2>$. We used Dirac-Fock
wave functions for free ions and atoms to calculate the
electron densities at the Te nucleus for different Te
configurations[4]. A more appropriate electron density could
be obtained for matrix-isolated molecules if self consistent
cluster calculations were available. Since such calculations are not available for TeF_6, $TeCl_4$ and Te_2 we have to
use the free ion Dirac-Fock calculations. The $\rho(o)$ for the
configuration $[Kr]4d^{10}5s^2$ will be a good approximation for
Te in $TeCl_4$ because of the ionic character of this compound.
Combining the IS value for Ar-$TeCl_4$ (1.0 ± 0.1) mm/sec. with
the IS value for the Te° monomer (0.0 ± 0.2) mm/sec. we
obtained using Eq. 2. $\Delta<r^2>$ = (2.8 ± 0.5) x $10^{-3} fm^2$. This
value is almost identical to the one obtained by comparison
of isoelectronic compounds[25]. In Fig. 7 the IS is plotted
for the different matrix isolated tellurium species versus
electron density differences. If TeF_6 were a completely
ionic compound, the Te electron configuration would be
$[Kr]4d^{10}$; on the other hand, the configuration 5s $5p^3$ $5d^2$

is predicted for Te in TeF_6 from the point of view of coordination chemistry[26]. From the IS calibration of Fig. 7 the IS of TeF_6 is more in agreement with the electron density expected for the 5s $5p^3$ $5d^2$ configuration. The Te_2 molecule has a slightly more positive IS than the Te monomer which indicates that overlap of the p and s orbitals is taking place in this molecule, changing consequently the electron density at the nucleus. Self consistent cluster calculations of $TeCl_4$ are necessary for a more accurate IS calibration for ^{125}Te.

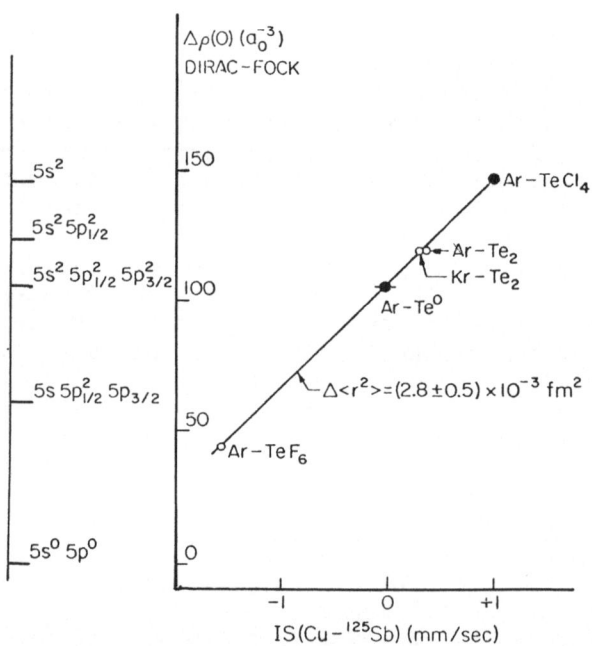

Fig. 7: Correlation between electron density differences $\Delta\rho(o)$ at the nucleus and the IS for ^{125}Te. Electron densities calculations are taken from Ref. 4.

REFERENCES

1) S. Bukshpan, C. Goldstein, and T. Sonnino, J. Chem. Phys. 49, 5477 (1968).

2) P. H. Barrett and T. K. McNab, Phys. Rev. Lett. 25, 1601 (1970).

3) H. Micklitz and P. H. Barrett, Phys. Rev. B5, 1704 (1972).

4) P.H. Barrett, P. A. Montano, H. Micklitz and J. B. Mann, Phys. Rev. B12, 1676 (1975).

5) Beat Mayer, Low Temperature Spectroscopy, American Elsevier Publishing Company, Inc., New York (1971). P. H. Kasai, Phys. Rev. Letters 21, 67 (1968).

6) Gmelins Handbuch Der Anorganischen Chemie, Tellur, edited by Erich Pietsch, Verlag Chemie G.M.B.H., Berlin (1940).

7) Tellurium, edited by W. Charles Cooper, Van Nostrand Reinhold Company, New York, Cincinnati, Toronto, London, Melbourne (1971).

8) D. P. Stevenson and Verner Schomaker, J. Am. Chem. Soc. 62, 1267 (1940).

9) N. N. Greenwood, B.P. Stranghan, and Anne E. Wilson, J. Chem. Soc. A, 1479 (1966).

10) Nikolaos Katsaros and John W. George, Chem. Commun. 21, 613 (1965).

11) R. N. Kuzmin, A. A. Opalenko, V. S. Shpinel, and I. A. Avenarius, Zh. Eksp. Teor. Fiz. 56, 167 (1969). (Sov. Phys. - Jetp 29, 94 (1969)).

12) Yu Kagan, Dokl. Akad. Navk. SSSR 140, 794 (1961) (Sov. Phys. - Dokl. 6, 881 (1962)).

13) S. V. Karyagin, Dokl. Akad. Nauk. SSSR 148, 1102 (1963).

14) P. A. Flinn, S. L. Ruby, and W. L. Kehl, Science 143, 1434 (1964).

15) R. N. Kuzmin, A. A. Opalenko and V. S. Shpinel, in Proceedings of the Conference on the Application of the Mössbauer Effect Tihany, 1969, edited by I. Dezsi (Akademiai Kiado, Budapest, 1971).

16) R. G. Barnes and W. V. Smith, Phys. Rev. 93, 95 (1954).

17) C. E. Violet, R. Booth and F. Wooten, Phys. Lett. 5, 230 (1963).

18) K. Rossman and J. W. Straley, J. Chem. Phys. 24, 1276 (1956).

19) W. Gordy, W. V. Smith and R. F. Trambarulo, Microwave Spectroscopy, John Wiley and Sons, Inc., New York (1953).

20) Charles A. Burrus, Jr. and Walter Gordy, Phys. Rev. 92, 274 (1953).

21) Linus Pauling, The Nature of the Chemical Bond, Cornell University Press, Ithaca, New York (1960).

22) P. Jung and W. Trifthäuser, Phys. Rev. 175, 512 (1968).

23) M. L. Unland, J. Chem. Phys. 49, 4514 (1968).

24) G. M. Kalvius, in Hyperfine Interaction in Excited Nuclei, edited by G. Goldring and R. Kalish (Gordon and Breach, New York, 1971), Vol. 2, p. 523.

25) G. M. Kalvius and G. K. Shenoy, Atomic Data and Nuclear Data Tables 14, 639 (1975).

26) S. L. Ruby and G. K. Shenoy, Phys. Rev. 186, 326 (1969).

THE ISOMER SHIFT IN ^{151}Eu

E. A. Samuel
Center for Surface and Coatings Research
Lehigh University, Bethlehem, PA   18015

and

W. N. Delgass
School of Chemical Engineering
Purdue University, West Lafayette, IN   47907

ABSTRACT

Molecular orbital calculations in the Löwdin formalism for five compounds, EuO, Eu_2O_3, Eu_3O_4, EuF_2 and EuF_3, give a value of $3.2 \pm 0.7 \times 10^{-4}$ for the isomer shift calibration constant of ^{151}Eu. The M.O. calculations are also used to predict the variation of the isomer shift with coordination number and bond distance in the $Eu^{3+} - O^{2-}$, $Eu^{2+} - O^{2-}$, $Eu^{3+} - F^-$ and $Eu^{2+} - F^-$ systems, and to give the values -18.7 and -8.3 mm/sec for the isomer shift of a free Eu^{2+} ion and a free Eu^{3+} ion with respect to Eu_2O_3. The relativistic enhancement of the free-ion non-relativistic electron density at an Eu nucleus is discussed and the variation of free-ion isomer shifts with electron configuration presented.

1.   INTRODUCTION

The isomer shift is a characteristic parameter of Mössbauer spectroscopy and arises as a result of the finite size of the nucleus and the departure of the electron

potential within the nucleus from the r^{-1} dependence that
electrons just outside the nucleus experience. In a model in
which the nuclear charge, Ze, is uniformily distributed in
a sphere of radius, R, this difference in potential is (1),

$$V'(r) = Ze^2(-3/(2R)+r^2/(2R^3)+1/r) \qquad (1)$$

in electrostatic units. Such a difference in potential leads
to a shift in the resonance energy, E_γ, given in cm/sec, by:

$$\Delta E = |4\pi Ze^2R^2c/(5E_\gamma)][\sum_A |\Psi_i(0)|^2(1-3a_iR^2/7)$$

$$-\sum_S |\Psi_i(0)|^2(1-3a_iR^2/7)][\Delta R/R] \qquad (2)$$

In equation 2, the three factors in brackets are arranged
from left to right in increasing order of uncertainty. The
first factor; in which R, the nuclear radius, is taken as (2)

$$R = 1.123A^{1/3} + 2.352A^{-1/3} - 2.070A^{-1}, \qquad (3)$$

A being the total number of nucleons (A=151 in the case of
^{151}Eu) and E_γ the unperturbed resonance energy of the
isomeric transition (21.6 keV in the case of ^{151}Eu); is known
very accurately. The second factor is the difference in the
electron density at the nucleus between absorber and source.
In keeping with the numerical relativistic Dirac-Fock calculations of Mann (3), we assume that the electron density of
$s_{1/2}$ and $p_{1/2}$ electrons varies near the nuclear origin as
$|\Psi_i(0)|^2(1-a_ir^2)$. The difficulty in estimating the electron
density at a nuclear site in the solid is due to the considerable departure of the electron density from its free-ion
value when the ions are incorporated in the solid state.
There is an added difficulty of estimating the free-ion
electron density itself when only non-relativistic Hartree-
Fock calculations of the s-radial functions of the free ions

are available. Shirley (1) has pointed out that the non-relativistic electron density at a nuclear site may be corrected for the large relativistic effects near the nucleus by multiplying it by the Racah-Breit-Rosenthal-Bodmer enhancement factor, S'(Z), which is constant for all electronic configurations of an atom as long as the nuclear charge remains the same. However, while the Racah, Breit-Rosenthal and Bodmer theories (4,5,6) predict the same correction factor for all ns orbitals of a given configuration, numerical relativity factors derived by comparing relativistic and non-relativistic densities of Eu configurations show different values for different orbitals within the configurations. An average $S_s'(Z)$ could still be defined numerically to give the total relativistic s-density from the non-relativistic density. A similar factor $S_{sp}'(Z)$ could also be defined to give the total relativistic s and p density from the total non-relativistic density. We see in the course of this paper that, for Eu ions, both these numerical factors, while being very nearly constant for different configurations, are lower by about 6 to 10 % than the theoretical value predicted by the Racah-Breit-Rosenthal-Bodmer theories. It appears that non-relativistic free-ion densities of Eu ions corrected for relativity effects by the theoretical S'(Z) tend to be an overestimation. Kalvius (7) has reviewed previous comparisons of non-relativistic and relativistic densities. He observes that Hafmeister (8) found the numerical ratios of non-relativistic to relativistic densities evaluated at the nuclear surface to be within 3 % of the theoretical S'(Z) values for a series of elements ranging from K(Z = 19) to Cs(Z = 55). On the other hand, Dunlap et al. (9) have found that the numerical factor for Np(Z = 93) was about 12 % lower than the

theoretical factor when the non-relativistic and relativistic densities were evaluated at the nuclear origin.
We derive our numerical relativity factors for Eu(Z = 63) by
comparing the non-relativistic and relativistic densities
evaluated at the nuclear origin. We can then estimate
separately the correction to the isomer shift arising from
the variation of the relativistic density over
the nuclear volume. The third factor in equation
2 is the relative change in the nuclear radius due to the
isomeric transition. It is a non-dimensional parameter
characteristic of an isomeric pair. Of the three factors in
equation 2 this factor is the least certain. This is because
ΔR, the change in nuclear radius between excited and ground
states, cannot be calculated with reasonable accuracy from
considerations of nuclear structure alone (7,10). Spectroscopists call the third factor the calibration constant and
in any evaluation of it through measurements of the isomer
shift are inherent the uncertainties in the electron density
at the nuclear site (term 2).

In this paper we consider the calibration of the isomer
shift of ^{151}Eu by analyzing the isomer shift in five compounds
of europium: EuO, Eu_2O_3, Eu_3O_4, EuF_2 and EuF_3. We estimate
the electron density at ^{151}Eu sites in these solids using the
Löwdin theory of molecular orbitals which prescribes a systematic method for correcting free-ion densities for cation-
ligand overlaps. The Löwdin theory (11), as outlined by
Flygare and Hafmeister (12) in their analysis of the isomer
shift of ^{129}I, is presented in section 2. Section 3 develops
the numerical procedure for obtaining overlap corrected
electron densities. Overlap integrals are calculated in a

way similar to that used by Flygare and Hafmeister. Two sets
of non-relativistic and relativistic calculations are available for Eu free-ion wave functions. The first set consists
of non-relativistic Hartree-Fock and relativistic Dirac-Fock
calculations by Mann (2,13). The second set, which includes
a wide variety of electronic configurations of Eu ions, consists of non-relativistic Hartree-Fock calculations by Wilson
(14) and relativistic Hartree-Fock calculations by Coulthard
(15). All three authors evaluate the exchange terms without
approximation, but Mann also includes correlation terms in
an approximate way in his Dirac-Fock routine (13). The
electron density at the nuclear origin may be obtained directly from Mann's Dirac-Fock routine because he assumes the
nucleus to be of finite size with a Fermi distribution of
charge. On the other hand, Coulthard (15) approximates the
nucleus with a point charge. In this case the electron density diverges at the nuclear origin and is taken as equal to
that at the nuclear surface. In our calibration of the
isomer shift of ^{151}Eu we use the electron densities obtained
from Mann's Dirac-Fock calculations. We use the Coulthard
and Wilson densities for obtaining a map of the variation of
the electron density at the nucleus with electronic configuration to aid us in chemically interpreting our calculations.
Section 4 contains a table summarizing the important structural characteristics of the compounds studied. We present
the results of our calculations in section 5 and discuss the
errors in our value of the calibration constant arising from
solid state effects other than overlap distortion. This
section also contains the chemical interpretation of our calculations relating to the isomer shift in europium oxides,
fluorides, europium metal and europium intermetallic compounds.

2. THEORY

Contribution of Nuclear Deformation to the Isomer Shift Calibration Constant of ^{151}Eu

Shirley (1) has pointed out that a nuclear deformation may be as important to the isomer shift calibration constant as an actual change in nuclear radius. The electric quadrupole moment is a measure of nuclear deformation and nuclei in the region beginning with ^{151}Eu and ending with ^{193}Ir display quadrupole moments that are several times higher than the single particle values expected on the basis of the shell model. Shirley has suggested that for ^{151}Eu, which lies on the edge of the collective zone, the calibration constant is

$$\Delta R/R = (\Delta(\varepsilon^2)/2)_{core} + (\Delta R/R)_{sp} \qquad (4)$$

the core being taken as the nucleus with all its nucleons but the odd nucleon and the single particle being taken as the odd nucleon (odd proton). The surface of the deformed core is taken as the ellipsoid $r = R[1+\varepsilon P_2(\cos\theta)]$, where $P_2(\cos\theta)$ is a Legendre polynomial. Rainwater's model of the nucleus (16,17) may be invoked to estimate $\Delta(\varepsilon^2)/2$, and if the contribution of the single particle is ignored, this estimate may be shown to be a lower limit on $\Delta(\varepsilon^2)/2$ (18). Assuming that the loss of a gamma photon results in a decrease in the deformability, ε, and using the results of Bohr and Wheeler (19) regarding the change on the liquid drop energy of a nucleus on deforming, $\Delta(\varepsilon^2)$ may be written as:

$$\Delta\varepsilon^2 = (E_\gamma/931000.0)/[(2/5)(0.0141)A^{2/3}$$
$$-(1/5)(0.000627)(Z^2/A^{1/3})], \qquad (5)$$

which for ^{151}Eu(Z = 63, A = 151) is 3.62 x 10^{-4}. An upper
estimate may be obtained for $\Delta(\varepsilon^2)$ using experimental values
for the quadrupole moments of ^{151}Eu in its ground and excited
states and comparing them with the expectation that the
quadrupole moment of a uniformly charged ellipsoid with
charge Z is:

$$Q = (6/5)\varepsilon ZR^2. \qquad (6)$$

This yields 11 x 10^{-4} for $\Delta(\varepsilon^2)$, and the calibration constant
for ^{151}Eu in which the effects of the core predominate may
thus be expected to be in the range:

$$1.8 \times 10^{-4} \leq \Delta R/R \leq 5.5 \times 10^{-4} \qquad (7)$$

The range given in 7 is also the range in which presently
available calibration constants lie (2,20,21,22). From this
simple analysis it appears that a nuclear deformation alone
is sufficient to explain the range of calibration constants
derived from experimental observations of the isomer shift.
There are, however, two calculations, among others (7), which
disagree with this conclusion. Speth (23,24), on the basis
of Migdal's theory of finite Fermi systems, has obtained a
value for the change in the second moment of the nuclear
charge distribution in ^{151}Eu as 10.99 x 10^{-29} cm^2. This
value, which includes both the single particle and the deformation contributions, yields for the calibration constant,
$\Delta R/R$, the value 1.34 x 10^{-4} when the nuclear radius is given
by equation 3 (6.40 x 10^{-13} cm in the case of ^{151}Eu). The
significance of the single particle contribution is more
strongly stated by Shirley (1) and Shirley and Barret (25).
These authors obtain a deformation that is actually lower for
the excited state than for the ground state. Thus the question

of the relative importance of nuclear deformation and an
actual change in the nuclear radius to the isomer shift
calibration constant does not appear to be completely
resolved.

Löwdin Theory

Löwdin (11) has pointed out that the total wave function
of an ion in a molecule may still be expressed as a Slater
determinant of one electron orbitals, not of the free-ion
orbitals, however, but of one electron orbitals constructed
from the free-ion orbitals by orthogonalizing them with
ligand orbitals to include the effect of overlap of metal
orbitals with ligand orbitals. According to the Löwdin
formalism (11), the orthogonalized ligand orbitals $|p_\ell)'$ and
the orthogonalized metal orbitals $|ns)'$ (assuming here for
simplicity that the ligand p orbitals form only σ bonds with
the metal ns orbitals) are given by,

$$|ns)' = |ns) - 1/2\sum_\ell |p_\ell)S(p_\ell|ns) + 3/8\sum_{\ell m}|p_\ell)S(p_\ell|p_m)S(p_m|ns)$$

$$+ 3/8 \sum_{n's\ell}|n's)S(n's|p_\ell)S(p_\ell|ns) - \cdots$$

$$|p_\ell)' = |p_\ell) - 1/2\sum_m|p_m)S(p_m|p_\ell) - 1/2\sum_{ns}|ns)S(ns|p_\ell)$$

$$+ 3/8 \sum_{ns}\sum_m|ns)S(ns|p_m)S(p_m|p_\ell) - \cdots \qquad (8)$$

where ℓ runs over all the ligand ions and $S(p_\ell|ns)$ and
$S(p_\ell|p_m)$ are elements of the overlap matrix defined by
Löwdin. The modified s-electron density at the site of the
nucleus of the metal ion is,

$$|\Psi(0)|^2 = \sum_{ns} |ns)'(ns|' + \sum_{\ell} |p_\ell)'(p_\ell|'$$

$$= \sum_{ns} |\Psi_{ns}(0)|^2 + \sum_{\ell} \sum_{ns} S^2(ns|p_\ell)|\Psi_{ns}(0)|^2$$

$$+ \sum_{\ell} \sum_{ns} \sum_{\substack{n's \\ \neq ns}} S(ns|p_\ell)S(p_\ell|n's)|\Psi_{ns}(0)||\Psi_{n's}(0)|-\cdots \tag{9}$$

In equation 9, only two-fold overlaps are shown. Inclusion of three-fold overlaps in equation 9 brings it in conformity with the similar result of Simanek and Sroubek (26) derived on the basis of a Gram-Schmidt orthogonalization procedure, while the inclusion of four-fold overlaps in equation 9 shows a departure from the Simanek and Sroubek formula. The two-fold overlap formula of Flygare and Hafmeister (12) (equation 9) and the three-fold overlap formula of Simanek and Sroubek become identical when ligand-ligand overlaps are neglected. We have applied the Löwdin orthogonalization procedure including only two-fold overlaps to analyze the isomer shift in EuO, Eu_2O_3, Eu_3O_4, EuF_2 and EuF_3. We have estimated that the neglect of ligand-ligand overlaps in these compounds leads to overestimating the overlap correction to the electron densities at ^{151}Eu sites in these compounds by less than 5 %. We have also used the Löwdin procedure to predict the variation of the isomer shift of Eu^{3+} and Eu^{2+} ions four and six coordinated to O^{2-} and F^- ions as a function of Eu-O and Eu-F bond distances.

3. NUMERICAL PROCEDURE

Free-Ion Electron Densities in the Non-Relativistic Hartree-Fock Approximation

In the non-relativistic approximation, the radial function

P for an electron in an orbital (n, ℓ) is written for small r, measured in atomic units, as (3)

$$P = Ar^{\ell+1}(1 - Zr/(\ell+1) + \alpha r^2 - \beta r^3 + \ldots) \qquad (10)$$

where the nucleus is approximated as a point charge, Ze, and the coefficients A, α and β are available from Mann's non-relativistic Hartree-Fock routine (3). The electron density near the origin is

$$\rho(r) = \omega P^2/(4\pi r^2) \qquad (11)$$

where ω is the occupation number of the (n, ℓ) orbital. In the non-relativistic approximation, only s electrons have non-zero electron density at the origin. Ignoring the spatial variation of the electron density over the nuclear volume, the electron density may be written as

$$|\Psi(0)|^2 = \sum_{ns} \omega_{ns} A_{ns}^2/(4\pi) \qquad (12)$$

by keeping only the leading term in the series expansion for the electron density near the origin as given by equations 10 and 11. Inclusion of first order terms in r shows that for a point nucleus the radial dependence of the electron density is of the form $|\Psi(0)|^2(1-2Zr)$. This gives the electron density at the nuclear surface as $|\Psi(0)|^2(1-2ZR)$. The average electron density within the nucleus is thus approximately $|\Psi(0)|^2 \cdot (1-ZR)$. Because of the radial dependence, the isomer shift contribution from each s orbital has to be corrected by a factor (1-ZR). Since this correction factor for the radial dependence of the non-relativistic electron density is the same for all s orbitals of all Eu ions, it may be factored out of the total electron density in both absorber and source in equation 2 for the isomer shift. For the [151]Eu nuclear radius of about 1.2×10^{-4} atomic units the factor ZR is of the order of 10^{-2}. Thus the error in the isomer shift arising from the neglect of the radial dependence of the electron density in the non-relativistic limit is about 1 % and correction for the radial dependence of the electron density at the nucleus is not necessary in this case. The non-relativistic Hartree-Fock electron densities at the nuclear origin of Eu ions, calculated from Mann's starting coefficients and equation 11, are tabulated in Table 1. For comparison non-relativistic electron densities calculated using the Fermi-Segrè-Goudsmit formula (27) along with Burns' shielding rules (28) are given in Table 2.

Free-Ion Electron Densities in the Relativistic Dirac-Fock Approximation

In the Dirac-Fock approximation, the major component of the radial function, P, and the minor component, Q, may be written for small r as (13,29)

$$P = PA \; r^{\ell+1} \; (1 - \alpha r^2 + \beta r^4 - \ldots) \qquad (13)$$

$$Q = PA \; r^{\ell} \; (q_0 + q_1 r^2 + q_2 r^4 + \ldots) \qquad (14)$$

where q_0 is zero for $s_{1/2}$, $p_{3/2}$, $d_{5/2}$ and $f_{7/2}$ electrons. The starting coefficients, PA, q_0, q_1, q_2, α and β are available from Mann's Dirac-Fock routine (13). The electron density at a distance r from the nuclear origin is

$$\rho(r) = (P^2 + Q^2)\omega/(4\pi r^2) \qquad (15)$$

In the relativistic limit, $s_{1/2}$ and $p_{1/2}$ electrons have non-zero electron densities at the nuclear origin. The total electron density at the nucleus, assuming it to be constant

Table 1. Non-relativistic Hartree-Fock electron densities of
          of $Eu^0(4f^76s^2)$, $Eu^{2+}(4f^7)$, and $Eu^{3+}(4f^6)$. From
          Mann (3).

ORBITAL	Eu^0		Eu^{2+}		Eu^{3+}	
	a_0^{-3}	10^{26} CM.$^{-3}$	a_0^{-3}	10^{26} CM.$^{-3}$	a_0^{-3}	10^{26} CM.$^{-3}$
1s	78222.105	5278.819	78222.105	5278.819	78221.726	5278.793
2s	8445.0959	569.9173	8445.0803	569.9162	8444.3389	569.8662
3s	1787.8871	120.6556	1787.8394	120.6523	1787.7058	120.6433
4s	406.54732	27.43585	406.42583	27.42765	408.92115	27.59604
5s	58.831618	3.970257	58.681120	3.960095	63.627360	4.293892
6s	3.7989726	0.256374	-	-	-	-
Total s	177848.53	12002.11	177840.26	12001.55	177852.64	12002.39

Table 2. Non-relativistic electron densities of Eu ions,
          using Burns' shielding rules and the Fermi-Segrè-
          Goudsmit formula. $Eu^{1+}(4f^76s)$, other configura-
          tions same as in Table 1.

ORBITAL	Eu^0		Eu^{1+}		Eu^{2+}		Eu^{3+}	
	a_0^{-3}	10^{26} CM.$^{-3}$	a_0^{-3}	10^{26} CM.$^{-3}$	a_0^{-3}	10^{26} CM.$^{-3}$	a_0^{-3}	10^{26} CM.$^{-3}$
1s	76588.7	5168.59	76588.7	5168.59	76588.7	5168.59	76588.7	5168.59
2s	8115.6	547.68	8115.6	547.68	8115.6	547.68	8115.6	547.68
3s	1633.7	110.25	1633.7	110.25	1633.7	110.25	1633.7	110.25
4s	351.6	23.73	351.6	23.73	351.6	23.73	355.8	24.01
5s	53.2	3.59	53.8	3.63	54.4	3.67	56.8	3.83
6s	5.0	0.34	6.2	0.42	-	-	-	-
Total s	173495.7	11708.36	173498.1	11708.52	173488.0	11707.84	173501.1	11708.72

over the nuclear volume, is

$$|\Psi(0)|^2 = \sum_{ns_{1/2}} \omega_{ns_{1/2}} PA^2_{ns_{1/2}} /(4\pi) \qquad (16)$$

for $s_{1/2}$ electrons, and

$$|\Psi(0)|^2 = \sum_{np_{1/2}} \omega_{np_{1/2}} (PA_{np_{1/2}} q_{0,np_{1/2}})^2 /(4\pi) \qquad (17)$$

for $p_{1/2}$ electrons. The electron density including radial dependence to order r^2 is given by $|\Psi(0)|^2[1-(2\alpha-q_1)r^2]$ for $s_{1/2}$ electrons and by $|\Psi(0)|^2[1+(2q_1/q_2+1/q_0^2)r^2]$ for $p_{1/2}$ electrons. Mann's Dirac-Fock calculations for the four Eu ions, Eu^{+0}, Eu^{+1}, Eu^{+2} and Eu^{+3}, show that the coefficients α and q_1 are very nearly the same for all $s_{1/2}$ electrons and that the coefficients q_0, q_1 and q_2 are very nearly the same for all $p_{1/2}$ electrons. These coefficients are also found to be very nearly the same for all four Eu ions. According to Mann (13), for all four Eu ions: $\alpha = 6.3 \times 10^6$ and $q_1 = 2.0 \times 10^3$ for all $s_{1/2}$ orbitals; $q_0 = -4.8 \times 10^{-4}$, $q_1 = 3.0 \times 10^3$ and $q_2 = -7.2 \times 10^{10}$ for all $p_{1/2}$ orbitals, all coefficients being given in atomic units. Over the ^{151}Eu nuclear radius of about 1.2×10^{-4} atomic units the correction factor $(1 - 3a_1R^2/7)$ in the expression for the isomer shift (formula 2) is about 0.92 for $s_{1/2}$ electrons and about 1.02 for $p_{1/2}$ electrons. We thus see that the radial dependence of the electron density is more pronounced in the relativistic limit than in the non-relativistic limit, and we are no longer justified in the relativistic limit to ignore its effect on the isomer shift.

Electron densities, uncorrected for the effect of their radial dependence on the isomer shift and calculated from

Mann's starting coefficients using equations 16 and 17, are
set in Table 3. Table 4 gives two sets of ratios of relativistic to non-relativistic electron densities, uncorrected for
their radial dependence, for the $ns_{1/2}$ orbitals of Eu ions
in different electronic configurations. The first set is
from Wilson's non-relativistic H-F calculation (14) and
Coulthard's relativistic H-F calculation (15). The second
set is from Mann's non-relativistic H-F and relativistic
Dirac-Fock calculations (3,13). We see from Table 4 that
Mann's density ratios are about 5% higher than the Wilson-
Coulthard ratios. Theoretically we expect the relativistic
correction, $S'(Z)$, to be 3.51 (1) for all ns orbitals on the
basis of the Racah-Breit-Rosenthal-Bodmer theories. We see
from Table 4, however, that numerical H-F and Dirac-Fock calculations in the relativistic approximation suggest that the
relativity enhancement is different for different ns orbitals
of the same electronic configuration. The average ratio,
$S'_s(Z)$ (total relativistic s-density/total non-relativistic
density) and the average ratio, $S'_{sp}(Z)$ (total relativistic
s,p density/total non-relativistic density) are both seen to
remain nearly constant for different ionic configurations.
Close examination of this data shows, however, that caution is
advised in the recovery of relativistic densities from the non-
relativistic electron densities. The Coulthard-Wilson $S'_s(Z)$
and $S'_{sp}(Z)$ constants are lower than the theoretical value by
about 10% while the Mann constants are lower by only about 6%.
We take this difference as due to the error in the relativistic
Hartree-Fock density arising from approximating the nucleus
with a point charge.

Table 5 contains the relativistic densities corrected
for their radial dependence over the nuclear volume. The

Table 3. Relativistic Dirac-Fock electron densities for Eu ions. From Mann (3), configurations same as in Tables 1 and 2.

ORBITAL	$Eu^{.0}$ a_0^{-3}	$Eu^{.0}$ $\frac{10^{26}}{CM.^{-3}}$	Eu^{1+} a_0^{-3}	Eu^{1+} $\frac{10^{26}}{CM.^{-3}}$	Eu^{2+} a_0^{-3}	Eu^{2+} $\frac{10^{26}}{CM.^{-3}}$	Eu^{3+} a_0^{-3}	Eu^{3+} $\frac{10^{26}}{CM.^{-3}}$
1s	249316.32	16825.11	249316.35	16825.11	249316.35	16825.11	249315.25	16825.04
2s	32476.223	2191.658	32476.197	2191.656	32476.160	2191.654	32473.808	2191.495
3s	6955.3897	469.3845	6955.3276	469.3803	6955.2159	469.3727	6954.4365	469.3201
4s	1591.1093	107.3760	1590.9180	107.3631	1590.5634	107.3392	1597.5621	107.8115
5s	240.79646	16.25015	240.30629	16.21707	239.90977	16.19031	256.35143	17.29988
6s	15.785419	1.065279	21.576234	1.456072	-	-	-	-
2p	1325.8536	89.47523	1325.8565	89.47542	1325.8625	89.47583	1325.7423	89.46772
3p	313.82071	21.17819	313.82236	21.17830	313.82588	21.17854	313.79971	21.17677
4p	69.396030	4.683191	69.395744	4.683172	69.395627	4.683164	69.777112	4.708908
5p	8.8841931	0.599550	8.9121345	0.601435	8.9662693	0.605089	9.8236536	0.662949
Total s	581191.25	39221.69	581201.34	39222.37	581156.39	39219.34	581194.81	39221.93
Total p	3435.9091	231.8723	3435.9734	231.8767	3436.1006	231.8852	3438.2856	232.0327
Total s, p	584627.16	39453.56	584615.74	39452.79	584592.49	39451.22	584633.10	39453.96

Table 4. Numerical relativity factors for Eu. Upper set from Coulthard (15) and Wilson (14), lower set from Mann (3,13).

ION	CONFIGURATION	$S_s'(Z)$ 1s	2s	3s	4s	5s	6s	Average $S_s'(Z)$	Average $S_{sp}'(Z)$
Eu^0	[Xe] $4f^7\,6s^2$	3.04848	3.67760	3.71929	3.73529	3.89949	3.95870	3.12546	3.14395
	$4f^7\,6s\,6p$	3.04848	3.67760	3.71930	3.73532	3.90099	3.91419	3.12544	3.14393
	$4f^7\,5d\,6s$	3.04848	3.67760	3.71930	3.73561	3.91534	4.12958	3.12544	3.14393
	$4f^7\,5d\,6p$	3.04848	3.67759	3.71926	3.73561	3.91574	-	3.12542	3.14392
	$4f^6\,5d\,6s^2$	3.04848	3.67767	3.71911	3.72794	3.85319	3.99570	3.12545	3.14394
Eu^{1+}	[Xe] $4f^7\,6s$	3.04848	3.67761	3.71928	3.73528	3.90053	3.87674	3.12544	3.14393
	$4f^7\,5d$	3.04849	3.67760	3.71929	3.73564	3.91409	-	3.12543	3.14392
	$4f^6\,5d\,6s$	3.04848	3.67768	3.71913	3.72796	3.85460	3.88438	3.12543	3.14393
Eu^{2+}	[Xe] $4f^7$	3.04848	3.67760	3.71931	3.73503	3.89460	-	3.12542	3.14391
	$4f^6\,5d$	3.04849	3.67768	3.71911	3.72781	3.85021	-	3.12540	3.14390
	$4f^6\,6s$	3.04849	3.67770	3.71913	3.72732	3.83810	3.76667	3.12544	3.14394
Eu^{3+}	[Xe] $4f^6$	3.04848	3.67767	3.71913	3.72722	3.83075	-	3.12539	3.14390
	AVERAGE	3.04848	3.67763	3.71922	3.73217	3.88064	3.96014	3.12543	3.14392
Eu^0	[Xe] $4f^7\,6s^2$	3.18729	3.84557	3.89029	3.91371	4.09297	4.15518	3.26790	3.28722
Eu^{2+}	[Xe] $4f^7$	3.18729	3.84557	3.89029	3.91354	4.08836	-	3.26786	3.28718
Eu^{3+}	[Xe] $4f^6$	3.18729	3.84563	3.89015	3.90677	4.02895	-	3.26784	3.28718
	AVERAGE	3.18729	3.84559	3.89024	3.91134	4.07009	-	3.26787	3.28719

Table 5.   Corrected relativistic electron densities of Eu
           ions, from Mann (13). Configurations same as in
           Tables 1 and 2.

ORBITAL	Eu^0		Eu^{1+}		Eu^{2+}		Eu^{3+}	
	a_0^{-3}	10^{26} $CM.^{-3}$	a_0^{-3}	10^{26} $CM.^{-3}$	a_0^{-3}	10^{26} $CM.^{-3}$	a_0^{-3}	10^{26} $CM.^{-3}$
1s	229542.59	15490.68	229542.62	15490.68	229542.62	15490.68	229541.61	15490.62
2s	29891.429	2017.223	29891.405	2017.221	29891.372	2017.219	29889.206	2017.073
3s	6401.5108	432.0060	6401.4536	432.0021	6401.3510	431.9952	6400.6335	431.9468
4s	1464.3890	98.82429	1464.2130	98.81241	1463.8866	98.79039	1470.3279	99.22508
5s	221.61825	14.95591	221.16712	14.92546	220.80218	14.90084	235.93434	15.92203
6s	14.528187	0.980435	19.857792	1.340103	-	-	-	-
2p	1362.6314	91.95718	1362.6343	91.95737	1362.6405	91.95779	1362.5169	91.94946
3p	322.53013	21.76595	322.53183	21.76606	322.53545	21.76630	322.50855	21.76449
4p	71.322181	4.813177	71.321887	4.813158	71.321766	4.813149	71.713839	4.839608
5p	9.1307873	0.616191	9.1595045	0.618129	9.2151417	0.621884	10.096324	0.681350
Total s	535072.13	36109.34	535081.43	36109.97	535040.06	36107.18	535075.41	36109.56
Total p	3531.2289	238.3050	3531.2950	238.3094	3531.4257	238.3183	3533.6713	238.4698
Total s,p	538603.36	36347.65	538592.86	36346.94	538571.48	36345.50	538609.09	36348.03

correction factors $(1-3a_iR^2/7)$, as obtained from Mann's
Dirac-Fock calculations, averaged 0.920647 for $s_{1/2}$ and
1.027742 for $p_{1/2}$ orbitals. Only one set of factors is
necessary for all four Eu ions: Eu^{+0}, Eu^{+1}, Eu^{+2} and Eu^{+3},
since, as we observed earlier, the correction factors depend
primarily on the nuclear charge Z and show only very slight
variation with changes in electronic configuration. We see
from Tables 4 and 5 that the difference in the total electronic density at the nucleus between Eu^{3+} and Eu^{2+} ions is
reduced from 2.74 x 10^{26} cm^{-3} to 2.54 x 10^{26} cm^{-3} due to the
effect of the variation of the charge density over the
nuclear volume.

Calculation of Overlap Integrals

Non-relativistic Hartree-Fock radial functions for $n\ell$

orbitals of Eu ions are available in digital form from Mann's program (3). Watson (34) has given analytical radial functions of the O^{2-} ion in the H-F approximation and Clementi (35) has given similar analytical radial functions for the F^- ion. The overlap between a metal ion orbital, $\Psi_{n\ell m}(a)$, and a ligand ion orbital, $\Psi_{n'\ell'm'}(b)$, is defined as,

$$S(a;b) = \int_0^{2\pi}\int_0^{\pi}\int_0^{\infty} R_{n\ell}(r_a)\, Y_m^{\ell *}(\theta_a,\phi_a)\, R_{n'\ell'}(r_b)$$

$$Y_{m'}^{\ell'}(\theta_b,\phi_b)\, \sin\theta_a\, dr_a\, d\theta_a\, d\phi_a \qquad (18)$$

where

$$\Psi_{n\ell m}(a) = R_{n\ell}(r_a)\, Y_m^{\ell}(\theta_a,\phi_a) \qquad (19)$$

and

$$\Psi_{n'\ell'm'}(b) = R_{n'\ell'}(r_b)\, Y_{m'}^{\ell'}(\theta_b,\phi_b). \qquad (20)$$

From the adjacent figure it can be seen that:

$$r_b = (r_{ab}^2 + r_a^2 - 2r_a r_{ab}\cos\theta_a)^{1/2} \qquad (21)$$

$$\cos\theta_b = (r_{ab} - r_a\cos\theta_a)/(r_{ab}^2 + r_a^2 - 2r_{ab}r_a\cos\theta_a)^{1/2} \qquad (22)$$

$$\sin\theta_b = r_a\sin\theta_a/(r_{ab}^2 + r_a^2 - 2r_{ab}r_a\cos\theta_a)^{1/2} \qquad (23)$$

where r_{ab} is the inter-nuclear separation. Since radial wave functions are usually given in the form $f(r) = rR_{n\ell}(r)$, the overlap integrals take the form shown below for $S(ns_a, n's_b)$:

$$S(ns_a;n's_b) = \frac{1}{2}\int_0^{\pi}\int_0^{\infty} \frac{f_s(r_a)f_s(r_b)\sin\theta_a\, r_a}{(r_{ab}^2+r_a^2-2r_{ab}r_a\cos\theta_a)^{1/2}}\, dr_a\, d\theta_a, \qquad (24)$$

where r_b is given by equation 21.

Mann's radial functions are available only digitally, with the distribution of radial points according to the system employed by Herman and Skillmann (36). This system sets up a radial mesh consisting of blocks of 41 points each, with 40 equal increments in r within each block. The increment is doubled in each succeeding block. Although the radial variable could have been expressed in Bohr units directly, Mann introduces, for consistency with Herman and Skillmann, a change in variable to x = r/U. The function U comes from the Thomas-Fermi theory and is equal to $(9\pi^2/128Z)^{1/3}$, which is 0.88534138 for hydrogen and 0.22250029 for europium.

In this work overlap integrals were calculated using Simpson's rule. For each radial point r_a, starting from the origin and moving outward, the integration over θ_a was performed using 100 equally spaced increments, calculating r_b according to equation 21 and evaluating $f_b(r_b)$ from the analytical expression for it. When the b ion wave-functions were known only digitally, $f_b(r_b)$ was evaluated by Lagrange interpolation using 5 points. The integration routine, when checked with Slater type orbitals, reproduced Mulliken's analytical results (37) within one part in 10^5. As a further check we have calculated overlaps for $Fe^{2+}-F^-$ and $Fe^{3+}-O^{2-}$ combinations using Clementi's (35) analytical wave functions for Fe^{2+} and Fe^{3+}. The results of these calculations agree within a few percent with the calculations of Simanek and Sroubek (26) and of Simanek and Wong (38) using Watson's (39) analytical wave-functions for Fe^{2+} and Fe^{3+}. The slight differences between our overlaps and those of Simanek and collaborators are to be expected due to the different wave-

functions used. The comparison thus serves to show that the
differences in overlaps using Hartree-Fock wavefunctions obtained by different numerical methods is small. Each integration took about 1 min. to evaluate on a PDP-10 computer
for each internuclear distance. The total overlap corrected
electron density for Eu-O or Eu-F combinations thus took
about 5 min. of CPU time for each internuclear distance.

4. STRUCTURES OF EUROPIUM COMPOUNDS

Structural characteristics of the five compounds of
europium; EuO, Eu_2O_3, Eu_3O_4, EuF_2 and EuF_3, for which overlap
corrected electron densities were calculated are tabulated in
Table 6. Both of the divalent compounds, EuO and EuF_2, are
cubic. Of the trivalent compounds only Eu_2O_3 is cubic. The
trivalent Eu ions in both Eu_2O_3 and Eu_3O_4 ($Eu_2O_3 \cdot EuO$) are
found in two crystallographically inequivalent positions and
Table 6 shows both coordinations in each of these compounds.
The divalent Eu ion in Eu_3O_4 is 9-coordinated with nearly
axial symmetry. The trivalent Eu ion in EuF_3 is also 9-coor-
dinated.

5. RESULTS AND DISCUSSION

Table 7 shows the overlap corrections to the free-ion
electron densities at the sites of ^{151}Eu nuclei in the compounds EuO, Eu_2O_3, Eu_3O_4, EuF_2 and EuF_3. For each overlap
n was summed from 2 to 5. Hybridization of the valence
electrons was not included. Table 7 shows that the total
correction due to direct overlaps involving the $Eu^{m+}(np)$
orbitals is, in addition to being small (10% of the total
correction), roughly equal for Eu^{2+} and Eu^{3+} in the compounds

Table 6.   Structural characteristics of Eu Compounds.

COMPOUND	STRUCTURE	COORDINATION OF Eu ION	BOND DISTANCES, Å	REFERENCE
EuO	Rock Salt		$Eu^{2+}- O =$ 2.572	Sinha (30)
Eu_2O_3	Bixbyite		$Eu^{3+}- O =$ 2.329	Pauling and Shappell (31)
Eu_3O_4	Eu_2SrO_4		$Eu^{3+}(1)- O = 2.236,$ 2.301, 2.386, 2.430 $Eu^{3+}(2)- O = 2.295,$ 2.297, 2.345, 2.388 $Eu^{2+}- O = 2.638,$ 2.720, 2.722, 2.955	Rau (32)
EuF_2	Fluorite		$Eu^{2+}- F^- =$ 2.519	Sinha (30)
EuF_3	Orthorhombic		$Eu^{3+}- F =$ 2.274, 2.391, 2 614 2.316, 2.407, 2.350	Zalkin and Templeton (33)

Table 7.   Overlap corrections.

	OVERLAP CORRECTION TO ELECTRON DENSITY, 10^{26} CM^{-3}						
OVERLAP	EuO	Eu_2O_3	$Eu(1)_3O_4$	$Eu(2)_3O_4$	$Eu(3)_3O_4$	EuF_2	EuF_3
$Eu^{m+}(ns) ; X(2p)$	1.02	1.67	1.67	1.62	0.99	0.75	1.15
$Eu^{m+}(ns) ; X(2s)$	0.08	0.19	0.20	0.27	0.06	0.06	0.12
$Eu^{m+}(np) ; X(2p)$	0.08	0.12	0.10	0.11	0.08	0.07	0.10
$Eu^{m+}(np) ; X(2s)$	0.02	0.04	0.04	0.05	0.02	0.02	0.03
TOTAL OVERLAP CORRECTION	1.20	2.02	2.01	2.05	1.15	0.90	1.40

studied. Thus the main objective of our calculations,
namely to obtain the overlap corrected electron density dif-
ference $\{|\Psi(0)|^2_{Eu^{3+}} - |\Psi(0)|^2_{Eu^{2+}}\}$ when the ions are in solids,
appears to depend only slightly on $Eu^{m+}(np)$; $X(2p)$ and
$Eu^{m+}(np)$; $X(2s)$ overlaps. Table 8 shows this electron density
difference for selected pairs of compounds along with their
observed isomer shift differences (40,41,20). The total
electron density difference was obtained by adding the overlap correction difference between Eu^{3+} and Eu^{2+} in the two
compounds to the free-ion electron density difference of
2.54×10^{26} cm^{-3} obtained from Table 5. The calibration
constant corresponding to the electron density differences
in Table 8 was then evaluated using equation 2, which can be
rewritten as

$$I.S. = 0.130(\Delta R/R \times 10^4)[\Delta|\Psi(0)|^2 \times 10^{-26}] cm/sec. \quad (25)$$

We see from Table 8 that the calibration constant obtained
from the oxides is significantly lower than that obtained from
the fluorides. This corresponds to the overlap correction in
the oxides being more than in the fluorides. The higher overlap corrections in the oxides may be partly due to errors in
the O^{2-} wavefunctions used. Watson (34), noting that the free
O^{2-} ion is unstable, has obtained the wavefunctions of the O^{2-}
ion in the solid state by assuming the free ion to be stabilized by a potential field from its neighbors. This external
stabilizing potential may have the effect of giving the O^{2-}
radial functions longer tails than in the actual solid, which
in turn would tend to overemphasize the overlap correction.
According to this argument the calibration constant derived
from the fluorides may be more reliable. However, the difference between the values of the calibration constant

Table 8.   The calibration constant, $\Delta R/R$.

PAIR OF TRIVALENT AND DIVALENT COMPOUNDS	TOTAL OVERLAP CORRECTED DENSITY DIFFERENCE $\lvert\psi(0)\rvert^2_{Eu^{3+}} - \lvert\psi(0)\rvert^2_{Eu^{2+}}$ 10^{26} CM.$^{-3}$	EXPERIMENTAL ISOMER SHIFT DIFFERENCE, I.S.(Eu^{3+}) - I.S.(Eu^{2+}) CM. /SEC.	CALIBRATION CONSTANT, $\frac{\Delta R}{R}$ X 10^4
Eu_2O_3, EuO	3. 36	1. 21	2. 77
$Eu(1)_3O_4$, $Eu(3)_3O_4$	3. 40	1. 31	2. 96
$Eu(2)_3O_4$, $Eu(3)_3O_4$	3. 44	1. 31	2. 93
EuF_3, EuF_2	3. 04	1. 36	3. 44
AVERAGE CALIBRATION CONSTANT = 3. 2 X 10^{-4} ± 0. 7 X 10^{-4}			

obtained from the oxides and the fluorides is also attribut-
able to the effects of hybridization (i.e., the mechanism of
potential distortion considered by Walch and Ellis (42)).
While the directional characteristics of bonding in EuO can
be accounted for by overlap of Eu p orbitals with oxygen
orbitals, in general hybridizations of the europium orbitals,
involving combinations of 5s, 6s, 5p, 6p, 5d and 4f orbitals,
must be considered to account for directional bonding. Thus,
the calibration constants derived from the pair of oxides,
EuO and Eu_2O_3, as well as from Eu_3O_4 contain errors due to
both the uncertainty of the 0^{2-} wavefunctions and the neglect
of hybridization. The calibration constant derived for the
pair of fluorides, EuF_2 and EuF_3, can be expected to be more
accurate because of the greater precision of the F⁻ wave-
functions and the decreased importance of hybridization due
to the higher ionicity of the fluorides. The largest difference among the calibration constants in Table 8, namely
the difference 3.44 - 2.77 between the (EuO, Eu_2O_3) value and

the (EuF_2, EuF_3) value, is thus taken as an estimate of the
error in the calibration constant calculated from considerations of overlap distortion alone. This difference accounts
for 20% of the mean of the fluoride and average oxide values,
and we give our value of the isomer shift calibration constant
of ^{151}Eu as 3.2 x 10^{-4} ± 0.7 x 10^{-4}.

Using 3.2 x 10^{-4} for $\Delta R/R$, we have calculated the isomer
shift with respect to Eu_2O_3 for Eu^{2+} and Eu^{3+} ions four, six
and twelve coordinated to O^{2-} and F^- ions. The results are
illustrated in Figures 1 and 2. An interesting feature of
this calculation is the prediction of isomer shifts with
respect to Eu_2O_3 of -8.3 mm/sec for the free Eu^{3+} ($4f^6$) ion
and -18.7 mm/sec for the free Eu^{2+} ($4f^7$) ion. Observed
isomer shifts in Eu^{3+} compounds vary between -1.0 and 0.3
mm/sec (20), consistent with the small range of Eu^{3+}-O^{2-} bond
distances observed in the Eu^{3+} oxides. The smallest value of
the isomer shift observed for the Eu^{2+} ion ($\sim$ -14.5 mm/sec)
is closer to the free ion value of -18.7 mm/sec than the
smallest Eu^{3+} shift is to the free Eu^{3+} value of - 8.3 mm/sec.
This suggests that Eu^{2+} tends to be more ionic than Eu^{3+} in
its bonding character with oxygen and fluorine.

As a check on our calculations we have located the
isomer shifts of Eu^{2+} in EuO, Eu_3O_4, $EuTiO_3$ and $EuZrO_3$ in
Figure 1(a). The average bond distances in $EuTiO_3$ and $EuZrO_3$
are as calculated by Berkooz (43). The discrepancy between
I.S. for EuO and the curve for Eu^{2+} six-coordinated to O^{2-}
reflects our averaging of the calibration constants in Table
8 but is well within the 20% error previously estimated. The
isomer shift of Eu^{2+} nine-coordinated with O^{2-} in Eu_3O_4 is
within 10% of the expected value. The isomer shifts of

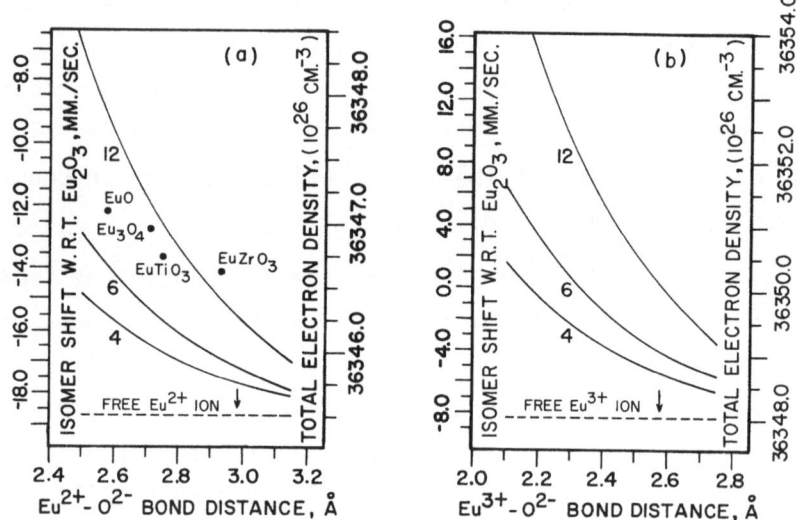

Figure 1.   Isomer shifts in europium oxides as a function of
            bond distance and coordination number for Eu^{2+},
            (a), and Eu^{3+}, (b).

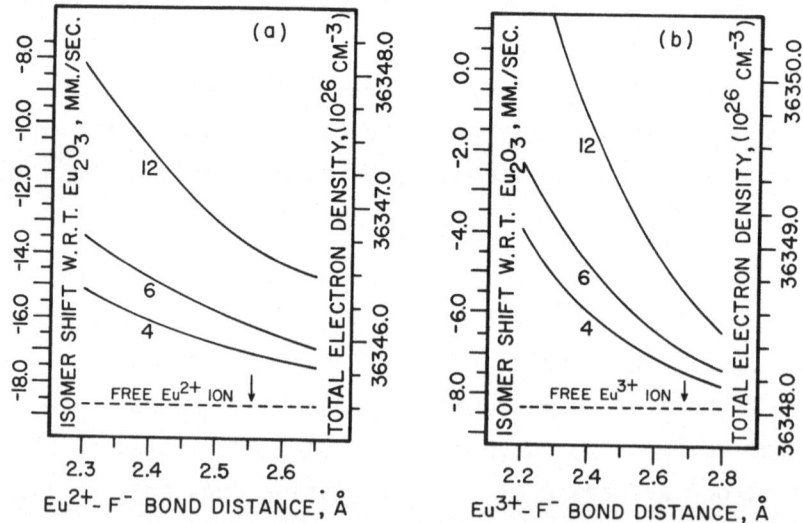

Figure 2.   Isomer shifts of europium fluorides as a function
            of bond distance and coordination number for
            Eu^{2+}, (a), and Eu^{3+}, (b).

EuTiO$_3$ and EuZrO$_3$ were not used in the calibration procedure and may be used to check our value of the calibration constant. The Eu^{2+} ions in both these perovskites are twelve-coordinated to oxygen ions. The isomer shifts of both compounds lie close to the twelve coordinated curve, the discrepancy between theory and experiment being of the order of 10%. This agreement may thus be taken to support the validity of our calculations.

Finally, we describe a procedure for obtaining a plot similar to the WWJ plot for ^{57}Fe (44) which maps isomer shifts of different free-ion configurations. Figure 3 is a plot (right ordinate) of the relativistic electron densities at the nucleus for the Eu free-ion configurations $4f^{6+q}5d^r 6p^t6s^u$, where q, r, t and u are integers, against x, where x = q + r + t + u. In this plot the electron densities, which are uncorrected for the effect on I.S. of their radial dependence over the nuclear volume, are obtained in most part from Coulthard (15) although our value of the calibration constant is based on Mann's electron densities (13). The wider range of configurations considered by Coulthard is the reason for this preference. The 5% discrepancy in the absolute electron densities as calculated by Coulthard and Mann will not affect the conclusions drawn below since they depend only on electron density differences. The configurations, $4f^65d^26s$, $4f^66s^26p$, $4f^75d^2$, $4f^76p$, $4f^66s^2$, $4f^65d^2$, $4f^8$, $4f^66p$ and $4f^66s6p$, not considered by Coulthard, were obtained by multiplying Wilson's non-relativistic densities for these configurations (14) by the average relativity factor $S'_{sp}(Z)$ = 3.143924 (see Table 4). We see from Figure 3 that the free neutral atom, Eu^0, and the free Eu^{3+} ion have nearly

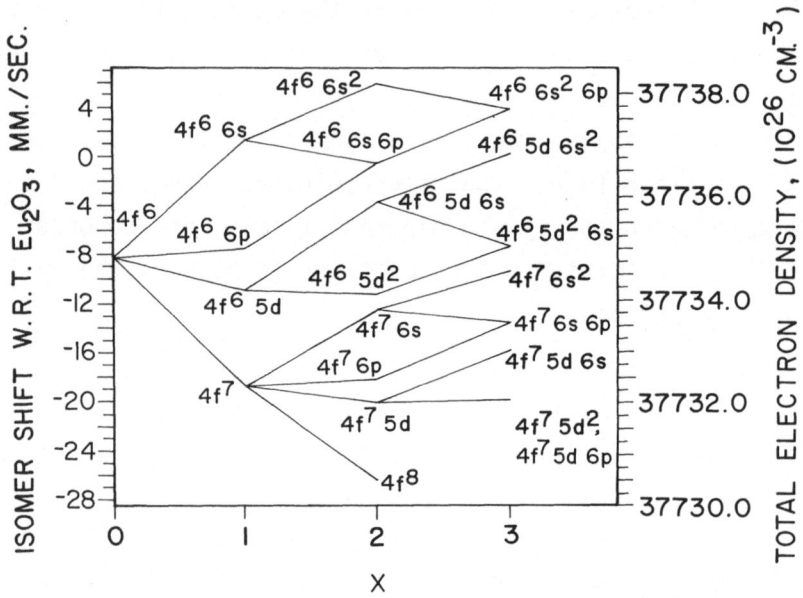

Figure 3. Map of I.S. and $|\Psi(0)|^2$ as a function of Eu free-
ion electron configuration. X is the number of
electrons added to the $[Xe]4f^6$ core.

the same electron density at the nucleus. There is experimental evidence that 5d covalency which strongly influences
the spin density at the nucleus (and thus the hyperfine field
experienced by the nucleus) is not reflected in the isomer
shift (45).

To fix the isomer shift scale in Figure 3 we assume our
calculated isomer shifts of the free Eu^{2+} and Eu^{3+} ions as
calibration points. Thus an electron density of 37732.32 x
10^{26} cm^{-3} corresponding to the free ion configuration, $4f^7$
(Eu^{2+}) is taken to have the isomer shift w.r.t. Eu_2O_3 of
-18.7 mm/sec and an electron density of 37734.80 x 10^{26} cm^{-3}
corresponding to the free-ion configuration, $4f^6$ (Eu^{3+}) is

taken to have the isomer shift of -8.3 mm/sec. This isomer
shift scale is shown on the left ordinate of Figure 3.

The variation of the isomer shift due to covalency in
Eu compounds may now be discussed in terms of _effective_ free-
ion configurations. Table 9 presents effective free-ion configurations for Eu ions in the solid state assigned by requiring that the calculated isomer shift of the free-ion
configuration be the same as the measured isomer shift of the
Eu ion in the solid and that the charge on the Eu ion be
reasonable. The physical meaning of the _effective_ configurations is difficult to assess because the clearly important
effect of ligand overlap is completely ignored. Nevertheless,
Table 9 strongly suggests that 6s electrons play an important
role in Eu bonding. In this light, it is interesting to note
that Sm^{1+}, produced in a KCl matrix, takes a $4f^6 6s^1$ rather
than a $4f^7$ configuration (46). For Eu metal Shirley (1) used
the proportionality between isomer shift and the hyperfine
field at 4.2K in Eu compounds to estimate the effective
electron configuration to be $4f^7 6s^1$. This result is based
on a value of 10 mm/sec for the change in the isomer shift
due to a single 6s electron outside the $4f^7$ core. Figure 3
gives approximately 5 mm/sec for this value and thus recon-
ciles the effective configuration $4f^7 6s^2$, shown for Eu metal
(I.S. = -8.2 mm/sec) in Table 9, with Shirley's results.
Pressing the analysis one step further leads to the association of the variation from -11.4 mm/sec to -8.8 mm/sec for the
the I.S. of Eu intermetallic compounds (45) with a change in
effective electron configuration from $4f^7 6s^1$ to $4f^7 6s^2$. This
result predicts that, in EuM_2, 6s electrons are increasingly
more effectively localized on Eu when M takes the order

Table 9. Effective free-ion configurations for Eu in solids.

ION	RANGE OF ISOMER SHIFT, MM./SEC. W.R.T. Eu_2O_3	EFFECTIVE FREE-ION CONFIGURATION
Eu^{3+}	-1.0 to 0.3	$4f^6 6s^x$, $0.7 \leq x \leq 1$
Eu^{2+}	-14.5 to -12.0	$4f^7 6s^x$, $0.7 \leq x \leq 1$
Eu^0	-11.4 to -8.2	$4f^7 6s^x$, $1 \leq x \leq 2$

Pt < Al < Pd < Zn < Cu. Except for Al, this series follows
expectations based on electronegativities of the M metals.

6. CONCLUSIONS

The calculations presented here show that ligand overlap
is an important factor in the understanding of the isomer
shift in ^{151}Eu. Correction for this effect has led to a
value of the calibration constant, $\Delta R/R$, of $3.2 \pm 0.7 \times 10^{-4}$
and shown large differences between Eu^{3+} and Eu^{2+} free-ion
I.S. and values observed for trivalent and divalent compounds.
Calculations based on the new value for $\Delta R/R$ show significant
variation in Eu I.S. with coordination number and bond distance in the oxides and fluorides and suggest 6s contributions
to covalent bonds with europium.

ACKNOWLEDGMENTS

Much of this work was carried out while the authors were
members of the Department of Engineering and Applied Science,
Yale University. Acknowledgment is made to the donors of the
Petroleum Research Fund, administered by the American Chemical
Society, and to Yale University for support of this work. We

thank B. D. Dunlap for helpful comments on the manuscript and are indebted to J. B. Mann for providing us with atomic structure calculations that have made this research possible.

REFERENCES

1. D. A. Shirley, Rev. Mod. Phys., <u>36</u>, 339 (1964).
2. G. K. Shenoy and G. M. Kalvius, Hyperfine Interactions in Excited Nuclei, Vol. 4, G. Goldring and R. Kalish (eds.), Gordon and Breach, New York (1971).
3. J. B. Mann, Los Alamos Scientific Laboratory Report 1-3691, Atomic Structure Calculations II, Unpublished.
4. G. Racah, Nature, <u>129</u>, 723 (1932).
5. J. E. Rosenthal and G. Breit, Phys. Rev. <u>41</u>, 459 (1932).
6. A. R. Bodmer, Proc. Roy. Soc. (Lon.), <u>A66</u>, 1041 (1953).
7. G. M. Kalvius, Hyperfine Interactions in Excited Nuclei, Vol. 2, G. Goldring and R. Kalish (eds.), Gordon and Breach, New York (1971).
8. D. W. Hafmeister, J. Chem. Phys., <u>46</u>, 5 (1967).
9. B. D. Dunlap, G. K. Shenoy, G. M. Kalvius, D. Cohen and J. B. Mann, Hyperfine Interactions in Excited Nuclei, Vol. 2, G. Goldring and R. Kalish (eds.), Gordon and Breach, New York (1971).
10. D. Shirley, Ann. Rev. Phys.Chem., <u>20</u>, 25 (1969).
11. P. O. Löwdin, J. Chem. Phys., <u>18</u>, 365 (1950).
12. W. H. Flygare and D. W. Hafmeister, J. Chem. Phys., <u>43</u>, 789 (1965).
13. J. B. Mann, J. Chem. Phys., <u>51</u>, 841 (1969) and private communication.
14. M. Wilson, J. Phys. B: Atom. Molec. Phys., <u>5</u>, 218 (1972).
15. M. A. Coulthard, J. Phys. B: Atom. Molec. Phys. <u>6</u>, 23 (1973).
16. J. Rainwater, Phys. Rev., <u>79</u>, 432 (1951).
17. M. G. Mayer and N. J. Jensen, Elementary Theory of Nuclear Shell Structure, Wiley, New York (1955).
18. E. A. Samuel, Ph.D. Thesis, Yale University (unpublished).
19. N. Bohr and J. A. Wheeler, Phys. Rev., <u>56</u>, 423 (1939).
20. G. Gerth, P. Kienle and K. Luchner, Phys. Letters, <u>27A</u>, 5 (1967).
21. S. Hüfner, P. Kienle, D. Quitman,and P. Brix, Z. Physik, <u>187</u>, 67 (1967).
22. P. Kienle, Rev. Mod. Phys., <u>36</u>, 372 (1964).
23. J. Speth, Phys. Letters, <u>28B</u>, 625 (1969).
24. J. Speth, Phys. Letters, <u>31B</u>, 513 (1970).

25. P. H. Barret and D. Shirley, Phys. Rev., 131, 123 (1963).
26. E. Simanek and Z. Sroubek, Phys. Rev., 163, 275 (1967).
27. E. Fermi and E. Segré, Z. Physik, 82, 729 (1933).
28. G. Burns, J. Chem. Phys., 41, 1521 (1964).
29. M. O. Faltons, Ph.D. Thesis, Lawrence Radiation Labor-
    atory, University of California, Report No. UCRL-18706
    (1969).
30. S. P. Sinha, Europium, Springer-Verlag, New York (1967).
31. L. Pauling and M. D. Shappell, Z. Krist., 75, 128 (1930).
32. R. C. Rau, Acta Cryst., 20, 716 (1965).
33. A. Zalkin and D. H. Templeton, Am. J. Chem., 75, 2453
    (1952).
34. R. E. Watson, Phys. Rev., 111, 1108 (1958).
35. E. Clementi, I. B. M. J. Res. Dev., 9, 2 (1965).
36. F. Herman and S. Skillman, Atomic Structure Calculations,
    Prentice-Hall, Englewood Clifs, New Jersey (1963).
37. R. S. Mulliken, C. A. Rieke, D. Orloff and H. Orloff,
    J. Che. Phys., 17, 1248 (1949).
38. E. Simanek and A. Y. C. Wong, Phys. Rev., 166, 348 (1968).
39. R. E. Watson, Solid State and Molecular Physics Group
    Technical Report No. 12, Massachusetts Institute of
    Technology, (1959), unpublished.
40. E. Stiechle, Z. Physik, 201, 331 (1967).
41. H. H. Wickman and J. Catalano, J. Appld. Phys., 39, 1428
    (1968).
42. P. F. Walch and D. E. Ellis, Phys. Rev., 7, 903 (1973).
43. O. Berkooz, J. Phys. Chem. Solids, 30, 1736 (1969).
44. L. R. Walker, G. K. Wertheim and V. Jaccarino, Phys.
    Rev., Lett., 6, 98 (1961).
45. H. H. Wickman, J. H. Wernick, R. C. Sherwood and C. F.
    Wagner, J. Phys. Chem. Solids, 29, 181 (1968).
46. F. K. Fong, J. A. Cape and E. Y. Wong, Phys. Rev., 151,
    299 (1966).

HIGH TEMPERATURE BEHAVIOR OF THE 6.2 keV MÖSSBAUER

TRANSITION OF TANTALUM-181 IN TANTALUM METAL

D. Salomon, W. Wallner and P.J. West

Institut für Atom- und Festkörperphysik

Freie Universität Berlin, D-1000 Berlin 33

The observation of the Mössbauer resonance of the 6.2 keV gamma rays of tantalum 181 in tantalum metal over the temperature range of 300 to 2400 K is reported here. This is the highest temperature range for any Mössbauer study up to now. Slight deviation of the linear dependence of the isomer shift is noted at higher temperatures. The increase and following decrease of the line-width with temperature due to absorption, diffusion and degassing of residual gases (oxygen and nitrogen) is considered.

INTRODUCTION

Earlier temperature measurements |1-3| of the 6.2 keV gamma transition, have been performed for sources of W-181 impurities diffused into various transition metal hosts over the range of 15 to 900 K. The measurements over the range of 300 to 900 K |1| showed a strong linear dependence of the isomer shift (IS), which varied from -32 (Ni host) to +8 (Pt host) times the expected high temperature limit of the second order Doppler (SOD) shift. At the low temperature range of 15 to 457 K, |2| a deviation of the IS from the linear dependence was noted for W-181(W) below 100 K, which was associated with the lattice expansion of tungsten. In order to extend these measurements to higher temperatures, the problem of activity diffusion had to be avoided. For this reason, the experiments were limited to tantalum metal absorbers heated resistivly in ultra high vacuum. Two variables, high current densities and residual gas pressure, were thus present. Since residual gas concentration in tantalum metal was shown by Sauer |4| to be inherit in the basic ex-

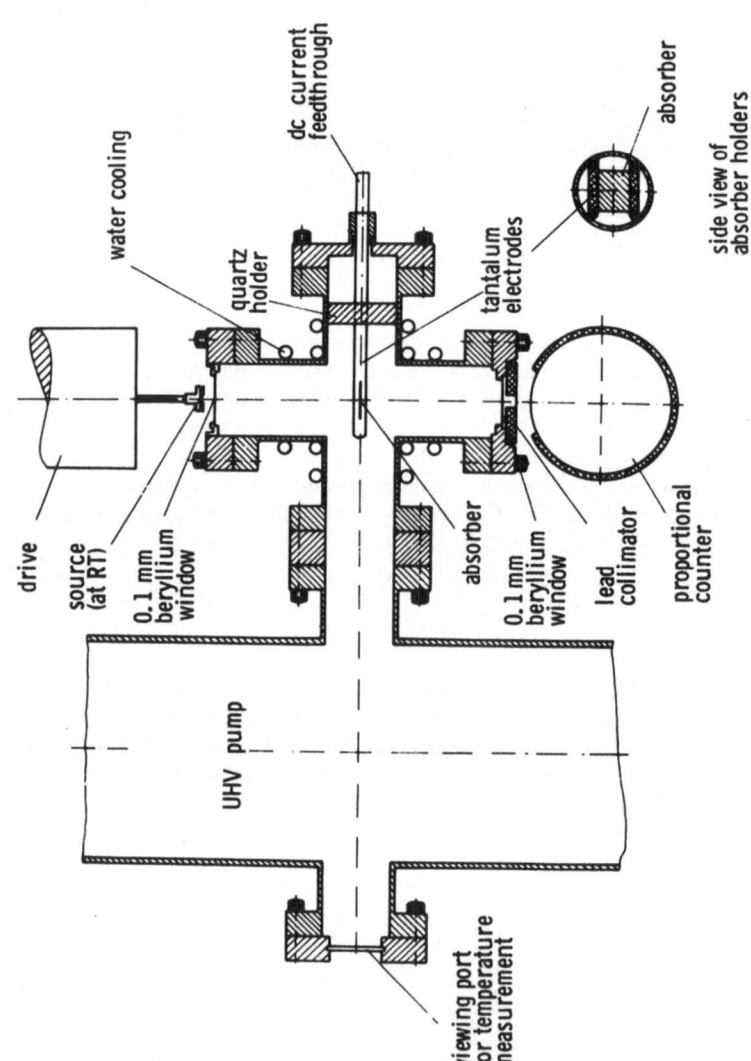

Fig. 1   Schematic drawing of the high temperature set up

perimentally broad linewidth (our best value Γ (FWHM) = 0.065(4) mm/s = 10 x Γ_o) and the heating method is similar to the usual absorber preparation, the study of sorption and desorption at low gas concentrations (0.04 at.%) as well as gas-metal lattice-dynamics was possible.

The low energy and high atomic mass, in addition to the high resolution of the tantalum resonance, should enable the observation of the Mössbauer absorption up to the melting point of tantalum metal (3269 K).

EXPERIMENTAL

Sources and absorbers were prepared in similar manner as has been described previously |3,5|. The W-181 activity was diffused onto a tungsten single crystal disc in high vacuum ($\sim 10^{-8}$ Torr) using r.f. heating at 2500°C for about 30 minutes. The Ta-foil absorbers were rolled to various thicknesses with periodic outgassing in ultra high vacuum ($\sim 10^{-9}$ Torr) from a 13μm thick foil of 99.996% purity. Final outgassing of the foils took place at temperatures of 1700°C for the 1.5μm to 2200°C for the 7μm thick foil. These temperatures were as high as possible before a breakdown occurred.

Since it is well known that tantalum metal has a large affinity to absorb gases |6| from about 200°C upwards, the metal acts, even at low pressures, as an absorption pump. Attempts to heat absorber foils indirectly have always led to a broader linewidth at room temperature, than the one obtained by resistivity heating. For this reason, the high temperature measurements can only be carried out in ultra high vacuum with direct current heating. It should be noted that the use of alternate current or an on-off cycle heating at 100 Hz, causes temperature fluctuations of about 40%. Thin foils of the type used in these experiments, have a large surface to volume ratios and thus have heating (or cooling) rate of several thousand degrees per second.

A schematic drawing of the high temperature set-up is shown in Fig. 1. The source was moved by a sinosoidal drive at room temperature outside the vacuum system. While the absorber was heated resistivly inside, a warm up at the source position of at most ten degrees centigrade was noted. This can readily be neglected. Absorber temperature was monitored via

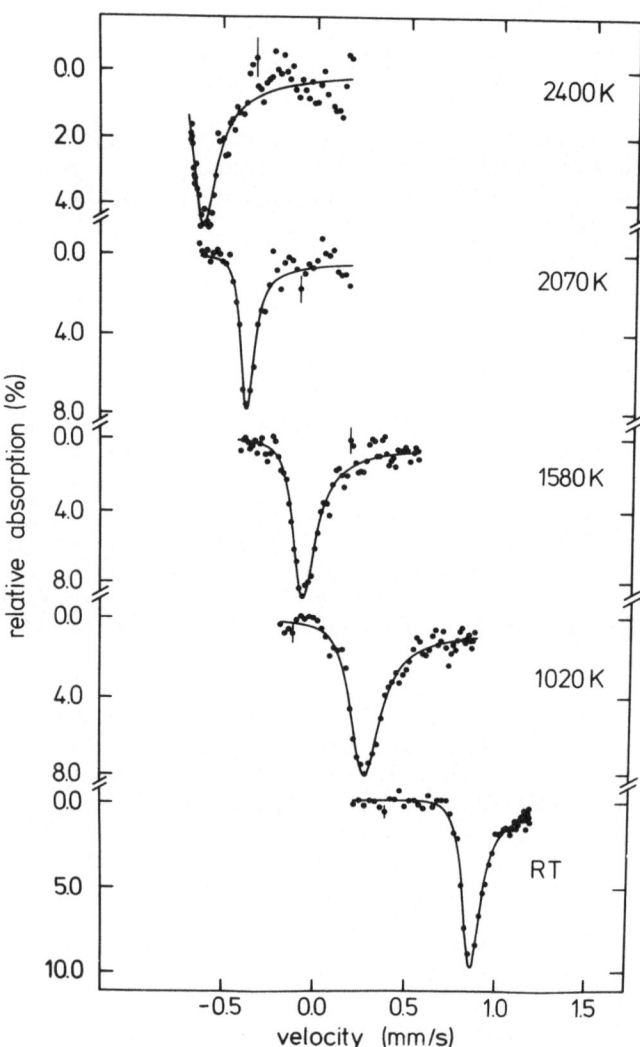

Fig. 2 Absorption spectra for a W-181(W) source at room
temperature, and Ta metal absorbers at various
temperatures.

the view window by an optical pyrometer. The distance between source and detector was about 14 cm, which provided for
a small solid angle. On the average a spectrum was accumulated over a period of 24 hours. It should be noted that this
apparatus, because of its tight geometry, where a residual
pressure of about 10^{-8} Torr was always present, was not used
for the initial foil preparations.

RESULTS AND DISCUSSION

Representative Mössbauer spectra at various temperatures
are shown in Fig. 2. The data were fitted with a single Lo-
rentzian line modified by a constant dispersion term $|7|$ (cor-
rected for absorber thickness). The resonance line, as expected from earlier source measurements, shifts toward lower
energies. The broadening of the line at medium temperature
range (500 - 1500 K) narrows ultimatly to its original room
temperature value at higher temperatures.

In Fig. 3 the IS data are summarized for three tantalum
metal absorbers as a function of temperature. The samples, of
thicknesses 2.4, 4.2 and 11.2 mg/cm^2, were all measured with
the same W-181($\underline{W}$) source which was moved at room temperature.
Below room temperature, the absorber was thermally cooled
in a cold-finger apparatus. Up to about 1300 K the data are
in agreement with the slope $(dS/dT)_p = -8.0(5)$ 10^{-4} mm/(s deg)
measured for a thermally heated W-181(Ta) source. A deviation toward higher transition energies occurred at the
higher temperature range. Three foils of increasing thickness, and hence (decreasing) current densities, were considered. Thus, at 1180 K, for example, these were 2.0, 1.6 and
1.0 x 10^4 A/cm^2 for the 2.4, 4.2 and 11.2 mg/cm^2 foils respectively. At 1700 K, the current densities increased to
4.0, 3.4 and 2.0 x 10^4 A/cm^2 respectively. A current dependent shift could be accounted for current densities above
2 x 10^4 A/cm^2, due to secondary s-d electron scattering.

The other prominent effect of the high temperature measurements is the first absorption and then degassing of residual oxygen and nitrogen. Absorption of these residual gases
occurs in the temperature range 500 - 1500 K, even though
the partial pressure is of the order of 10^{-8} Torr. This is
readily seen in the wider linewidth of the resonance. In
Fig. 4, Mössbauer spectra are shown at room temperature (RT)

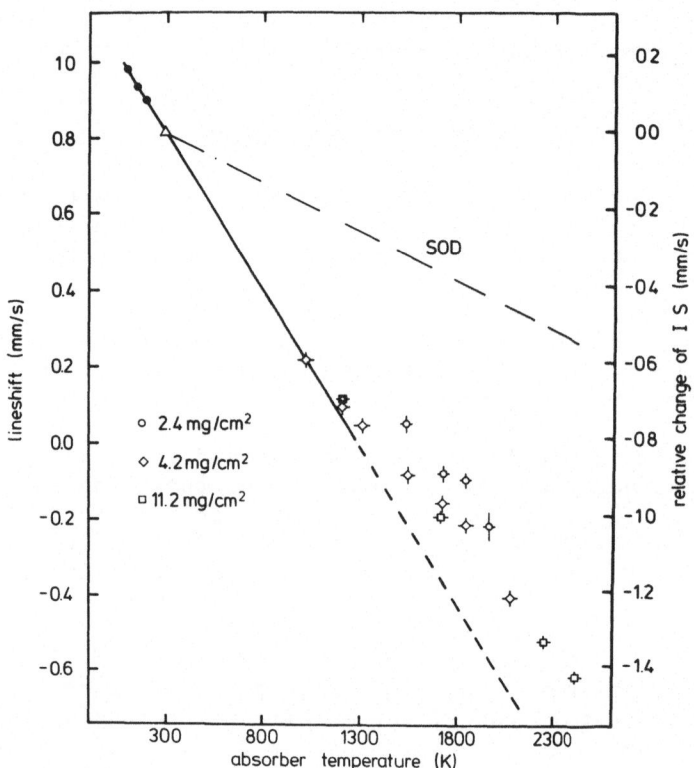

Fig. 3 Line positions (uncorrected for SOD) versus temperature for three different tantalum metal absorbers.
The highest temperature for each foil was the one obtained before they burned out. The dashed line is the
extension of the low temperature slope dS/dT=-8.0(5)
x 10^{-4} mm/(s·deg).

and three increasingly high temperatures each followed by a
RT measurement following a fast current shutdown. Assuming
that the high temperature and its RT sequent have the same
residual gas concentration, and using the result of Sauer
|4|, $\partial\Gamma/\partial c$=13.8 mm/s per at.%, relative oxygen and nitrogen
concentrations in the tantalum foil can be estimated. Thus
for example, for $\Gamma(exp)$=0.33 mm/s in Fig. 4, a concentration
of 0.023 at.% can be assigned. The IS dependence on concentration is about one fifth than that of the linewidth. Thus
slight correction has to be taken into account for the cases
of wide lines. The data shown in Fig. 3 has been corrected
accordingly.

Comparison of the high temperature and the corresponding
immediate RT spectra (the upper right in Fig. 4), shows that
a motional narrowing of the impurity gases is present at the
higher temperature. This is similar to the motional-narrow-
ing observed in the $\propto$ phase of Ta-H in the temperature range
230 to 450 K |8|. While at RT the diffusion of O and N is
15 orders of magnitude lower than that of H in tantalum me-
tal |9,10|, extrapolation of O and N diffusion values to
higher temperatures (above 1500 K) predicts equivalent mobility to H in Ta at RT. Because of the rather low concentrations of O and N involved in the present case, and their
strong dependence on temperature (as can be seen in the
spectra on the right-hand column in Fig. 4) estimates of the
activation energy are hard to make. At best one observes the
general trend to the Ta-H like system. On the other hand,
the hopes to observe a narrow linewidth (narrower that the
best RT value) by the motional narrowing of the impurities
at these very elevated temperatures, diminish, as at these
temperatures the Mössbauer nucleus would start to diffuse,
and thus would give rise to a new source of broadening |11|.

CONCLUSION

We have shown that Mössbauer spectroscopy of the 6.2
keV gamma resonance is feasible at very high temperatures.
Furthermore, linewidths comparable to best RT values, and
effects which are only fractionally smaller than RT absorption indicate that there are no basic handicaps for recoil-
free emission and absorption at these temperatures.

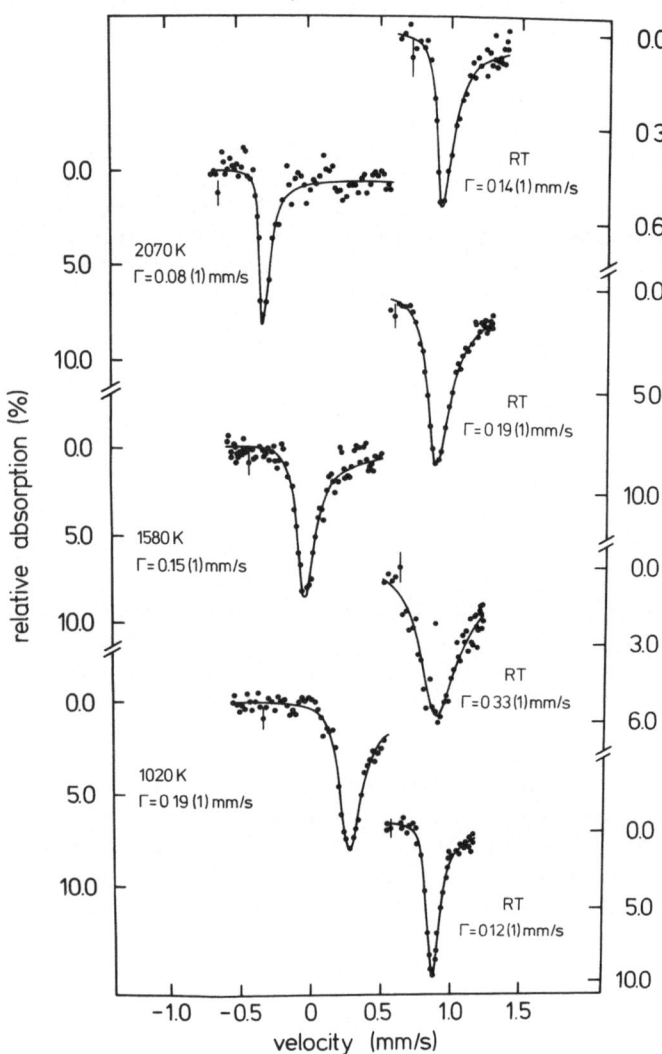

Fig. 4 Absorption spectra (from bottom to top) of a Ta metal
absorber at RT and three increasingly high temperatures, each followed by a RT measurement following a fast
current shut-down, using a single line W-181(W) source.
The experimental linewidth is Γ(FWHM).

ACKNOWLEDGEMENT

The authors would like to thank Prof. E.Matthias for his constant interest. The partial support of the Deutsche Forschungsgemeinschaft, SFB 161, is acknowledged.

REFERENCES

1. G. Kaindl, D. Salomon, Phys. Rev. Letters 30 (1973) 579
2. D. Salomon, B.B.Triplett, N.S. Dixon, P. Boolchand, S. S. Hanna, J. Physique C6 35 (1974) 285
3. G. Kaindl, D. Salomon, G. Wortmann, in: Mössbauer Effect Methodology 8, 211 edited by I.J. Gruvermann (Plenum Press, New York, 1973)
4. C. Sauer, Z. Physik 222 (1969) 439
5. G. Kaindl, D. Salomon, G. Wortmann, Phys. Rev. B 8, (1973) 1912
6. R.B. McLellan, W.A. Oates, Acta Met. 21 (1973) 181
7. G.T. Trammel, J.P. Hannon, Phys. Rev. 180 (1964) 337 and Yu. M. Kagán, A.M. Afanasev, V.K. Voltovetskii, JETP Letters 9, 91 (1969)
8. A. Heidemann, G. Kaindl, D. Salomon, H. Wipf, G. Wortmann, Phys. Rev. Letters, in press
9. R.W. Powars, M.V. Doyle, J. App. Phys. 30 (1959) 514
10. J. Völkl, G. Alfeld, in: Diffusion in Solids: Recent Developments, edited by A.S. Nowick and J.J. Burton (Academic Press, New York, 1975)
11. M.A. Krivoglaz and S.P. Repetskii, Sov. Phys. - Solid State 8 (1967) 2325

APPLICATIONS OF PARALLEL-PLATE AVALANCE COUNTERS IN

MÖSSBAUER SPECTROSCOPY

Gerd Weyer[+]

Institut für Atom- und Festkörperphysik der
Freien Universität Berlin, Berlin 33

INTRODUCTION

Transmission techniques are the dominating detection
techniques in Mössbauer spectroscopy. As illustrated in
Fig.1 the usual set-up consists of a (moving) source emitting Mössbauer γ-radiation which is partially absorbed in a
resonance absorber; the transmitted radiation is detected
in a suitable γ-detector as a function of a relative velocity
between source and absorber. A typical transmission-spectrum
shows a resonance absorption effect of a few percent. In
general this percentage cannot be increased due to Debye-
Waller factors f < 1 for source and absorber and due to
the necessity of thin absorbers to avoid undesired line
broadening. Moreover, the effect may be decreased by additionally detected background radiation. The alternative
method to measure resonantly scattered γ-radiation (γ') or
X-rays or conversion electrons emitted after resonance
absorption has been applied for special problems only. Here,
the background of non-resonantly scattered radiation can be
kept low so that large (>1) effect-to-background ratios are
obtainable for many cases. This advantage, however, is
restricted by a loss in intensity due to a comparably small
solid angle for the detection of the scattered radiation in
conventional experimental arrangements. For scattered γ-radi-
ation a possibly large conversion coeffient α may cause an
additional restriction. It has been shown by Debrunner [1]
that the detection of scattered γ-radiation is better than
transmission techniques for high energy Mössbauer transitions

[+]Present address: Institute of Physics, University of Aarhus,
DK-8000 Aarhus C, Denmark

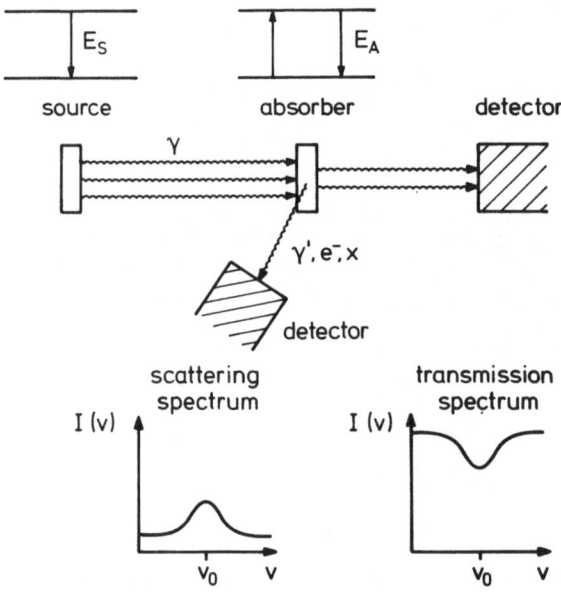

Fig.1: Schematic illustration of transmission and
       scattering experiments

having a low α and f. For cases with α $\gtrsim$ 1 the detection of
X-rays or conversion electrons may be favourable. However,
the low range in material of either low energy X-rays or
conversion electrons decreases the intensity considerably.
On the other hand the relation between range and energy for
electrons allows for depth selective surface studies in
connection with high resolution β-spectrometers |2,3,4|. The
disadvantage of a small solid angle for the detection of
conversion electrons or X-rays can be compensated for by the
construction of counters where the resonance absorber mate-
rial is incorporated in the counter |5|. In general, propor-
tional counters have been used for this method |e.g. 6,7|.
Large effect-to-background ratios can still be maintained by
suppressing the efficiency of the counter for non-resonant
γ-radiation. These resonance detectors have been used for
surface studies mainly. An interesting application in high
resolution studies was given by Mitrofanov et al.|5|.

    In this paper a conversion electron detector based on a
parallel-plate avalanche counter (PPAC) and its applications
to Mössbauer spectroscopy will be described.

EXPERIMENTAL ARRANGEMENT

Mechanical Construction of PPACs

Fig.2 shows a typical construction for a parallel-plate avalanche counter. Two parallel-plate systems are coupled in this counter. High-voltage is put on the middle electrode. Lucite plates or rings (A,B) bear mylar foils covered with thin graphite or aluminium layers sprayed or evaporated on, respectively. Resonance absorber material is deposited on the inner surfaces of these electrodes (C). The system is mounted in a housing (F) made of lucite or aluminium. The gasfilled counter volume is connected to a reservoir volume through a plastic tube on the gas inlet (G) for long time stability of the gasfilling. In many experiments thin beryllium discs (D) are used as entrance windows. Because of its low weight the counter can easily be moved by conventional Mössbauer drive systems. The dimensions of the plate-systems may be chosen in accordance with experimental requirements.

Operation of PPACs

The principles of operation and properties of PPACs have been given in detail by Christiansen |8| and Draper |9|; timing characteristics have been discussed by Krusche et al. |10|. The detection of a primary ionizing event in the gas between the electrodes is achieved by gas multiplication. In the homogeneous electric field of the plate electrodes, electrons and ions will cause avalanches moving towards the electrodes. The statistical mean value for the number of electrons in an electron avalanche n grows exponentially with the distance x from the point of primary ionization: $n(x) = \exp(\alpha x)$ (1); here, α is the first Townsend coefficient which depends on field strength, the type of gas and gas pressure. The distribution for the number of electrons in an avalanche is $w'(n,x) = (n_o)^{-1} \exp{-(n/n_o)}$ (2) with $n_o(x) = \exp(\alpha x)$ and $n_o, n \gg 1$. For a homogeneous distribution of the primary ionization in the counter gap (this can normally be assumed for conversion electrons) the distribution of the total number of electrons in an avalanche is approximated by

$$w(n,d) \approx (n \, \alpha \, d)^{-1} \exp(-n/n_o) \quad (3) \; ; \; n_o = \exp(\alpha \, d) \; |10|,$$

where d is the distance of the electrodes. Typical operating
values for α d are 10^{15}-10^{17} for gases like aceton vapor or
a 90% He/10% CH_4 mixture. Because of the statistical distribution of the primary ionization and the growth of the

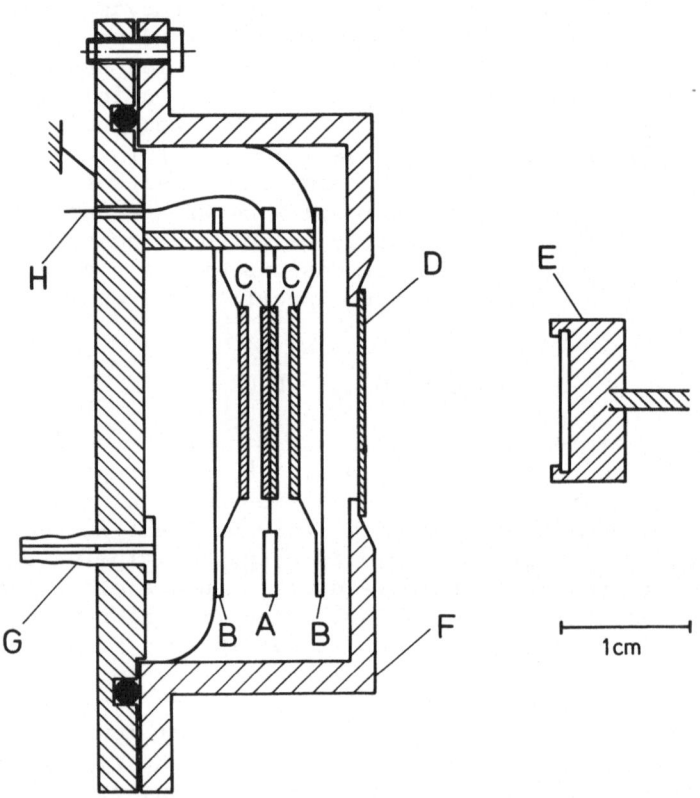

Fig.2: Parallel-plate avalanche counter. A,B- parallel-
plate system (aluminized circular lucite plates
or rings covered with mylar foils.), C- resonance
absorber material, D- thin entrance window, E-
Mössbauer source,F- counter housing (lucite or
aluminium), G- gas inlet, H- high voltage feed
through.

avalanches, no energy resolution can be obtained for conversion electrons. The efficiency for low energy conversion
electrons ($\lesssim$ 100 keV) can be made close to 1. The efficiency
for incoming (non-resonant) γ-radiation is mainly due to
electron producing interactions of the γ-radiation with the
wall or electrode materials. This is predominantly the photoeffect for the γ-energies of interest. The γ-efficiency can be
to suppressed $\lesssim 10^{-4}$ by choosing thin electrodes of low
atomic number (e.g. Be-discs). The background counting rate
is typically $\lesssim$ 1 count/min. Organic quenching vapors like
methan, aceton or ether are reasonable counting gases with
good counting characteristics but He/CH_4-mixtures are also
suitable if a low atomic number of the counting gas is of
importance. Typical operating pressures are between 20-1000
mb. Mössbauer material is normally deposited on the electrode plates. Different simple deposition techniques have
been found to give good results. Metal foils can directly
be used as electrodes, crystalline material can be crystal-
lized from a solution directly on the electrode plates or
foils, powder may be sedimented or brushed on from an alco-
holic suspension. For insulating materials the addition of
colloidal graphite serves as a glue and provides for electri-
cal conductivity. In surface studies of pieces of material
unsuitable as electrodes the cathode has been replaced by a
high transmission metal mesh so that the conversion electrons could enter the sensitive volume of the counter.

Characteristics of PPACs

It is useful to estimate the obtainable effect-to-background ratio ε_s for a counter with a thin layer of monoatomic
absorber material (Debye-Waller factor f_A, unsplit line).
For an idealized Mössbauer source emitting a single line only
(Debye-Waller factor f_S), this is given by the ratio of
detected conversion electrons to photoelectrons:

$$\varepsilon_s = \frac{n \, \beta \, d \, f_S \, f_A \, \alpha \, \sigma_o \, \Omega}{2 \, n \, d \, (\alpha+1) \, \sigma_{ph} \, \Omega} = \frac{\beta \, f_S \, f_A \, \alpha \, \sigma_o}{2 \, (\alpha+1) \, \sigma_{ph}} \qquad (4)$$

(n- density of atoms, β- abundance of the Mössbauer isotope,
α- internal conversion coefficient, σ_o, σ_{ph}- cross sections
for Mössbauer and photoeffect, respectively, d- absorber
thickness, given by the range of electrons if the absorber
thickness is larger than the range of the conversion electrons, Ω- relative solid angle for the detection of electrons

(normalized to 4π). It follows from equation (4) that in
emission spectroscopy with a given source with Debye-Waller
factor f_s, the value of ε_s can be maximized by choosing
highly enriched absorber material with large f_A. Any background γ- or X-rays from the source will diminish ε_s due to
the photoelectrons emitted from the absorber. For chemical
compounds as absorber material this contribution may be
minimized by selecting material with low atomic number.
Often additional background suppression is possible by an
appropriate filtering of the radiation of the source. In
practice, the limiting condition for ε_s (for low energy
Mössbauer transitions $\alpha \approx 1$, $f_A \approx 1$ at roomtemperature) is
given by the ratio σ_0/σ_{ph}. As can be seen from equation (4),
if the γ- and X-ray spectra emitted by different sources are
identical, relative Debye-Waller factors for these sources
can be determined with high accuracy for $\varepsilon_s \gtrsim 1$.

Experimentally observed values for ε_s^{exp} are listed in
table 1 normalized to a source with $f_s = 1$ together with
information about source spectra and absorber material.
Except for ^{57}Fe and ^{119}Sn no particular efforts were made to
maximize ε_s.

The efficiency of the counter for recoilfree emitted
γ-radiation may be optimized for experimental requirements
by the coupling of several absorption layers in a row. The
thickness of the individual layers should be smaller than
the range of the conversion electrons in the absorber material. With this technique efficiencies of the order of $\sim 0,1$
can be realized.

For a comparison of the resonance counter with transmission techniques with respect to the statistical accuracy
obtained in a given measuring time interval, the following
assumptions can be made for simplicity: The effect to back-
ground ratio $\varepsilon = (|N_0 - N_\infty|)/N_\infty$, with N_0, N_∞ the counting
rates on or off resonance, respectively, is large ($\varepsilon_s \gg 1$)
for the scattering experiment and small ($\varepsilon_t \ll 1$) for the
transmission experiment. The source emits single-line Mössbauer γ-radiation only. The transmitted radiation is detected
with a high efficiency (~ 1) in the transmission experiment.
The relative statistical errors of $|N_0 - N_\infty|$ can then be

Table 1:

Mössbauer isotope	^{57}Fe	^{57}Fe	^{119}Sn	^{119}Sn	^{151}Eu	^{161}Dy	^{181}Ta
source	^{57}Co	^{57}Co	^{119m}Sn	^{119m}Te	^{151}Sm	^{161}Tb	^{181}W
E_γ	14	14	24	24	22	26	6
I_γ	0.06	0.06	0.4	0.05	1.0	0.3	0.01
E_e	7	7	20	20	14	15	4
absorber	steel	PFC	SnO_2	SnO_2	Eu_2O_3	DyF_3	Ta
f_A	0.8	0.3	0.4	0.4	0.5	0.4	1.0
β	0.9	0.9	0.9	0.9	1.0	0.9	1.0
Γ_{exp}	2.0	1.1	2.3	2.3	2.0	19	14
$\alpha/(\alpha+1)$	0.9	0.9	0.8	0.8	1.0	0.7	1.0
σ_o/σ_{ph}	410	410	450	450	25	120	18
ε_s^{exp}	24	10	22	10	1.3	0.7	0.01

(E_γ, E_e: γ- and conversion electron energies in keV, I_γ- relative γ-intensity, Γ_{exp}- experimentally observed line width in units of $2\,\Gamma_o$, Γ_o- natural line width, β- isotope abundance of the Mössbauer isotope, PFC- $K_4Fe(CN)_6 \cdot 3H_2O$)

approximated by

$$\frac{\Delta|N_o^t - N_\infty^t|}{|N_o^t - N_\infty^t|} \approx \frac{\sqrt{2}}{\varepsilon_t \sqrt{N_\infty^t}} \qquad (5) \quad \text{and}$$

$$\frac{\Delta|N_o^s - N_\infty^s|}{|N_o^s - N_\infty^s|} \approx \frac{1}{\sqrt{N_o^s}} \qquad (6)$$

(t- for transmission and s- for scattering). N_0^s in the
scattering experiment is given by $N^s = N_\infty^t \varepsilon_t \Omega \alpha/(\alpha+1)$.
The condition for a better accuracy in a scattering than in
a transmission experiment is $2 \Omega \alpha/(\alpha+1) > \varepsilon_t$ (7). Typical
values for these parameters are $\alpha/(\alpha+1) \approx 0.8$, $\Omega \approx 0.4$ so that
$2 \Omega \alpha/(\alpha+1) \approx 0.6$ and condition (7) becomes $0.6 > \varepsilon_t$. Because
of Debye-Waller factors f_s, $f_A < 1$ and a considerable
decrease in resolution when thick absorbers are used ε_t will
hardly reach a value of 0.6 in any Mössbauer transmission
experiment. Therefore, especially when high resolution is
needed resonance detectors will be superior to transmission
techniques. A relative error smaller by a factor of 10 is
obtainable for ^{57}Fe and ^{119}Sn which results in a shortening
of the measuring time by a factor of 100 for a given relative error.

Conversion electron detectors based on proportional
counters do have energy resolution and thus should be superior to parallel-plate counters if a high background contribution is reducing ε_s. However, in practice this advantage
is restricted, because the energy of the conversion electrons is not sharp if the absorber thickness is of the order
of the range of these electrons. Furthermore, high energy
background electrons do not lose their entire energy in the
counting gas and thus cannot be completely discriminated. As
a rule it can be assumed that no particular advantage is
obtained from energy resolution if $\varepsilon_s \gtrsim 1$. For ^{57}Fe and
^{119}Sn, where proportional counters have been used, no larger
effect to background ratios have been observed as compared
with parallel-plate avalanche counters.

In the analysis of conversion electron Mössbauer spectra, the interference between photoeffect and internal
conversion has to be taken in account $|11,12|$. The shape of
these spectra is given by $N(v) = N_\infty[1+\varepsilon(1+\beta x)/(1+x^2)]$ (8)
with $x = 2 (v-v_0) / \Gamma$. The interference effect is known to
be observable in transmission spectra for E1-transitions
only; here the interference parameter β is given by

$$\beta_\gamma = 2[\alpha(2I_1+1)\sigma_{ph} / 3(\alpha+1) (2I_0+1)\sigma_0]^{1/2} \quad (9)$$

(I_0, I_1- spins of the ground and excited state, respectively). A comparison of a transmission and scattering spectrum
is shown in Fig.3 for the 26 keV E1-transition of ^{161}Dy. The
measured interference parameters are $\beta_\gamma = 0.06$ and $\beta_e = 0.08$
in accordance with the expectation from theory that $^e\beta_e$ is

larger than β_γ by a factor $(\alpha+1)/\alpha$, which is 1,3 for this case $|13|$. The largest value for β has been found for the 6 keV E1-transition of ^{181}Ta, $\beta_\gamma \approx \beta_e = 0.26$ (see Fig.10). For M1- and E2-transitions the interference effect vanishes in transmission but is observable for the detection of conversion electron $|14|$.However, in this case smaller values of β have been measured than for E1-transitions $|15,16|$.

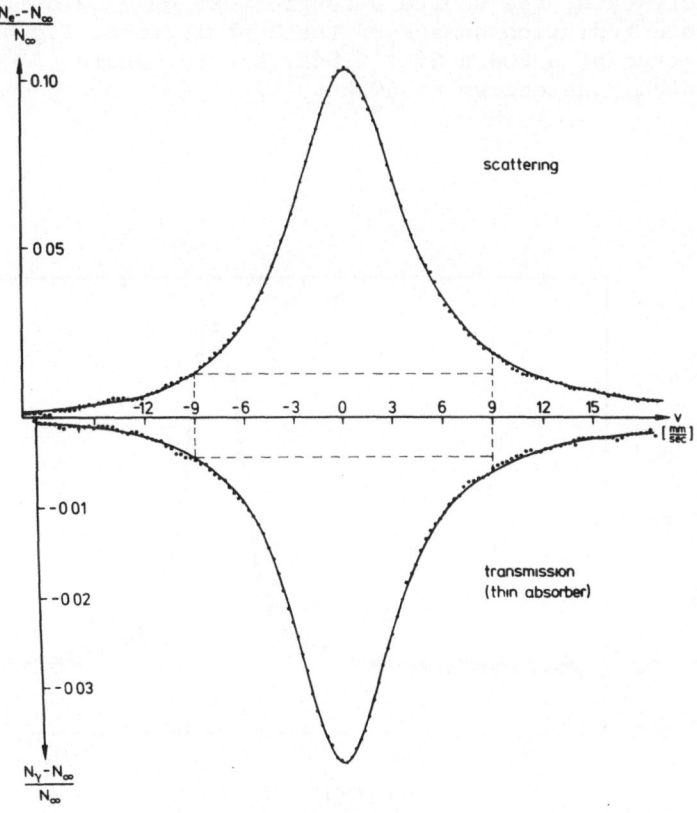

Fig.3: Mössbauer transmission and scattering spectra of the 26 keV-transition of ^{161}Dy. The scattering spectrum was measured with an angle of $0°$ between incoming γ-rays and detected conversion electron. Source: ^{161}Tb Dy F$_3$, absorber: ^{161}Dy F$_3$. From $|13|$.

EXAMPLES OF APPLICATIONS

Emission Spectroscopy

The high sensitivity of the resonance counter for recoilfree emitted Mössbauer γ-radiation can be utilized in emission spectroscopy of weak sources or of sources with extremely low Debye-Waller factors. In Mössbauer studies of isotope-separator implanted radioactive impurity atoms in host crystals, the source strength must often be quite low to avoid radiation damage of the host lattice. Fig. 4 shows a spectrum of a 600 n Ci ^{119m}SnSi source, where ^{119m}Sn was implanted at an energy of 60 keV [17].

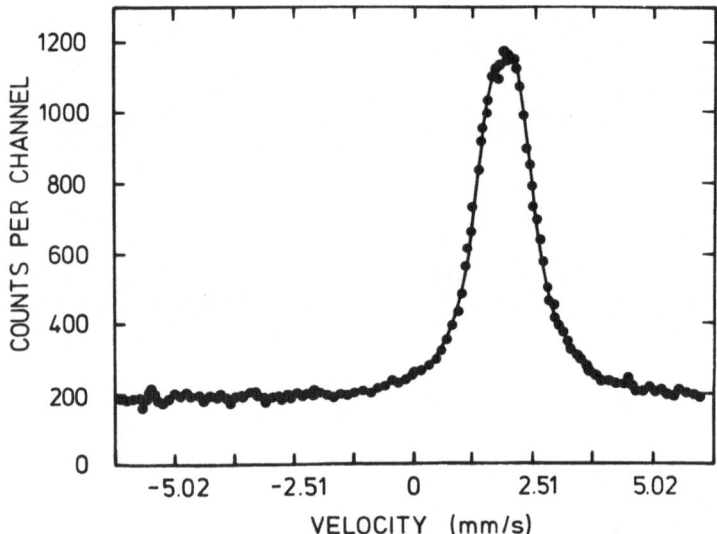

Fig.4: Mössbauer spectrum obtained with a SnO_2-resonance counter. Source: ^{119m}Sn implanted in silicon at 60 keV, activity 600 n Ci, f_S = 0.3. Measuring time 12 hours at room-temperature. From [17].

Mössbauer Polarimeter

The versatility in mechanical construction of PPACs is demonstrated in Fig.5, where an extremely flat type is shown operated in the gap of the pole shoes of an electromagnet. A thin iron foil (enriched in ^{57}Fe) serves as electrode (E). Without an applied magnetic field, this foil was polarized in a direction perpendicular to the normal of the

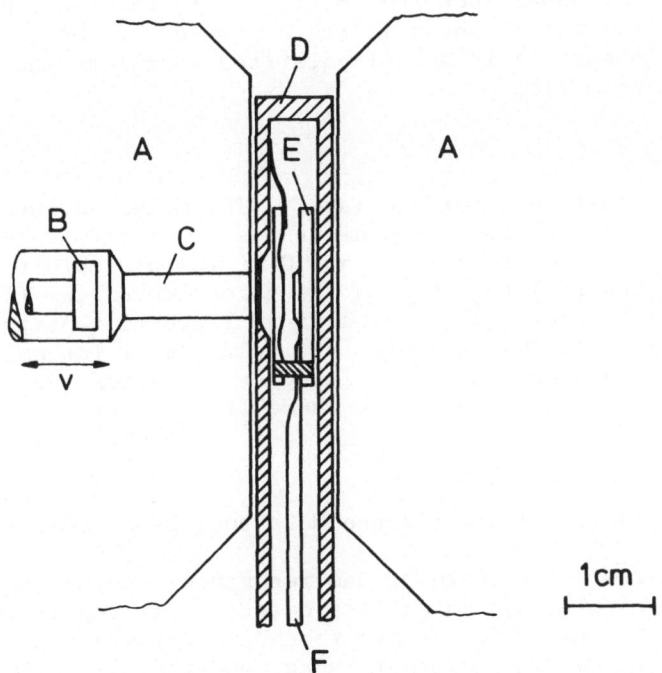

Fig.5: PPAC operated in the gap of the pole shoes of an electromagnet. A- pole shoe, B- vibrating source, C- boring through of the pole shoe, D- counter housing, E- electrode (cathode) covered with an iron foil (enriched in ^{57}Fe), F- high voltage connection and support of the parallel-plate system. From |18|

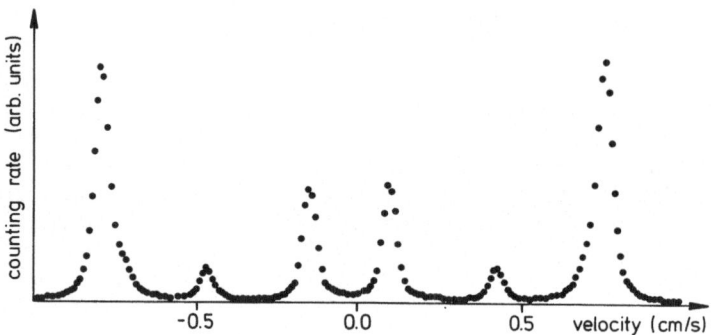

Fig.6: Mössbauer-spectrum obtained with the experimental set-up shown in Fig.5. Source: ^{57}CoPd,
absorber: iron foil polarized normal to the foil.
From |18|

foil (at least in a surface layer). The measured intensity
ratio of the hyperfine components was ≈ 3 : 4 : 1. In an
applied magnetic field of 24.5 k G it was possible to
polarize the foil to ≈ 90% in the direction of the field
normal to the foil as can be seen in Fig.6 (intensity ratio
≈ 3 : 0,3 : 1). Thus a convenient polarimeter for emission
studies of sources can be built-up by vibrating the source
in a hole through a pole shoe where the magnetic field is
low (≈ 1KG)(Fig.5).

In-Beam and Coincidence Mössbauer Experiments

The low sensitivity of the resonance counter for non-
resonant γ-radiation makes this detector particular suited
for in-beam experiments where the Mössbauer state is populated by a nuclear reaction. Background radiation from nuclear reactions often exceeds that from radioactive sources
by orders of magnitude, so that the counting rate in the
Mössbauer line contains large background fractions. Furthermore, the counting rate may be limited by a detector over-
load. These problems can be evaded by the application of
resonance counters where counting rates of > 10^6 sec^{-1} can
be utilized.

Additional reduction of the background has been
achieved by pulsed beam techniques so that delayed γ-radia-
tion is measured intermittently only |19|. This requires a
time resolution of the PPAC much shorter than the nuclear
lifetime of the Mössbauer level. The internal time resolution of PPACs has been shown to be independent of the electron energy and a resolution of < 1 ns has been measured
|10|. This resolution is superior to that of most competing
γ-detectors for energies of ≈ 10 keV. However, the deposition of resonance material on the electrodes and electronic
timing problems from the exponential pulse-height spectrum
may lower the resolution considerably.

Fig.7 shows the experimental arrangement for a delayed
coincidence experiment on the resonance scattering process
of ^{57}Fe. The 122 keV γ-transition preceeding the emission
of 14 keV Mössbauer γ-quanta from the source is detected in
a NaI(Tl)-detector and used as time zero signal. Conversion
electrons reemitted after resonance absorption of the 14 keV
γ-quanta are detected in the PPAC. The time distribution of
the electrons is displayed in Fig.8. A constant background
of accidental coincidences has been subtracted from the
data. From the peak in the spectrum (which is due to iodine
X-rays escaping from the NaI-detector) a time resolution of

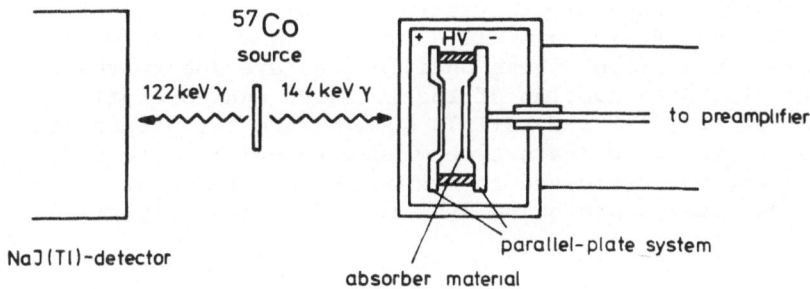

Fig.7: Experimental arrangement for a delayed coincidence
        experiment on ^{57}Fe

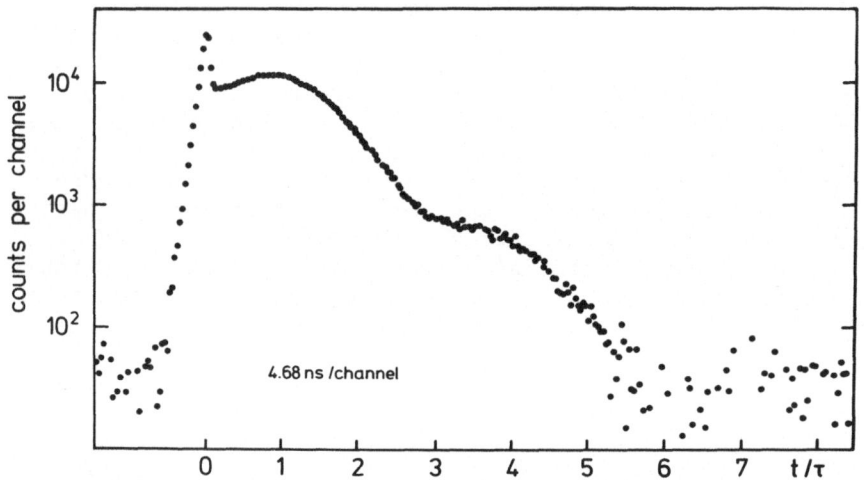

Fig.8: Time distribution of conversion electrons reemitted
from a Mössbauer absorber. Accidental coincidences
are subtracted. Time scale in units of nuclear lifetime. Source: ^{57}Co Pd, absorber: $K_4Fe(CN)_6 \cdot 3H_2O$
(enriched in ^{57}Fe), isomer shift between source and
absorber 2,4 Γ_o.

FWHM = 25 ns is inferred. The delayed spectrum contains an
exponential fraction due to photoelectrons released from the
14 keV γ-rays; the major fraction exhibiting characteristic
oscillations stems from the resonance absorption of the
14 keV γ-radiation. These oscillations are due to the incom-
plete resonance overlap of the emission and absorption lines
(see ref. 20,21 for details). Coincidence techniques allow
by using an appropriate time window to enlarge the effect
to background ratio and to narrow the experimentally
observed line width below 2 Γ_o in a Mössbauer experiment.

Surface Studies

    Finally, examples of the possibilities for surface
studies with these counters will be given. Fig.9 shows a
transmission and scattering spectrum of finely powdered

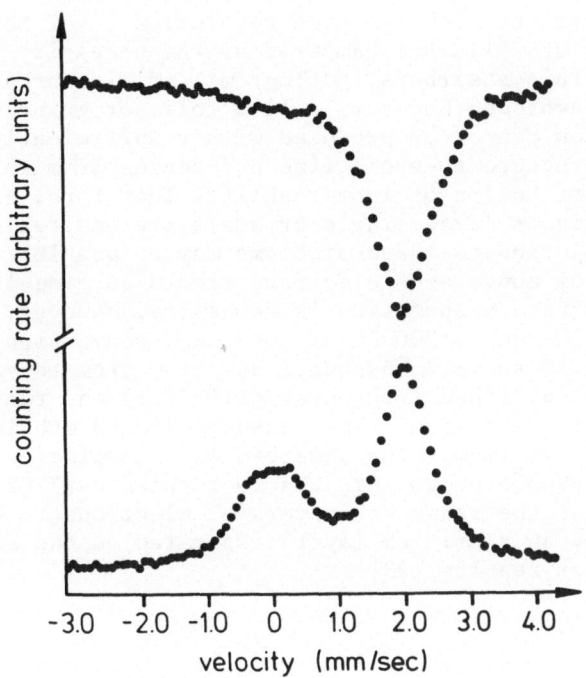

Fig.9: Transmission and scattering spectrum of α-tin
powder. Source: ^{119m}Sn(Ca Sn O$_3$)

α-tin. The quadrupole split lines centered about v=o from
the surface oxide are clearly resolved in the scattering
spectrum. The advantage of the application of conversion
electron detectors in surface corrosion measurements is
obvious from these data. Moreover, corrosion processes may
be followed in a time range of seconds or less because
sufficiently accurate spectra can be obtained in these times
with high activity sources.

Experiments with the high resolution 6 keV Mössbauer transition of ^{181}Ta are hampered by the necessity to use extremely thin absorbers ($\ll$ 1mg/cm^2) if line broadening has to be avoided. However, rolled foils or grinded compounds often cannot be produced with a sufficiently good crystal structure to avoid line broadening from an inhomogenous distribution of isomer shifts. Therefore, experiments with relatively large single crystals are desirable. A convenient solution to these problems may be possible by the detection of conversion electrons from a surface layer. Careful surface preparation is necessary, however, because of the low range ($\approx$ 200 Å) of the 4 keV conversion electrons. Fig.10 shows a Mössbauer spectrum from the surface of a highly polished (roughness < 10^{-5} cm) and tempered tantalum single crystal. The measured line-width is comparable to the minimum value observed in transmission experiments (extrapolated to zero absorber thickness) $|22|$. A reduction of the range of conversion electrons to $\approx$ 60 Å in tantalum by an aluminium layer evaporated on the surface gave similar results $|23|$.

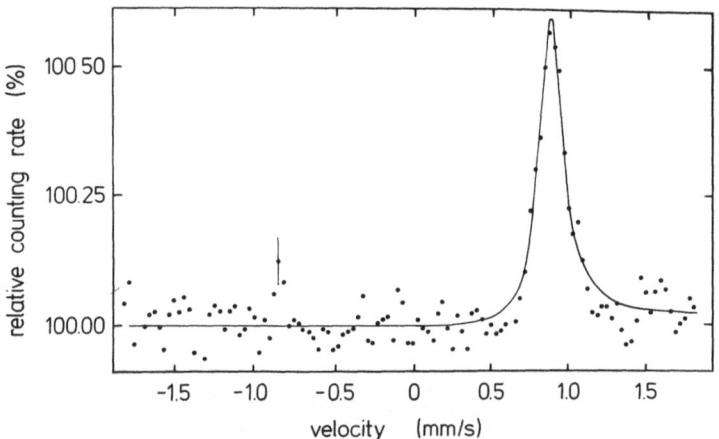

Fig.10: Mössbauer conversion electron spectrum of a tantalum single crystal surface ($\sim$ 200 Å). Source: ^{181}W($\underline{W}$). From $|22|$

CONCLUSION

PPACs have been shown to have several oustanding properties which can be utilized for applications in Mössbauer spectroscopy. Examples have been presented for different Mössbauer transitions and experimental requirements where advantage has been taken of different properties of these counters. In all cases the superiority of this solution for the experimental problems to conventional arrangements has been demonstrated. It is obvious that many other applications can be made in the future. Possible lines of extension of the method are: e.g. low temperature measurements which become possible by using pure noble gases and electronic quenching techniques for the counter, experiments of resonance absorption of synchrotron radiation |24| and applications in various high precision experiments.

ACKNOWLEDGEMENT

I am indebted to many colleagues with whom I cooperated in the different experiments related to this paper, especially to Prof. J. Christiansen who introduced PPACs in Mössbauer spectroscopy.

This work was supported in part by the Deutsche Forschungsgmeinschaft (SFB 161) and by the Dansk Forskningsrad.

REFERENCES

1 P. Debrunner, Mössbauer Effect Methodology (ed.I.J. Gruvermann) 1, 97 (1965)

2 K.P. Mitrofanov and V.S. Shpinel JETP 13, 686 (1961)

3 Zw. Bonchev, A. Jordanov and A. Minkova, Nucl. Instr. Meth. 70, 36 (1969)

4 U. Bäverstam, C. Bohm, T. Ekdahl, D. Liljequist and B. Ringström, Mössbauer Effect Methodology (ed.I.J. Gruvermann and C.W. Seidel and D.K. Dieterly) 9, 259 (1974)

5 K.P. Mitrofanov, N.V. Illarionova, and V.S. Shpinel Instr. Exp. Tech. 3, 415 (1963)

6 C.M. Yagnik, R.A. Mazak, and R.L. Collins Nucl. Instr. Meth. 114, 1 (1974)

7 J.J. Spijkerman,Mössbauer Effect Methodology (ed. I.J. Gruvermann) 7, 85 (1971)

8 J. Christiansen, Z. angew. Phys. 4, 326 (1952)

9 J.E. Draper, Nucl. Instr. Meth. 30, 148 (1964)

10 A. Krusche, D. Bloess and F. Münnich, Nucl. Instr. Meth. 51, 197 (1967)

11 G.T. Trammel, J.P. Hannon, Phys. Rev. 180, 337 (1969)

12 Yu. M. Kagan, A.M. Afanasev, V.K. Voitovetskii JETP Letters 9, 91 (1969)

13 P. Steiner and G. Weyer, Z. Phys. 248, 362 (1971)

14 A.M. Afanasev and Yu. Kagan, Phys. Lett. 31A, 38 (1970)

15 K.P. Mitrofanov, M.V. Plotnikova, N.I. Rokhlov and V.S. Shpinel, JETP Lett. 12, 60 (1970)

16 P. Steiner and G. Weyer, Phys. Lett. 36A, 201 (1971)

17 G. Weyer, J.U. Andersen, B.I. Deutch, J.A. Golovchenko and A. Nylandsted-Larsen, Rad. Eff. 24, 117 (1975)

18 P. Steiner and G. Weyer, unpublished

19 J. Christiansen, P. Hindennach, U. Morfeld, E. Recknagel, D. Riegel and G. Weyer, Nucl. Phys. A99, 345 (1967)

20 H. Drost, H.v. Lojewski, K. Palow, R. Wallenstein and G. Weyer, Proc. of the 5th Int. Conf. on Mössb. Spectr., Bratislava 1973, Ed. U. Hucl and T. Zemcik, page 713 (1975)

21 H. Drost,K. Palow and G. Weyer, Proc. of the Int. Conf. on the Appl. of the Mössb. Eff., Bendor 1974, Journ. de Phys. (Paris) 12, C 6 - 679 (1974)

22 P.J. West, E. Matthias, D. Salomon, W. Wallner and G. Weyer, Proc. of the Int. Conf. on Mössb. Spectr., Cracow 1975, Ed. A.Z. Hrynkiewicz and J.A. Sawicki, p. 457 (1975)

23 D. Salomon, P.J. West, G. Weyer and E. Matthias
Mössbauer conversion electron studies on tantalum metal
surfaces, to be published

24 S.L. Ruby, Proc. of the Int. Conf. on the Appl. of the
Mössb. Eff., Bendor 1974, Journ. de Phys. (Paris) 12,
C 6-209 (1974)

MÖSSBAUER STUDIES WITH THE RARE-GAS ISOTOPE ^{83}Kr*

Berend Kolk

Rutgers University, Dept. of Physics

Busch Campus, New Brunswick, N.J.08903

I INTRODUCTION

The isomer shift and other hyperfine-interaction parameters depend largely on the electron configuration of the Mössbauer atom in the host lattice, which in most cases is not sufficiently well known. One way to overcome this problem is to study isolated atoms trapped in rare-gas matrixes, where it is assumed that these isolated atoms have a free-atom electron configuration. Studies on rare-gas matrix isolated molecules of ^{125}Te compounds are presented at this symposium by Montano and coworkers.

An alternative method is to use a rare-gas atom as an impurity. Because of its inert character it is unlikely that a chemical bond is formed between the rare-gas impurity and the host atoms. Of the rare-gas isotopes suitable for Mössbauer work ^{83}Kr, whose decay scheme is shown in Fig.1, has various advantages. Compared with ^{129}Xe and ^{131}Xe its natural linewidth, $\Gamma_n = \hbar/\tau = 0.10$ mm/sec, is about thirty times smaller, and owing to the low transition energy (9.4 keV) its recoilless fraction is considerably larger.

*Work supported in part by NSF

Various methods of $^{83}Kr^m$ source preparation
are discussed in Sec.II. Special attention is
given in Sec.III to sources and absorbers of ^{83}Kr
implanted in iron and aluminum. Standard absorbers in ^{83}Kr Mössbauer spectroscopy are the Kr hydroquinone clathrate and solid krypton, which both
were extensively investigated[1-4]. A discussion
of the results is given in Sec.s IV to VI.

In the cage formed by the hydroquinone molecules in the clathrate β hydroquinone (HQ) the Kr
atom experiences an electric-field gradient.
From the observed quadrupole interaction the
nuclear quadrupole-moment ratio, Q^*/Q, of the
quadrupole moment of the 9.4 keV state to that of
the ground state of ^{83}Kr was derived[1] as well as
the location of the Kr atom in the cage (Sec.IV).
Contrary to the generally accepted view[5] that the
Kr atom moves freely in such a cage, we found that
the Kr atom is bound by dipole-induced dipole forces to one of the oxygen-hydrogen rings (Fig.4)
which form the bottom or top of the HQ cage.

Solid krypton is an ideal crystal for studying anharmonic effects on the Debye-Waller factor.
Rare-gas crystals in general are highly anharmonic
but the interatomic potentials are fairly well
known[6,7], so that the various anharmonic effects
can be well calculated. A brief discussion of the
quasi-harmonic approximation and of the *direct*
anharmonic effects arising from the coupling
between the normal modes is given in Sec.V.
Comparison of theory with experiment[2-4] shows that
the quasi-harmonic model alone is a poor approximation, and that *direct* anharmonic effects must be
taken into account.

Investigations[2] of solid-krypton sources at
4.2K, discussed in Sec.VI, show that the recoilless fraction of these sources is 40% lower than
that of a solid-krypton absorber at the same
temperature. This effect is caused by *after effects*
of the 32keV γ-ray decay preceding the resonant
γ-ray transition in $^{83}Kr^m$.

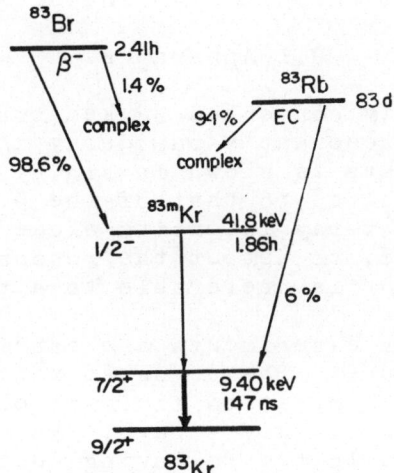

Fig.1. Electron-capture (EC) decay of ^{83}Rb and β^- decay of ^{83}Br to ^{83}Kr^m.

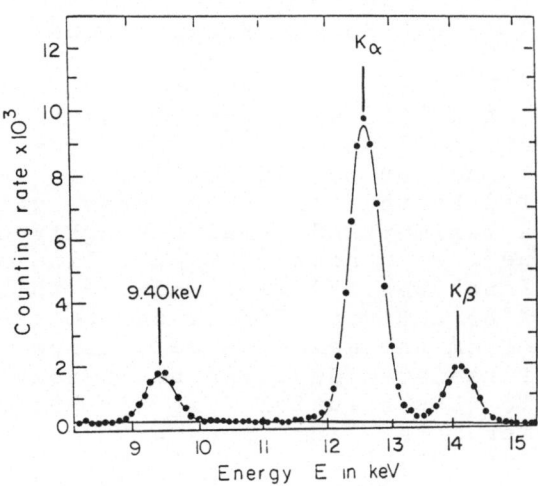

Fig.2. γ-Ray spectrum of a solid ^{83}Kr^m source taken with a Si(Li) detector.

II EXPERIMENTAL PROCEDURES

2.1 Apparatus

The highly converted 32 keV transition which preceeds the resonant γ-ray decay in ^{83}Kr produces Kα and Kβ x rays as shown in Fig.2. The x-ray energies are close to that of the 9.4 keV γ rays and the total x-ray intensity exceeds that of the γ rays. Hence, to detect the resonant γ rays a Si(Li) detector is preferable to a proportional counter.

Most ^{83}Kr experiments are carried out below room temperature. The cryostat windows must be of beryllium or mylar to minimize electronic absorption of the 9.4 keV γ rays. For the experiments discussed in the following sections special glass cryostats were constructed (see Ref.4), in which the source could be replaced very quickly, which was necessary because of the 2h half life of ^{83}Kr^m. A conventional electro-mechanical feedback system in combination with a multichannel analyzer was used for the Mössbauer drive system.

2.2 Source Preparations

The ^{83}Kr^m activity can be produced by irradiating enriched ^{82}Se with thermal neutrons. Via (n,γ) reactions ^{83}Se and ^{83}Se^m are formed which decay through ^{83}Br to ^{83}Kr^m (Fig.1). The ^{83}Kr^m activity fed by the ^{83}Br decay grows within two hours to a maximum and remains roughly constant over the next four hours. Single-line sources of ^{83}Kr^m in ZnSe (Sec.IV) were made in this way.

The ^{83}Br activity formed can be chemically separated from the selenium activities very quickly. The irradiated SeO$_2$ can be dissolved in 50 ml 1N KOH. To this solution 1 mg KBr has to be added as a carrier. The ^{83}Br$^-$ ions will precipitate when 1.5 mg AgNO$_3$, dissolved in 5N HNO$_3$, is added. The Mössbauer spectrum shown in Fig.3b was recorded with such an ^{83}Kr(AgBr) source. Another chemical separation suitable to produce alkali-bromide sources has been given by Pasternak and Sonnino.[8]

^{83}Kr^m can also be produced directly by thermal-neutron irradiation of krypton gas via the reaction ^{82}Kr$(n,\gamma)^{83}$Kr^m. However, the 11.6% stable ^{83}Kr in natural krypton gives rise to resonant self absorption which leads to significant line broadening and to reduction in the effective recoilless fraction.[2,4]

The disadvantage of the methods described above is that they produce very short living Mössbauer sources (2 to 4 hours). Therefore, ^{83}Rb with an 83 days half life (Fig.1) is of interest. This isotope can be made by the reaction ^{81}Br$(\alpha,2n)^{83}$Rb. Unfortunately also ^{84}Rb will be formed via the reaction ^{81}Br$(\alpha,n)^{84}$Rb. The K x rays of this nuclide give an increase in background of the resonant γ rays, and thus decrease the signal to noise ratio.

The formation of ^{84}Rb can be avoided by using the reactions ^{87}Sr$(p,5n)^{83}$Y and ^{88}Sr$(p,6n)^{83}$Y. The ^{83}Y decays via ^{83}Sr to ^{83}Rb. The Yttrium isotopes eventually produced with atomic numbers 84 to 88 decay to stable strontium and rubidium isotopes. The ^{83}Rb activity can be separated from the irradiated SrO with standard radio-chemical methods. Fairly strong ^{83}Rb sources were obtained after irradiating SrO with 20 µAh 60 MeV protons using the cyclotron of the Kernfysisch Versneller Instituut at Groningen, Netherlands. When SrO enriched in ^{84}Sr is used ^{83}Rb can also be formed via ^{84}Sr$(p,2n)^{83}$Rb using 30 MeV protons.

TABLE I

The recoilless fraction of ^{83}Kr in various hosts at 92K. The characteristic temperature $\Theta_a(-2;T)$ (Eq.4) is derived from $f(T)$ using Eq.8.

HOST	$f(92K)$	$\Theta_a(-2;92K)$
iron	0.8±0.1	150±50K
aluminum	0.33±0.03	55±6 K
β hydroquinone	0.58±0.04	82±5 K
solid krypton (theory) Sec.V	0.29	54K

III STUDIES ON KRYPTON-IMPLANTED SOURCES AND ABSORBERS

Most [83]Kr sources show significant line broadening as result of resonant self absorption (solid krypton[2]) or of after effects (alkali halides[1]). To date only [83]Se sources of the semi-conducting ZnSe show linewidths close to the natural one at higher temperatures.[9]

To avoid resonant selfabsorption and after effects we implanted [82]Kr with the Groningen isotope separator over the entire area of both sides of 1×1 cm iron and aluminum foils.[4] Six foils of either metal were stacked together. This stack formed the [83]Kr[m] source after a one hour thermal neutron irradiation in the high flux reactor of the Reactor Centrum Nederland ($\phi = 5\times10^{13}$ n/cm^{2}sec). To keep electronic absorption of the 9.4 keV γ rays low and also to minimize the background due to the activated host material, the foil thicknesses had to be about 10^{-4} cm. The sources could be reactivated several times.

To get sufficiently strong sources high doses ($\sim3\times10^{16}$ ions/cm^2) of 60 kV Kr$^+$ ions were implanted.[10] At such a high dosage the implanted atoms are not substitutional but form clusters.[11-13] With an electron microscope we observed Kr bubbles in iron with a size of ~130 Å.

A typical Mössbauer spectrum of such an [83]Kr[m](Fe) source, taken with a KrHQ absorber, is shown in Fig.3a. The absorption line is asymmetric and quite broadened. Correction for the intrinsic broadening in the KrHQ absorber yielded the source linewidth $\Gamma_s = 4.5\ \Gamma_n$. The asymmetry and the broadening may be due to magnetic and quadrupole interactions of Kr atoms at sites associated with vacancies or at grain boundaries and dislocations in iron.[12,13] A least-squares fit of the [83]Kr[m](Fe) spectrum with a single Lorentzian and a superimposed magnetic hyperfine splitting yielded a field of ~200 kOe. This result gives only an order of magnitude since the Kr atoms may occupy sites corresponding to a wide range of fields. Magnetic hyperfine fields at nuclei of rare-gas atoms in iron have been attributed to overlap polarization.[14]

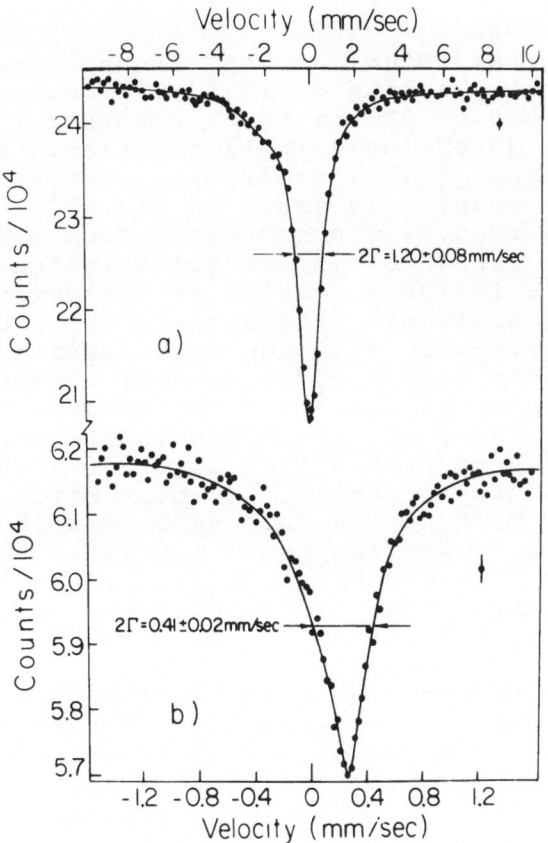

Fig.3. Mössbauer spectrum of a) ^{83}Kr^m(Fe) *source* taken with a βHQ absorber, b) ^{83}Kr(Al) *absorber* taken with an ^{83}Kr^m(AgBr) source, both recorded at 92K. Note the difference in the upper and lower velocity scales.

      The recoilless fraction $f(92K) = 0.8\pm0.1$ of ^{83}Kr^m in iron is the highest value observed for any ^{83}Kr source or absorber at that temperature to date (see Table I).

      The linewidth of the ^{83}Kr^m(Al) source, $\Gamma_s = 2\Gamma_n$, is a factor 2 smaller than that of ^{83}Kr^m in iron.

However, the recoilless fraction is also decreased by about the same factor.

An ^{83}Kr(Al) *absorber* containing 10 μg ^{83}Kr/cm^2 was made by implanting 60 kV ^{83}Kr$^+$ ions over a 1×1 cm area of a 0.6×10^{-4} cm thick aluminum foil and by stacking 30 of these foils together. The Mössbauer spectrum of this absorber, taken with an ^{83}Kr^m(AgBr) source, is shown in Fig.3b.

Of interest is the presence of an isomer shift δ = + 0.25±0.04 mm/sec which indicates a ~2% increase in the s density at the krypton nucleus in aluminum.[15] This appears to be consistent with our conversion-electron measurements.[16]

The conversion coefficients of electron subshells are proportional[17],[18] to the density close to the nucleus of the electrons in the subshells. Since the 9.4 keV transition is predominantly converted in the s shells, its conversion-coefficient ratio α_N/α_M is in good approximation equal to the ratio $\rho_{4s}(o)/\rho_{3s}(o)$. The value[16] 0.163 ± 0.024 which we observed for α_N/α_M of an ^{83}Kr^m(Al) source is larger than the free-atomic value[19] 0.123 for $\rho_{4s}(o)/\rho_{3s}(o)$. This result thus also indicates an increase in $\rho_{4s}(o)$.

This effect may be explained[4] by the renormalization of the 4s wavefunction of the Kr atom due to overlap with those of the Al host atoms or other Kr atoms, when Kr bubbles are formed (In these bubbles the Kr atom may undergo pressure).

TABLE II

Values obtained from the least-squares fit of the thickness-corrected KrHQ spectrum (Fig.5a).

Q^*/Q		2.05 ± 0.15
$e^2qQ(C/E_\gamma)$	mm/sec	+1.00 ± 0.12
δ	mm/sec	−0.15 ± 0.03
Γ_n	mm/sec	0.13 ± 0.03

IV THE LOCATION OF Kr IN β HYDROQUINONE

βHydroquinone (HQ) is an unusual compound, where the molecules form two interpenetrating three-dimensional networks which enclose approximately spherical cages of 8Å diameter (Fig.4a) with a ratio of one cage to every three HQ molecules.[20] Many types of molecules can be enclosed in such a cage if their size is not too large or too small. This structure has been considered to be an ideal case for the study[5,21-24] of the cell model, which is used in statistical thermodynamics to describe the behavior of liquid and solid solutions. In the cell model the guest atom moves freely in the cage and its vibrations are completely decoupled from the dynamics of the host lattice as well as from other trapped atoms in adjacent cells.

The KrHQ clathrate is one of the few absorber materials available for [83]Kr Mössbauer spectroscopy. It has the advantage that it can be easily made and handled and that its recoilless fraction is high.[25]

The clathrate can be made by dissolving 3 g of hydroquinone in 25 ml amyl acetate at 80°C and cooling this solution to room temperature over 15h under a krypton pressure of 30 to 40 atm. This method yields roughly the maximum possible amount of Kr that can be trapped in HQ, which is 20% by weight.[1,4] To avoid possible orientation effects the KrHQ crystals were powdered.

A Mössbauer spectrum of a KrHQ absorber taken with an [83]Kr^m(ZnSe) source at 92K is shown in Fig. 5b. This spectrum was corrected for absorber thickness using the Fourier deconvolution method of Ure and Flinn[26] (For a detailed discussion of the analysis of the KrHQ spectrum see Ref.1 and 4). The corrected spectrum (Fig.5a) was fitted with the following free parameters: the isomer shift δ, the absorber linewidth Γ_a, the quadrupole moment ratio Q^*/Q, and the interaction constant $B = e^2qQ(c/E_\gamma)$, where Q^* and Q are the quadrupole moments of the 9.4 keV and ground state, eq is the electric-field gradient (EFG) along the z axis of the cage, and E_γ is the transition energy.

The results of the least-squares fit are given in Table II. The value found for the linewidth, $\Gamma_a = 0.13\pm0.03$ mm/sec, is very close to the natural

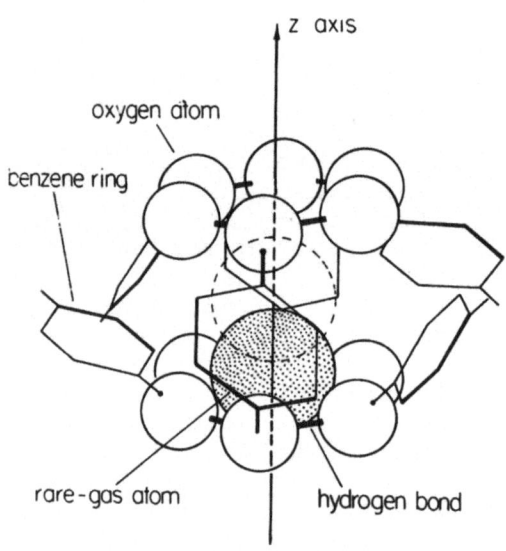

Fig.4a. Structure of βHQ cage containing a krypton
atom. Dotted circle represents a free Kr atom in
the cell approximation, shaded circle a Kr atom in
its actual position.

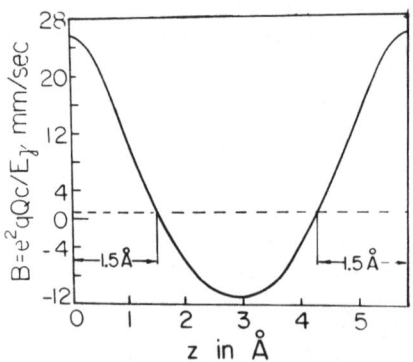

Fig.4b. Quadrupole-interaction constant B, calculated along the z axis of a βHQ cage. Center of
lower oxygen-hydrogen ring in Fig.4a corresponds
to z = 0.

linewidth, Γ_n= 0.10 mm/sec. The quadrupole-moment
ratio $Q*/Q$ = 2.05±0.15 agrees very well with the
recently revised results[1,27,28] from KrF$_2$ Mössbauer
spectra, $Q*/Q$ = 1.98±0.05.

 The unique value of B indicates that the Kr
atoms in βHQ all occupy equivalent sites. To determine the location of the krypton atom in the cage
the EFG, eq_0, was calculated along the z axis.[29]
In Fig.4b the values of the interaction constant
$B = e^2q_0(1-\gamma_\infty)Qc/E_\gamma$ are shown as function of z,
with the Sternheimer shielding factor[30] γ_∞ = -84,
and the quadrupole moment of the ground state[31],
Q = 0.26 barn.

 Because the observed value of B is positive
the krypton atom cannot be in the center of the
cage, but must be located close to the top or bottom ring (Fig.4b).

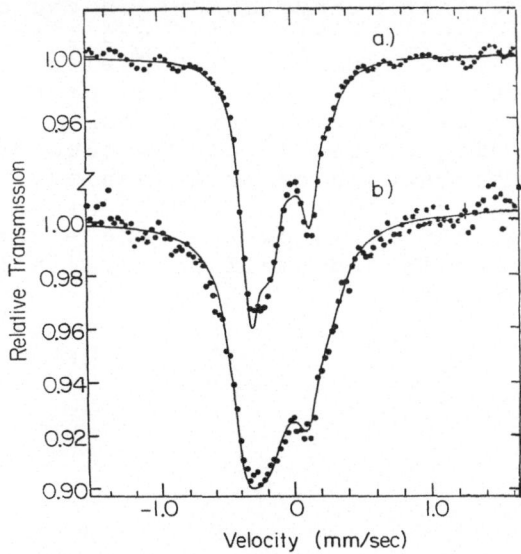

Fig.5 Lower figure represents the measured spectrum
of a KrHQ absorber taken with an ^{83}Kr^m(ZnSe) source
at 92K. Number of counts in first channel is
12.0×10^4. Upper figure shows the spectrum corrected for thickness using the Fourier deconvolution
method of Ure and Flinn.

From the position of the krypton atom with
respect to the center of the oxygen-hydrogen ring,
$z_0 = 1.5$ Å, it follows that the distance between
the krypton nucleus and the oxygen nuclei is 3.1 Å,
which is about the sum of their atomic radii.

The position of the krypton atom can be understood as follows: the oxygen-hydrogen ring induces
a dipole moment in the krypton atom. The induced
dipole is attracted toward the ring by the effective electric field of the oxygen atoms. The results
obtained here show that the generally accepted cell
model for guest atoms in βHQ is invalid in this case.

Of interest is the small isomer shift, $\delta =$
$- 0.15\pm0.03$ mm/sec, which cannot be explained by
second-order Doppler shift. The value of δ corresponds to a ~1% *decrease*[15] in contact density for a
krypton atom in βHQ.

To explain the quadrupole splitting in βHQ
and its high recoilless fraction, Hazony and Herber[32] suggested that a krypton-oxygen bound might
exist. The quadrupole splitting would then be
attributed to the transfer of 0.1 electronic
charge from the oxygen atoms to the empty 4d
($m = \pm1$) orbitals of the krypton atom. Qualitatively the shielding of the 4s and 3s electrons could
explain the decrease in contact density. However,
actual calculations[15] of the effect of 4d shielding
on the contact density show that the effect of
0.1 4d electrons is too small to explain the observed isomer shift.

IV ANHARMONIC EFFECTS IN SOLID KRYPTON

Solid krypton provides a good opportunity to study anharmonic effects on the recoilless fraction. Measurements of absolute $f(T)$ values of ^{83}Kr in solid krypton yielded, however, unrealistically low values.[33] This was due to the fact that an incorrect value of the maximum resonant cross section was used and that *after effects* in the solid krypton source (see Sec.V) were not taken into account.[2] The revised results (see Ref.2 and 4) are shown in Fig.6. They will be dealt with after a brief discussion of the quasi-harmonic approximation (QHA) and *direct* anharmonic effects.

In the QHA the motions of the N individual atoms in a crystal are described by the superposition of $3N$ normal modes, where the volume dependence of the frequencies is given by

$$\omega_i^{qh} = \omega_i^h \left(1 - \gamma_i \frac{\Delta V}{V}\right) \tag{1}$$

Here γ_i is the Grüneisen parameter of the ith mode, and $\Delta V/V$ the relative increase in volume due to lattice expansion.

In the QHA the normal modes are considered to be *independent* which in an actual crystal, however, is not completely true.[7,34] Because of a (small) coupling between the normal modes a Fourier analysis of the motion of each mode will give a peak with linewidth $\Gamma(\omega_i)$ instead of a sharp line as expected in the QHA. Further, the mean frequency of this peak will be shifted by $\delta\omega_i^a$ relative to the quasi-harmonic frequency ω_i^{qh}. The effects arising from the interaction between the normal modes will be called *direct* anharmonic effects to distinguish them from volume effects described by the QHA.

The shift and the linewidth are temperature-dependent. According to Barron[34,35] a useful approximation is $\delta\omega_i^a = \tau_i E_V(T)/3Nk_B = \tau_i e(T)$, where $E_V(T)$ is the vibrational energy of the lattice and k_B the Boltzmann constant. Hence, the total shift in the normal frequency ω_i^h due to anharmonicity is[7,36]

TABLE III

Lattice-dynamical parameters of solid krypton

$\Theta(-3)$	72 K	Ref.42
$\Theta(-2)$	66 K	Ref.42
$\Theta(-1)$	65 K	Ref.42
$\gamma(-3)$	2.5	see Eq.6
$\gamma(-2)$	2.6	see Eq.6
$\gamma(-1)$	2.7	see Eq.6
τ	1.1×10^{-3} K^{-1}	Ref.41

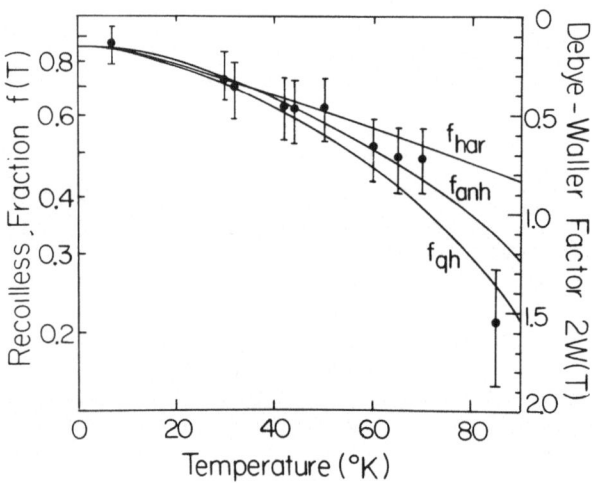

Fig.6. Data reevaluated[2-4] from work of Gilbert and Violet.[33] The smooth curves represent the recoilless fraction of ^{83}Kr in a solid-krypton absorber in harmonic approximation, $f_{har}(T)$, in quasi-harmonic approximation, $f_{qh}(T)$, and in totally anharmonic approximation, $f_{anh}(T)$, which includes direct anharmonic effects.

$$\omega_i^a = \omega_i^h (1 - \gamma_i \Delta V/V + \tau_i e(T)) \tag{2}$$

At higher temperatures $e(T) \to T$ so that

$$\omega_i^a/\omega_i^h = 1 - (\gamma_i \beta - \tau_i)T \tag{3}$$

where β is the coefficient of volume expansion.

To describe the effect of anharmonicity on various physical properties of a solid it is convenient to define characteristic effective temperatures.[37] For $n > -3$ and $n \neq 0$

$$\Theta_a(n;T) = \hbar/k_B \left\{ (n+3)/3 \sum_{i=1}^{3N} (\omega_i^a)^n/3N \right\}^{1/n} \tag{4}$$

and for $n = -3$ and $n = 0$ the limit of $\Theta_a(n,T)$ has to be taken. $\Theta_a(n,T)$ can be written to first order in γ_i and τ_i as[36]

$$\Theta_a(n;T)/\Theta_h(n) = 1 - (\gamma(n)\beta - \tau(n))T \tag{5}$$

where
$$\gamma(n) = \sum \gamma_i (\omega_i^h)^n / \sum (\omega_i^h)n$$
and
$$\tau(n) = \sum \tau_i (\omega_i^h)^n / \sum (\omega_i^h)^n \tag{6}$$

and $\Theta_h(n)$ is given by Eq.4 with ω_i^a replaced by ω_i^h. $\Theta_h(n)$ is temperature *independent*.

Maradudin and Flinn[38] found that for an anharmonic fcc lattice like solid krypton the recoilless fraction is given in very good approximation by $f_{anh}(T) = \exp \{ -k^2 <x_a^2(T)> \}$, where $<x_a^2(T)>$ is the mean square displacement of the γ-ray absorbing or emitting atom along the direction of the γ-ray wave vector $\vec{k}$. Pathak and Deo[39] showed that $<x_a^2>$ can be evaluated from expressions for $_h<x^2>$ in the harmonic approximation by replacing ω_i^h by ω_i^a. Using this result the recoilless fraction, $f_{anh}(T)$, can be expressed in terms of the characteristic effective temperatures, $\Theta_a(n;T)$. In the low-temperature limit[40]

$$2W_{anh}(T) = - \ln f_{anh}(T) =$$

$$= \frac{3E_R}{2k_B} \left\{ \frac{1}{\Theta_a(-1;T)} + \frac{2\pi^2}{3} \frac{m}{M} \frac{T^2}{\{\Theta_a(-3;T)\}^3} \right\} \quad (7)$$

where $2W_{anh}(T)$ is the Debye-Waller (DW) factor, $E_R = E_\gamma^2/2mc^2$ the recoil energy and m and M the masses of atom and unit cell. At moderate and high temperatures[40] $(T > \Theta(-2)/2\pi)$

$$2W_{anh}(T) = - \ln f_{anh}(T) =$$

$$= \frac{6E_R T}{k_B \{\Theta(-2;T)\}^2} \left\{ 1 + \left(\frac{\Theta_a(-2;T)}{6T} \right)^2 \right\} \quad (8)$$

The influence of the anharmonic effects on the recoilless fraction can be readily seen by combining Eq.s (5) and (8) for $T > \Theta(-2)$:

$$2W_{anh}(T) \simeq 2W_{har}(T) \left[1 + 2T\{\gamma(-2)\beta - \tau(-2)\} \right] \quad (9)$$

Obviously the quasi-harmonic contribution in Eq.9, $2T\gamma(-2)\beta$, gives rise to a *decrease* in the recoilless fraction. The *direct* anharmonic contribution, however, leads to an *increase* or *decrease* of the recoilless fraction, depending on whether $\tau(-2)$ is positive or negative. The sign of $\tau(-2)$ is determined by the cubic ϕ_3 and quartic ϕ_4 terms in the vibrational Hamiltonian.[34,36] Since ϕ_3 is an odd function of the atomic displacement its first-order contribution to most crystal properties vanishes. Its second-order terms are of the same order of magnitude as the first-order terms of ϕ_4. Hence, both anharmonic contributions have to be considered; $\tau(n) = \tau_3(n) + \tau_4(n) + ..$, where τ_4 is positive and τ_3 tends to be negative.

For solid krypton Feldman and Horton[41] showed that τ_4 is dominant, and they calculated $\tau = 10^{-3} K^{-1}$. Hence, in this case the *direct* anharmonicity will cancel partly the decrease of the recoilless fraction due to the quasi-harmonic contribution. To investigate this interesting effect we calculated

the recoilless fraction of ^{83}Kr in solid krypton
and compared the results with experiment (Fig.6).

In the harmonic approximation $f_{har}(T)$ was derived from Eq.s (7) and (8) replacing $\Theta_a(n,T)$ by
$\Theta_h(n)$. The values of $\Theta_h(n)$ (Table III) were obtained from the neutron-diffraction measurements
of solid krypton at 10K of Skalyo et al.[42] In the
temperature region from 10 to 20K, where neither
Eq.7 nor Eq.8 is valid, a smooth transition between
the low and moderate-temperature curve in Fig.6
was made by hand.

The recoilless fraction in QHA, $f_{qh}(T)$, was
determined from Eq.s (7) and (8) using
$\Theta_a(n;T) = \Theta_h(n)\{1-\gamma(n)\beta T\}$ where $\gamma(n)$ for $-3 < n < 0$
was derived from[43]

$$\gamma(n) = \int_0^\infty \frac{\beta(T)V(T)}{\chi(T)} T^{n-1} dT \bigg/ \int_0^\infty C_V(T) T^{n-1} dT \quad (10)$$

Here $V(T)$ is the molar volume, $\chi(T)$ the isothermal
compressibility and C_V the specific heat. From
experimental data[44] on β, χ, and C_V of solid krypton the values of $\gamma(n)$ in Table III were obtained.

Nearly all the values of f_{qh} represented in
Fig.6 are systematically lower than the experimental ones. An excellent agreement with the experimental data, however, is found, when the direct
anharmonic effects are taken into account by using
the value $\tau = 10^{-3}$K^{-1} of Feldman and Horton to
calculate $\Theta_a(n;T)$ (Eq.5).

Actual calculations[39] of the anharmonic effects
on the DW factor have been mostly carried out in
the QHA. The results obtained here indicate, however, that the direct anharmonic effects arising
from the coupling of the normal modes are substan-
tial and have to be taken into account.

VI AFTER EFFECTS IN SOLID KRYPTON AT 4.2K

To investigate after effects of the 32 keV
decay which preceeds the resonant γ-ray decay in
^{83}Kr^m, Mössbauer spectra of solid-krypton sources
were taken with a Kr HQ absorber at 4.2K (Fig.7).
From a detailed analysis, given in Ref.2, of the
resonant absorption area, in which resonant self
absorption and the quadrupole interaction in the
absorber were taken into account, the recoilless
fraction for a γ-ray *emitting* atom, f_{se}, was obtained. The result f_{se}=0.53±0.06 agrees very well
with f_{se}= 0.56±0.04 derived from the data of Gilbert and Violet[33] and $f_{se}\leq$ 0.70 determined from resonant self absorption data of Brown[45] (see Ref.2).
The value of f_{se}, however, is 40% smaller than the
recoilless fraction of a γ-ray *absorbing* atom in a
solid-krypton source, f_{sa}= 0.9±0.1, and in the absorber, f_{abs}= 0.87±0.08. The values of f_{sa} and
f_{abs} are in good agreement with theory (see Fig.6).
The observed effect implies that the mean square
displacement of the γ-ray emitting atom is increased by a factor four.

The fact that $f_{sa}\simeq f_{abs}$ rules out poor quality
of the krypton crystals used as sources, or an increase in source temperature due to dissipation of
decay energy, as the origin of the observed effect.
The result $f_{se}< f_{sa}\simeq f_{abs}$ indicates that the lattice is disturbed locally around the γ-ray emitting
atom, which can be caused by after effects as discussed below.

Most ^{83}Kr^m nuclei decay to the 9.4 keV state
by internal conversion (IC). In this process the
released nuclear energy is carried away by a core
electron of the atom. The hole in an inner shell
will be filled by an electron from a higher shell
on a fast time scale ($\sim 10^{-15}$ sec) with emission
of either an x ray or an Auger electron. The latter process dominates in this case. As a result of
the IC process and the following Auger cascade the
krypton atom is highly ionized. The fact that no
Kr$^+$ line has been observed in the spectra of Fig.7
indicates that the Kr atom picks up electrons from
the 4p valence band within 10^{-9} sec. A similar
result has been observed by Trifthäuser *et al.*[46,47]

from delayed-coincidence Mössbauer measurements on ^{57}Co in various insulators.

Hence, while the ^{83}Kr nucleus is in the 9.4 keV state (hereafter denoted by ^{83}Kr*) various holes in the 4p band of solid krypton are produced. These holes will be trapped close to the ^{83}Kr* atom since the mobility of holes in the rare-gas solids at 4.2K is very low.

Druger and Knox[48] showed that a hole can be trapped by two rare-gas atoms R, forming an $[R_2]^+$ center. In a simple model such a center can be regarded as two rare-gas atoms sharing a $4p_z$ hole. The recoilless γ rays emitted from an ^{83}Kr atom which is part of a $[Kr_2]^+$ center will hardly contribute to the central absorption line because of the large quadrupole splitting arising from the $\frac{1}{2}p_z$ hole. The data in Fig.7, however, show no evidence for such a quadrupole splitting.[2]

Most holes probably recombine within 10^{-9} sec with Auger electrons or with electrons produced by inelastic collisions of the Auger electrons with the krypton atom, because of the high mobility of electrons in rare-gas solids at this low temperature.[48] When a recombination between a hole and an

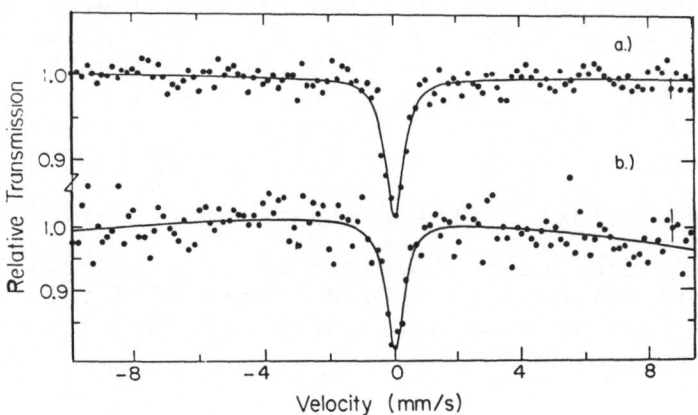

Fig.7 Mössbauer spectrum of a solid ^{83}Kr^m source versus a Kr HQ absorber at 4.2K. Spectrum a) contains 7.3×10^3 counts in the first channel; spectrum b) 2.5×10^3 counts. The curvature of the background in spectrum b) is caused by solid-angle effects.

electron occurs, roughly 10 eV is released. This
energy is sufficient to displace one or both Kr
atoms of the $[Kr_2]^+$ center from their lattice sites,
since the interatomic forces in solid krypton are
very weak (The formation energy for a vacancy in
solid krypton[49] is 0.11 eV). The remaining part
of the released energy will excite lattice vibrations which can yield a local temperature increase
far beyond the melting point (115K). Another contribution to the local temperature increase arises
from the energy of the Auger electrons dissipated
in the surroundings of the $^{83}Kr^*$ atom. The local
temperature will drop to its initial value within
10^{-9} sec, but as result of thermal processes the
lattice will be locally deformed. Because of the
very weak restoring forces in solid rare gases we
assume that the inelastic deformation exists longer than 10^{-7} sec. The lattice constant a_0 in the
deformed region is then increased on the average
by Δa, where $\Delta a / a_0$ will be of the order of magnitude of the relative expansion of a Kr lattice
brought from 4K to the melting temperature, i.e.
$\Delta a / a_0 \simeq 0.10$.[44]

An estimate of $\Delta a / a_0$ can be obtained in the
framework of the theory given in the previous section. The characteristic temperature correspond-
ing to a γ-ray emitting atom, $\Theta_{se}(-1)$, can be related to that of the absorber, $\Theta_{abs}(-1)$, using the
QHA. Effects from the coupling between the normal
modes can be neglected at this low temperature.
From Eq.s (2), (4) and (7) it follows that

$$\ln f_{abs} / \ln f_{se} = \Theta_{se}(-1) / \Theta_{abs}(-1) = 1 - 3\gamma(-1)\Delta a / a_0 \quad (11)$$

Inserting the observed values of f_{abs} and f_{se} in
Eq.11 and using $\gamma(-1) = 2.72$ (Table III) the value
$\Delta a / a_0 = 0.08$ is obtained, which has the expected
order of magnitude.

ACKNOWLEDGEMENTS

I am much indebted to Prof. G.K. Horton,
Prof. R.H. Herber and Dr. G. Collins for their
helpful discussions, and to Prof. Noémie Koller
for her stimulation and encouragement.

REFERENCES

1 B. Kolk, Phys. Rev. B $\underline{12}$,1620 (1975)
2 B. Kolk, Phys. Rev. B $\underline{12}$, 4695 (1975)
3 B. Kolk, Phys. Lett. $\underline{35A}$, 83 (1971)
4 B. Kolk, thesis, Rijks Universiteit Groningen,
   Netherlands (1974)
5 J.H. van der Waals and J.C. Platteeuw,Adv. in
   Chem. Phys. Vol. II, Interscience Publ., London,
   1959, p.1
6 G.L. Pollack, Rev. Mod. Phys. $\underline{36}$, 748 (1964)
7 G.K. Horton, Am. J. Phys. $\underline{36}$, 93 (1968)
8 M. Pasternak and T. Sonnino, Phys. Rev. $\underline{164}$, 384
   (1967)
9 S. Bukshpan, C. Goldstein and T. Sonnino, Phys.
   Lett. $\underline{27A}$, 372 (1968)
10 Recently low-dose implantations of ^{83}Rb in iron
   have been carried out by S. Bukshpan and H. de
   Waard, private communication
11 R.S. Nelson, Proc. Roy. Soc. $\underline{A311}$, 53 (1969)
12 H. de Waard, Physica Scripta $\underline{11}$, 157 (1975)
13 H. Bernas, Physica Scripta $\underline{11}$, 167 (1975)
14 D.A. Shirley, Phys. Lett. $\underline{25A}$, 129 (1967)
15 The density at the Kr nucleus was calculated for
   various electron configurations with the Herman-
   Skillman atomic-structure program (Ref. 19).
   The calibration constant $C = 0.521(mm/sec)a_o^3$ in
   the isomer-shift relation

   $$\delta = -C\left\{|\psi_{source}(o)|^2 - |\psi_{abs}(o)|^2\right\} \qquad \text{was}$$

   derived from the KrF$_2$ data (see Ref.s 1 and 28)
16 B. Kolk, F. Pleiter and W. Heeringa, Nucl. Phys.
   $\underline{A194}$, 614 (1972)
17 I.M. Band, L.A. Sliv and M.B. Trzhaskovskaya,
   Nucl. Phys. $\underline{A156}$, 170 (1970)
18 F. Pleiter and H. de Waard, to be published
19 F. Herman and S. Skillman, Atomic Structure Calculations (Prentice-Hall Inc., New Jersey, 1973)
20 D.E.Palin and H.M.Powell, J.Chem.Soc., 208 (1947)
21 N.R. Grey, N.G. Parsonage and L.A.K. Staveley,
   Mol. Phys. $\underline{4}$, 153 (1961)
22 B.Barnett and Y.Hazony, J.Chem.Phys. $\underline{43}$,3462(1965)
23 G.A. Neece and J.C. Poirier, J. Chem. Phys. $\underline{43}$,
   4282 (1965)
24 J.C. Burgiel, H. Meyer and P.L. Richards, J. Chem.
   Phys. $\underline{43}$, 4291 (1965)

25 Y. Hazony and R.L. Ruby, J. Chem. Phys. 49, 1478 (1968)

26 M.C.D. Ure and P.A. Flinn, Mössbauer Effect Methodology, Vol.7, (New Engl. Nucl. Corp., 1971) p. 245

27 V.M. Krasnoperov, A.N. Murin, N.K. Cherezov and I.A. Yutlandov, Sov. Phys. Doklady 14, 458 (1969)

28 S.L. Ruby and H. Selig, Phys. Rev. 147, 348 (1966) and Erratum, ibid. B 12 1991 (1975)

29 B. Kolk, to be published

30 F.D. Feiock and W.R. Johnson, Phys. Rev. 187, 39 (1969)

31 Mössbauer Data Index 1973 (IFI/Plenum, NY-London)

32 Y. Hazony and R.H. Herber, J. Inorg. Nucl. Chem. 33, 961 (1971)

33 K.G. Gilbert and C.E. Violet, Phys. Lett. 28A, 285 (1968) and K.G. Gilbert, thesis, Un. of California, 1968, UCRL- 50474

34 T.H.K. Barron, Disc. Faraday Soc. 40, 69 (1965)

35 T.H.K. Barron, in Lattice Dynamics (Pergamon Press Inc., NY, 1965) p. 247

36 L.S. Salter, Adv. in Phys. 14, 1 (1965)

37 M.P. Tosi and F.G. Fumi, J. Phys. Chem. Solids 23, 395 (1962)

38 A.A. Maradudin and P.F. Flinn, Phys. Rev. 129, 2529 (1963)

39 K.N. Pathak and B. Deo, Physica 35, 167 (1967)

40 R.H. Nussbaum, B.G. Howard, W.L. Nees and C.F. Steen, Phys. Rev. 173, 653 (1968)

41 J.L. Feldman and G.K. Horton, Proc. Phys. Soc. 92, 227 (1967)

42 J. Skalyo Jr., Y. Endoh and G. Shirane, Phys. Rev. B 9, 1797 (1974)

43 T.H.K. Barron, A.J. Leadbetter and J.A. Morrison, Proc. Roy. Soc. 279, 62 (1964)

44 D.L. Losee and R.O. Simmons, Phys. Rev. 172, 944 (1968)

45 J.B. Brown Jr., in Mössbauer Effect Methodology Vol.9 (New Engl. Nucl. Corp., 1974) p. 23

46 W. Triftshäuser and P.P. Craig, Phys. Rev. 162, 274 (1967)

47 W. Triftshäuser and D. Schroeer, Phys. Rev. 187, 491 (1969)

48 S.D. Druger and R.S. Knox, J. Chem. Phys. 50, 3143 (1969)

49 R.M. Cotterill and M. Doyama, Phys. Lett. 25A, 35 (1967)

LIST OF CONTRIBUTORS

P. H. Barrett, Department of Physics, University of California, Santa Barbara, California 93106

F. J. Berry, Birkbeck College, University of London, London WC1E 7HX, England

C. A. Clausen, III, Department of Chemistry, Florida Technological University, Orlando, Florida 32816

P. Debrunner, Department of Physics, University of Illinois at Urbana-Champaign, Urbana, Illinois 61801

W. N. Delgass, School of Chemical Engineering, Purdue University, West Lafayette, Indiana 47907

J. A. Dumesic, University of Wisconsin, Department of Chemical Engineering, Madison, Wisconsin 53706

Floyd E. Farha, Phillips Petroleum Company, Bartlesville, Oklahoma

R. L. Garten, Corporate Research Laboratories, Exxon Research and Engineering Company, Linden, New Jersey 07036

M. L. Good, Department of Chemistry, University of New Orleans, New Orleans, Louisiana 70122

Peter R. Gray, Phillips Petroleum Company, Bartlesville, Oklahoma

G. P. Huffman, U.S. Steel Research Laboratory, Monroeville, Pennsylvania 15146

C. H. W. Jones, Simon Fraser University, British
    Columbia, Canada V5A 1S6

Berend Kolk, Rutgers University, Department of
    Physics, Busch Campus, New Brunswick, New
    Jersey 08903

G. Lang, Department of Physics, The Pennsylvania
    State University, University Park,
    Pennsylvania 16802

H. Leidheiser, Jr., Center for Surface and Coatings
    Research, Lehigh University, Bethlehem,
    Pennsylvania 18015

Yu. V. Maksimov, Institute of Chemical Physics,
    Academy of Sciences of the Soviet Union,
    Moscow, USSR

W. R. McWhinnie, The University of Aston in Birmingham,
    Birmingham B4 7ET, England

H. Micklitz, Department of Physics, Technische
    Universität München, 8046 Garching, Germany

P. A. Montano, Department of Physics, West Virginia
    University, Morgantown, West Virginia 26506

E. Münck, Freshwater Biological Institute/University
    of Minnesota, Navarre, Minnesota 55392

H. H. Podgurski, U. S. Steel Research Laboratory,
    Monroeville, Pennsylvania 15146

D. Salomon, Institut für Atom- und Festkörperphysik,
    Freie Universität Berlin, D-1000 Berlin 33

E. A. Samuel, Center for Surface and Coatings
    Research, Lehigh University, Bethlehem,
    Pennsylvania 18015

C. Schulz, Department of Physics, University of
    Illinois at Urbana-Champaign, Urbana,
    Illinois 61801

G. W. Simmons, Center for Surface and Coatings
    Research, Lehigh University, Bethlehem,
    Pennsylvania 18015

K. Spartalian, Department of Physics, The Pennsylvania
    State University, University Park,
    Pennsylvania 16802

I. P. Suzdalev, Institute of Chemical Physics,
    Academy of Sciences of the Soviet Union, Moscow,
    USSR

W. Wallner, Institut für Atom- und Festkörperphysik,
    Freie Universität Berlin, D-1000 Berlin 33

P. J. West, Institut für Atom- und Festkörperphysik,
    Freie Universität Berlin, D-1000 Berlin 33

Gerd Weyer, Institute of Physics, University of Aarhus,
    DK-8000 Aarhus C, Denmark

R. Zimmermann, University of Erlangen-Nürnberg,
    Germany

INDEX